The Illustrated Encyclopedia of
Astronomy and Space

Editor Ian Ridpath

Thomas Y. Crowell Company
New York
Established 1834

PHOTOCREDITS: Aerofilms; Aldus Books; Anglo-Australian Observatory; Australian News and Information Bureau; Big Bear Solar Observatory; British Insulated Callender's Cables Ltd; Camera Press; A. C. Cooper Ltd; Heather Cooper; Cornell University; Dominion Astrophysical Observatory; European Southern Observatory; Hale Observatories; Jet Propulsion Laboratory; Kitt Peak National Observatory; Lick Observatory; Mcdonnell Douglas; Mount Wilson and Palomar Observatories; Museum of the History of Science, Oxford; NASA; National Film Archive; National Radio Astronomy Observatory; National Research Council; Novosti Press Agency; P. Popper Ltd; Rainforth Collection; I. Ridpath; Rockwell International; Royal Astronomical Society; Royal Observatory, Edinburgh; R. Scagell; The Science Museum, London; Science Research Council; U. K. Schmidt Telescope Project; Westerbork Radio Observatory; Yarkes Observatory.

3

SBN 333 21365 3

First published in Great Britain 1976 by
Macmillan London Limited
London and Basingstoke
Associated companies in New York, Toronto,
Dublin, Melbourne, Johannesburg and Delhi.

Published in the U.S.A. 1976 by
Thomas Y. Crowell Company Inc. New York, N.Y.
Published simultaneously in Canada by
Fitzhenry & Whiteside Limited, Toronto.

Phototypeset by Tradespools Ltd., Frome, Somerset
Printed in the United States of America

The Encyclopedia of Astronomy and Space

Editor **IAN RIDPATH**

Major contributors

HEATHER A. COUPER
Department of Astrophysics, University of Oxford

Dr. PAUL C. W. DAVIES
Department of Mathematics, King's College,
University of London

Dr. JOHN GRIBBIN
Science Policy Research Unit; formerly
assistant editor, *Nature*; author of *Galaxy Formation*,
Astronomy For The Amateur
and *Our Changing Universe*

NIGEL HENBEST
Mullard Radio Astronomy Observatory, Cambridge

Dr. KEITH HINDLEY
Director, Meteor Section,
British Astronomical Association

JAMES MUIRDEN
Manufacturer of astronomical instruments
and science writer

Dr. PAUL MURDIN
Royal Greenwich Observatory;
Anglo-Australian Observatory

IAN RIDPATH
Science author and editor, astronomy and
astronautics contributor to *New Scientist*, author of
Worlds Beyond

COLIN A. RONAN
Editor, *Journal of the British
Astronomical Association*

Dr. JOHN ROSS
Science Correspondent,
The Times Higher Education Supplement

Professor IAN W. ROXBURGH
Department of Mathematics, Queen Mary College,
University of London

ROBIN S. SCAGELL
Assistant editor, *How It Works* encyclopedia

Introduction

Astronomy is both the oldest and the most modern of the sciences. Certainly the skies were the first domain into which man brought order, but this was long before the all-embracing era of modern science. To begin with he merely classified the stars, grouping them into constellations, and made studies of the odd behavior of the planets. Simple though these investigations were, they enabled him to formulate theories and then refine them by examining the skies and seeing whether his calculations of future behavior coincided with what actually happened. Thus he discovered the regular seasonal movements of the Sun and Moon, was led to devise a calendar, and developed his mathematics and his techniques of precise measurement. Yet for the first four thousand years and more of his investigation of the skies, all man could do was to study the positions of the celestial bodies. He was quite unable to determine the nature of the stars or the planets, or to study our own Sun or Moon in detail.

The arrival of the telescope in the early 1600s changed all this. Now it was possible to see that the Moon, like the Earth, had mountains, valleys, and plains. The Sun could be seen to be disfigured with spots that came and went spasmodically. The planets were found to have disks, some to possess satellite systems, and all to be objects with features that showed they rotated on their axes. Moreover, the telescope showed thousands of stars too dim to be seen by unaided eyes, and the presence in space of hazy nebulous patches of light. Clearly, the Universe was a more curious and a more complex place than anyone had previously imagined. Yet an observational breakthrough was not all that happened in the 17th century. Owing to the efforts of men like Kepler, Galileo, and Newton, a mathematical and physical theory of planetary motion was evolved. This theory, with its concept of universal gravitation, completed the old astronomy, which had spent almost all its efforts trying to account for planetary movements, and left the astronomer free to use his telescope for probing deep into space.

The astronomical revolution of the 17th century gradually changed the face of science. The Universe was found not only to be vaster than previously imagined, but also to operate in a way that could be described in purely mechanical terms. In the metaphor of the time, the Universe was no more than a giant piece of celestial clockwork. However far he peered into space with telescopes of ever increasing size, the astronomer found that there was evidence of the existence of gravity and the same laws for the behavior of stars orbiting around one another as for planets orbiting the Sun. In 1838, using the telescope as a precision measuring instrument, it became possible at last to measure the distances of the stars and gain some insight into the true immensity of the Universe. Twenty years later, the scientific spirit of the new astronomy showed its enormous potential when it became possible to identify the chemical elements in the stars using the spectroscope. Here was a totally new technique, which taken together with the 20th century research into the structure of the atom, led to an understanding of what makes the stars shine, explode and die.

In our own decade astronomy has once more entered a revolutionary phase. The picture of the Universe of even twenty years ago has changed to something broader and more comprehensive than it has ever been before. With radio telescopes a new eye has opened on the universe—cold clouds of gas, never observable before, can now be charted and examined in detail, objects too distant in space to be seen visually can be picked out, and a radiation background detected, a background whose very existence has had the profoundest effect on man's ideas of how the Universe began. Yet although the radio telescope plays a powerful role in modern astronomy, it is not the sole new way of probing space. We are now aware, as earlier astronomers were not, that celestial bodies emit radiation of all kinds; not only light and heat, but also infrared radiation and short wave ultraviolet rays, X rays, and the very penetrating gamma rays. These all give information about the Universe never available before. But, except for some infrared and a little ultraviolet, none of these radiations can reach the ground; to observe them telescopes have to be used out in space, away from the blanketing effects of the Earth's atmosphere. The technology of space exploration has made it possible to design, build and launch orbiting astronomical observatories so that these vitally important observations can be made. What is more, manned exploration of space has meant not only that men have been able to land on the Moon, but also that they could carry out actual experimental work elsewhere in the Universe.

With computerized techniques for analyzing what telescopes and space probes detect, observational astronomy is now seeing as great an advance as that of nearly four hundred years ago when the telescope was first used as a scientific research tool. As in the 17th century, the revolution has not only been in observation. Theoretical astronomy has also leaped forward with new methods of analysis, the use of computerized models, advances in nuclear physics and relativity. Astronomers now routinely talk about objects like neutron stars, black holes, pulsars and quasars. They have brought into their discipline more scientific fields than ever before: geologists concerned with the crust of the Moon, geophysicists with the Moon and the Earth as bodies subject to physical changes in themselves and in their environment due to magnetic and other effects emanating from the Sun, biologists and other life scientists probing the possible existence of life elsewhere in the Universe, physicists, chemists, mathematicians, electronics engineers, information theory experts and computer technologists.

This concentration of scientific disciplines is not to be wondered at. Astronomy is a study of immense importance. Stars, nebulae, and galaxies, to say nothing of more esoteric objects like black holes, all act as extraterrestrial laboratories in which the very matter of the Universe is subjected to conditions of a degree and on a scale impossible to imitate in any laboratory on Earth. Studies of the Universe as a whole, and of its beginning and possible end, stretch the mind to the full and exert the most profound effects on our general outlook on matters that no thinking person can ignore. In 1543 when Copernicus' theory displaced man from his position at the hub of creation, no department of human thought, no art, no science, remained unaffected. Nothing could be the same again. The present revolution in astronomy—indeed, our whole outlook on the universe—seems about to have the same effect. Modern physics now appears due for fundamental changes because of new astronomical evidence, and the life sciences after further investigations into the presence of the molecular building blocks of living matter in the depths of space, will not emerge without change. Should it prove possible to detect the existence of other civilizations, there is nothing that will not be revitalized and re-oriented.

As we begin to glimpse a stranger and more dynamic Universe, we have reached the moment to take stock of present knowledge. This is the purpose of the *Encyclopedia of Astronomy and Space*—to gather together the myriad aspects of modern astronomy and space science and to present them in a readily accessible way to the interested layman. The variety of its coverage, the many illustrations, and the alphabetical arrangement should make it eminently suitable as a basic reference tool.

A note on units and symbols used in this Encyclopedia

The U.S. system of designating large numbers is used. Thus the U.S. billion equals the British thousand million, and the U.S. trillion equals the British billion. To save writing endless zeros in large numbers, scientists often use a power of ten notation. In this the figure 10 is given with a superscript which shows the total number of zeros to be written. Thus, 10^2 is 100, 10^6 a million (1,000,000), 10^9 a billion (1,000,000,000), and 10^{12} a trillion (one million million). This notation is sometimes used in the book. All units are given in standard and metric measure; the tons are metric (1,000 kg or 2,205 pounds). To the accuracy with which most astronomical figures are known, metric tonnes can be considered equivalent to American long tons or British tons.

When labeling bright stars the tradition is to use Greek letters. For convenience, we have spelled out the names of these letters. The complete Greek alphabet is given below.

α, A	alpha	ι, I	iota	ρ, P	rho
β, B	beta	κ, K	kappa	σ, Σ	sigma
γ, Γ	gamma	λ, Λ	lambda	τ, T	tau
δ, Δ	delta	μ, M	mu	υ, Y	upsilon
ε, E	epsilon	ν, N	nu	ϕ, Φ	phi
ζ, Z	zeta	ξ, Ξ	xi	χ, X	chi
η, H	eta	o, O	omicron	ψ, Ψ	psi
$\vartheta, \theta, \Theta$	theta	π, Π	pi	ω, Ω	omega

The 150-foot diameter radio telescope of the Algonquin
Radio Observatory has a paraboloidal surface accurate to
a few millimeters, and can therefore focus radio waves
as short as 2.8 centimeters on to the receiver mounted
in front. It is used extensively to monitor the changing
emission from quasars.

A

α
The symbol for the astronomical coordinate known as RIGHT ASCENSION.

Å
The symbol for the unit of measure of the wavelength of light called an ANGSTROM.

aberration

Any of a number of faults that mar the image in an optical instrument. In SPHERICAL ABERRATION, light from different parts of a lens or mirror is brought to different foci, causing a blurred image. Lenses suffer also from CHROMATIC ABERRATION, in which light of different colors is bent by differing amounts, as in a prism, producing color fringes around the object. In COMA, the images are progressively elongated toward the edge of the telescope field. ASTIGMATISM is the failure of an optical system to bring light rays from different planes, such as horizontal and vertical, to a common focus, causing elliptical images. A focal plane that is not flat produces so-called *curvature of field* and arises when different parts of the image are formed at different distances from the lens or mirror. Where the telescope's magnification varies across the field, the image is distorted so that straight lines look curved. So-called *barrel distortion* occurs where the telescope's magnification decreases toward the edge of the field, bowing straight lines outwards. Where magnification increases toward the edge of field, straight lines are bowed inwards in what is termed *pincushion distortion*.

aberration of starlight

An effect caused by the Earth's motion in orbit which produces slight changes in the apparent positions of stars. Aberration of starlight was discovered by JAMES BRADLEY in 1728, after careful observations showed a displacement in position of the star Gamma Draconis during the year. Bradley realized that the effect, which amounted to only about 20 seconds of arc, was caused by the motion of the Earth across the path of light rays from the star. This made the light rays appear to be coming in at a slant, in the same way that vertically falling raindrops seem to be moving at an angle as seen from a speeding vehicle. This effect gives a slightly false position for all stars, which must be corrected for in observations. The aberration of starlight was the first direct observational proof of the Earth's movement around the Sun.

absolute magnitude

A measurement used in comparing the total light output of stars. The absolute magnitude of a star is the brightness it would exhibit if it were at a standard distance of 10 parsecs (32.6 light-years). Absolute magnitude can usually be calculated from knowledge of the star's nature. By comparing its calculated absolute magnitude with the brightness with which it actually appears, astronomers can determine how far away it must be. (See APPARENT MAGNITUDE; DISTANCE MODULUS; MAGNITUDE.)

absolute zero

The coldest temperature possible. At absolute zero all thermal motion of molecules ceases because all heat

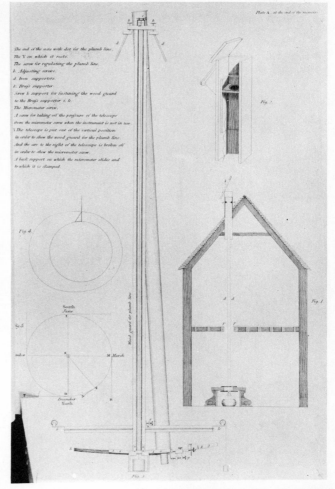

The zenith telescope with which James Bradley discovered the aberration of light. A weight held the telescope firmly against a micrometer screw on the scale at the bottom of the telescope, enabling accurate position measurements of Gamma Draconis and other stars to be made.

has been removed; the object is therefore totally cold. Absolute zero is equal to $-273.16°$C or $-459.67°$F.

absorption lines

Wavelengths in an object's spectrum which have been absorbed by cooler gas. The wavelengths of the absorption lines depend on the elements that cause them (see SPECTROSCOPY). The FRAUNHOFER LINES in the Sun's visual spectrum are caused by absorption of sunlight by the cooler gases in the Sun's outer layers. All stars show such dark absorption lines in their spectra, which allow astronomers to analyze their composition. Absorption lines are also caused by clouds of gas between the stars. Absorption can occur at any wavelength, including long wavelengths which can be observed only by radio astronomers.

achondrite

A stony meteorite lacking the inclusions called *chondrules* common in most other stony meteorites (see CHONDRITE). Achondrites are similar in composition to terrestrial basalt rocks.

achromatic

Term given to lenses that are specially corrected to prevent the appearance of color fringes around the image; this fault, common in cheap optics, is called CHROMATIC ABERRATION. Achromatic lenses are made from more than one piece of glass, each piece canceling out the chromatic effect of the other. The first achromatic lenses were made in 1733 by the English amateur optician Chester Moor Hall (1703–1771), after a study of the human eye had convinced him that achromatic lenses were possible in spite of the well-known prism experiments of Isaac Newton, which had seemed to show that refraction of light must inevitably cause rainbow-colored images. However, Hall never published his discovery, and the achromatic lens is usually credited to John DOLLOND.

Adams, John Couch (1819–1892)

British mathematician and astronomer, best known for his calculations that predicted the existence of the planet Neptune. Adams graduated with highest honors from Cambridge University in 1843, having already begun to investigate irregularities that were being observed in the motion of Uranus, the outermost planet of the solar system known at that time. Adams calculated that these irregularities could be caused by the gravitational effect of a hitherto unknown planet, and predicted where it was likely to be found. In 1845 he presented his calculations to the astronomer royal, Sir George Airy, who took no action. Only after the French mathematician Urbain LEVERRIER had published similar results was a search ordered. By then, however, the German astronomer Johann Galle was hot on the trail at the Berlin Observatory, and on September 23, 1846, he found Neptune close to the position calculated by Adams and Leverrier. Although the honors at first went to the Frenchman, Adams eventually won recognition for having been first to arrive at the solution.

Adams, Walter Sydney (1876–1956)

American astronomer, from 1923 to 1946 director of the Mount Wilson Observatory, known for his important spectroscopic studies of stars and planets. In 1914 he and Arnold Kohlschütter found that the nature of the dark lines in a star's spectrum shows whether it is a dwarf or giant star, thus allowing its ABSOLUTE MAGNITUDE to be deduced. Comparing a star's absolute magnitude with its observed brightness gives its distance (see DISTANCE MODULUS). This valuable distance-measuring technique developed by Adams is termed the *method of spectroscopic parallaxes*. In 1915 Adams examined the small, faint companion star of Sirius and found from its spectrum that its surface is white-hot; this was the first so-called WHITE DWARF to be recognized. In 1925 Adams found that this star's light shows a slight red shift, known as the Einstein shift, caused by its high gravity. Later, in 1932, he showed that the atmosphere of Venus consists of carbon dioxide; he worked with Theodore Dunham, Jr. (b. 1897), who himself made important spectroscopic studies of planetary atmospheres.

aerolite

Technical name for a stony meteorite, the most common type of meteorite that falls to Earth (see CHONDRITE).

Agena

An upper stage used on American launch vehicles, introduced in 1959. The Agena's engine, of 16,000-lb. (7,260-kg) thrust, can be stopped and started at will. Agenas were used with Atlas and Thor first stages to launch probes toward the Moon and planets, as well as to put into orbit numerous Earth satellites. The Agena is 5 feet (1.5 m) in diameter and approximately 23 feet (7 m) long. Agena stages were used as orbiting rendezvous targets during the Gemini series of manned space flights.

Airy, Sir George Biddell (1801–1892)

English astronomer royal from 1835 to 1881; previously director of the Cambridge University Observatory, where he set up the 12-inch (30-cm) Great Northumberland refractor later used in the search for Neptune. Airy totally reorganized the Royal Greenwich Observatory, installing a new series of instruments he himself had designed. Most famous of these was the transit telescope used to determine time from the passage of stars across the meridian. The position of this instrument defines 0° longitude, the Greenwich meridian, established by international agreement in 1884. Airy supervised the analysis and publication of all lunar and planetary observations made at Greenwich between 1750 and 1830, which astronomers still use, and which allowed John Couch ADAMS to predict the existence of Neptune. Airy failed, however, to act promptly on Adams' calculations. The British government used Airy as a consultant on numerous scientific and technical matters.

Aitken, Robert Grant (1864–1951)

American astronomer, discoverer of over 3,000 double stars. In 1932 he published a major listing of double stars, the *New General Catalogue of Double Stars within 120° of the North Pole*, containing measurements of 17,180 double stars made by him and William Joseph Hussey (1862–1926) with the 36-inch (91-cm) telescope at Lick Observatory. This superseded the *General Catalogue of Double Stars*, published in 1906 by the American double-star observer Sherburne Wesley Burnham (1838–1921), who had discovered a total of 1,274 pairs. From 1930 to 1935 Aitken was director of the Lick Observatory in California.

Albategnius (c.858–929)

The greatest Arab astronomer, also known as al-Battani. Using the most accurate astronomical instruments of his day, Albategnius redetermined the length of the year, timed the occurrence of the spring equinox to within a few hours, improved the value for precession, and accurately measured the tilt of the Earth's axis relative to its orbit. His tables of lunar, solar, and planetary positions were better than those in Ptolemy's *Almagest*. Albategnius also studied the changing apparent diameter of the Sun during the year, caused by the eccentricity of the Earth's orbit. He found that the point where the Sun's apparent diameter is smallest (and thus the Earth–Sun distance greatest) had moved from its position at the time of Ptolemy. Modern astronomy has since confirmed that the Earth's farthest point from the Sun does indeed move.

albedo

A measure of the proportion of light reflected by a

non-shiny surface. Dark objects have a lower albedo than light-colored ones. An object with an albedo of 1 would reflect all the light that hits it, making it brilliant white; an object with zero albedo would be totally black. The albedo of a planet or satellite is therefore a guide to the nature of its surface.

Aldebaran
The brightest star in the constellation Taurus, the bull, of magnitude 0.86. Aldebaran, also called Alpha Tauri, is a red giant star, held to represent the bull's glinting eye. It is 64 light-years away, and about 36 times the diameter of the Sun, giving out as much light as over 100 Suns.

Aldrin, Edwin Eugene (b. 1930)
Lunar module pilot of the Apollo 11 crew, which made the first lunar landing on July 20, 1969. Aldrin, a qualified engineer, was selected as an astronaut in 1963. He first flew in space on the Gemini 12 mission in November 1966, during which he made a record space walk of over two hours. On the Apollo 11 flight, Aldrin joined crew commander Neil Armstrong on the lunar surface for a walk of some two hours. Aldrin practiced various ways of moving about under the low lunar gravity, and helped set up experiments and collect rocks. Like the other crew members of Apollo 11, Aldrin left the astronaut corps after his return to Earth.

Alfvén, Hannes Olof Gösta (b. 1908)
Swedish physicist who received the Nobel Prize for physics in 1970 for his work on ionized gases and their interaction with magnetic fields. This research has been vital in attempts to produce controlled nuclear fusion, since powerful magnetic fields must be used to bottle up the hot gases in which the fusion reactions take place. It has also contributed to our understanding of the origin of the solar system, by showing that the fast rotation of the young Sun could have been transferred by magnetic fields to the tenuous gas cloud around it from which the planets formed, explaining why most of the spin, or angular momentum, of the solar system is present today in the planets rather than in the Sun. This work has also aided understanding of the outer regions of the Earth's magnetic field called the MAGNETOSPHERE; in 1939 Alfvén published a theory that linked solar storms with the activity of particles in the Earth's magnetosphere that cause aurorae. Alfvén's work ranges over a wide field, from studies of asteroid groupings which suggest that the asteroid belt is a failed attempt at forming a small planet, to the field of cosmology, in which he has strongly supported the view that equal amounts of matter and antimatter must exist in the Universe (see ANTIMATTER COSMOLOGY).

Algol
Famous star in the constellation Perseus, also called Beta Persei. Algol appears to vary in brightness between magnitudes 2.2 and 3.5 every 2.87 days. The variations are in fact caused by a fainter companion that periodically eclipses the main star of Algol, and the system was the first such ECLIPSING BINARY to be discovered. The English amateur John Goodricke suggested the correct explanation for Algol's behavior in 1782. Later astronomers discovered that Algol is actually a triple star, although the third companion does not take part in the eclipses. In 1971 Algol was identified as a radio source; some of the radio emission is believed to be caused by transfer of gas between the two main stars. Algol is 82 light-years away.

Algonquin Radio Observatory
Site of the largest radio telescope in Canada, 150 feet (46 m) in diameter, opened in 1966. The observatory, at Algonquin Park, Ontario, is operated by the National Research Council of Canada. The 150-foot telescope works at wavelengths down to 3 centimeters. It has been used to pioneer the technique known as very long baseline interferometry (VLBI), by comparing signals it receives with those received by an 84-foot (26-m) telescope at the associated observatory at Penticton, British Columbia. The Algonquin dish has also been linked with the radio telescope at Parkes, Australia, giving a baseline almost equal to the diameter of the Earth, for distinguishing small details in objects such as quasars.

Alouette satellites
Two Canadian scientific satellites, launched by the United States, designed to study the Earth's ionosphere. Successor to Alouette is the ISIS (International Satellites for Ionospheric Studies) series, a joint Canadian–U.S. program.

Satellite	Launch date
Alouette 1	September 28, 1962
Alouette 2	November 28, 1965 (launched with Explorer 31, which made supporting studies)
ISIS 1	January 30, 1969
ISIS 2	April 1, 1971

Alpha Centauri
A triple-star system (also called Rigil Kent), containing the nearest stars to our Sun, 4.3 light-years away. To the naked eye, Alpha Centauri appears as one star, the third-brightest in the entire sky, of magnitude −0.27. A small telescope shows two stars, of magnitudes 0 and 1.4; it is their combined light that makes Alpha Centauri appear so bright. These stars orbit each other once every 80 years. The main star is very similar to our Sun; its mass is 1.1 times the Sun's, and its radius 1.23 times that of the Sun. The second star is slightly smaller and cooler. Bigger telescopes reveal a magnitude 10.7 red dwarf star in the Alpha Centauri system. This third star is called PROXIMA CENTAURI, because in part of its orbit around the other two it comes closer to us than either of the main stars. The name Alpha Centauri when used without qualification usually refers to all three stars. The Alpha Centauri system is approaching us at about 15.5 miles (25 km) per second. In about 28,000 years it will be 3.1 light-years away, and 0.7 magnitudes brighter.

Altair
Brightest star in the constellation Aquila, also called Alpha Aquilae. It appears of magnitude 0.77. Altair is far hotter than the Sun, and is several times more massive. It is 16 light-years away, and 1.65 times the Sun's diameter.

altazimuth

The simplest form of mounting for telescopes. As the name implies the instrument is free to move independently in altitude (up and down) and in azimuth (left to right). Continual adjustments must thus be made in both axes to follow a star across the sky. EQUATORIAL MOUNTING allows a telescope to follow an object in one movement.

altitude

The angle between a celestial object and the horizon, measured at right angles to the horizon.

Ambartsumian, Viktor Amazaspovich (b. 1908)

Soviet (Armenian) astrophysicist who in 1947 discovered the existence of STELLAR ASSOCIATIONS, vast fields of hot, young stars in our Galaxy, which have apparently been born together. Their existence showed that large-scale star formation still continues in our Galaxy's spiral arms. In 1955 Ambartsumian suggested that the radio-emitting galaxies then being discovered had suffered explosions at their centers, and he proposed that galaxies grow by ejecting material; subsequent observations apparently showing just such a process in operation have helped substantiate his view (see RADIO GALAXY). In 1946 Ambartsumian founded the BYURAKAN ASTROPHYSICAL OBSERVATORY in Soviet Armenia.

Ames Research Center

NASA scientific and engineering establishment, adjoining the U.S. Naval Air Station at Moffett Field, California. It conducts studies on the possibility of life elsewhere in space, as well as on the effects of spaceflight on humans. Aeronautical studies at Ames include research on reentry into the Earth's atmosphere of space capsules and the winged SPACE SHUTTLE. The Ames Research Center has directed the PIONEER series of space probes. The Center was set up in 1940 by NASA's forerunner, the National Advisory Committee for Aeronautics (NACA), and is named for former NACA president Joseph Sweetman Ames (1864–1943).

anastigmat

A wide-aperture or "fast" photographic lens whose image quality does not deteriorate away from the center of the field. This deterioration, known as ASTIGMATISM, is cured by using several lenses in combination. The first modern anastigmat, the Zeiss Tessar, was introduced in 1902 and worked at $f/4.5$.

Anaxagoras (c.500–c.428 B.C.)

Greek philosopher, who maintained that the Earth and the objects in the sky were made of essentially the same substance. He believed that the Sun was a large, red-hot stone, and that the Moon was a dark, solid body like the Earth, illuminated only by light from the Sun. This led him to the true explanation of eclipses: that they are caused by the blocking off of light from the Sun.

Anaximander (c.610–c.546 B.C.)

Greek philosopher and astronomer, a pupil of THALES and sometimes called the father of astronomy. Anaximander put forward the first known cosmological theory, in which he described the Earth as a cold, hard body surrounded by luminous heavenly objects like wheels of fire in space. Realizing that the sky rotates around the pole star, he visualized the heavens as a spinning sphere—thus setting an unfortunate precedent, for astronomers until the time of Kepler assumed that there were real spheres in the sky. Anaximander could see that the Earth's surface was curved, but imagined our planet as a squat cylinder, floating in space at the center of the Universe.

Anders, William Alan (b. 1933)

American astronaut who flew in Apollo 8, the first manned spacecraft to orbit the Moon. Anders photographed the lunar surface during Apollo 8's 10 orbits, showing that the selected landing sites would be safe for later Apollo missions to touch down on, and improving astronomers' knowledge of lunar features. Anders, a qualified nuclear engineer, was selected as an astronaut in 1963.

Andromeda

Constellation in the northern hemisphere of the sky, named for a princess of Greek mythology; it is best placed for viewing during the northern hemisphere autumn. Andromeda can be easily located, lying next to the famous square of Pegasus; one corner of the square is actually the star Alpha Andromedae. The planetary nebula NGC 7662 in Andromeda is 5,000 light-years distant. At the heart of the constellation lies its most famous feature—the ANDROMEDA GALAXY.

Andromeda galaxy

The most distant object in space visible to the naked eye, also known to astronomers by its catalog numbers of M31 and NGC 224. It is a separate spiral galaxy in space, like our own Milky Way but apparently twice the diameter. It is sometimes incorrectly called the Andromeda nebula. The galaxy appears as a fuzzy oval patch to the naked eye, in the center of the constellation Andromeda, but large telescopes reveal that it is composed of individual stars; one estimate puts their number at 300 billion. Observations of CEPHEID VARIABLE stars in the Andromeda galaxy have shown that it is about 2.2 million light-years away, making it the nearest major galaxy to our own. It shows, like our own Galaxy, areas of dark and light gas, clusters of stars, and exploding stars. The Andromeda galaxy has two small bright companions, similar to the MAGELLANIC CLOUDS that accompany our own Galaxy. Recent investigations have found at least one other midget companion to the Andromeda spiral. The Andromeda galaxy, along with our own Galaxy and several others, is part of what is termed the LOCAL GROUP.

angstrom

Unit of measure of the wavelength of light, equal to one ten-billionth (10^{-10}) of a meter, symbol Å. It is named for the Swedish physicist Anders Jonas Ångström (1814–1874), a pioneer of spectroscopy who made studies of the solar spectrum and in 1862 discovered hydrogen in the Sun. In 1868 he mapped the Sun's spectrum, using what became known as the angstrom unit; it was officially named for him in 1905. Ten thousand angstroms equal 1 micron (μ).

angular diameter

The apparent diameter of an object measured in

degrees, or fractions of a degree. As seen from the
Earth, for instance, the angular diameters of the Sun
and Moon are very similar—about half a degree each.
Their actual diameters, of course, are very different.

angular distance
The apparent separation of two points or objects in the
sky, measured in degrees or parts of a degree.

Anik satellites
Satellites for communications within Canada; also
called Telesat, after the Canadian company that owns
and operates them. Anik 1 (the name is the Eskimo for
"brother") was launched by the United States on
November 9, 1972, followed by Anik 2 on April 20,
1973, and Anik 3 on May 7, 1975. The satellites are in
SYNCHRONOUS orbit, so that they appear to hang
stationary in the sky above the Earth's equator; they
make up the world's first domestic synchronous
satellite system. The satellites are stationed at 104°,
109°, and 114° west longitude, with their transmission
and reception antennae covering the area of Canada.
Each satellite can carry the equivalent of 12 color-
television channels or up to 5,760 telephone circuits.

Antares
The star alpha in the constellation Scorpius. Its name,
meaning "rival of Mars," was given it because of its
distinctive red tint, much like the color of the planet.
Antares is a red supergiant star of magnitude 1.08,
with a diameter about 285 times that of the Sun. It has
a peculiar hot companion star of magnitude 6.8. The
luminosity of Antares is equal to about 5,000 Suns,
which allows it to appear the brightest star in Scorpius
even though it is 430 light-years away. In 1971 its

companion star was discovered to give out radio
emission, due to gas streaming between the two stars or
to matter being ejected from the companion.

antimatter
Matter consisting of so-called antiparticles, which are
the same as the elementary particles of ordinary
matter, such as protons and electrons, but have the
opposite electric charge. For example, an antielectron,
or *positron*, has a positive, not a negative charge, and
the antiproton has a negative, not positive charge. If
matter and antimatter come into contact, they
annihilate each other, releasing vast amounts of energy.

antimatter cosmology
Theory which maintains that equal amounts of matter
and antimatter exist, and always have existed, in the
Universe. According to standard cosmology, on the
other hand, the Universe began with slightly more
matter than antimatter; all the antimatter was
destroyed by annihilation together with most of the
matter, leaving behind our present Universe, made up
entirely of matter. Since matter and antimatter
annihilate, they could coexist only if the Universe were
to contain separate regions of matter and antimatter,
with annihilation taking place at the boundary between
them. Antimatter cosmologists believe this separation
occurred in the BIG BANG, and as the Universe

expanded, regions with ordinary galaxies and regions with antimatter galaxies developed. The energy released by annihilation between the regions should produce gamma rays that could be observed as a gamma-ray background radiation. At present we cannot measure such radiation accurately enough to know if the Universe really is symmetrical.

Antlia (the air pump)
A faint constellation in the southern hemisphere of the sky, south of Hydra. It was named by Nicolas Louis de Lacaille in the 1750s.

Antoniadi, Eugène M. (1870–1944)
French astronomer born of Greek parents, one of the greatest visual observers of the planets. He is best known for his painstaking and expertly drawn maps of the planets Mars and Mercury. Many of his finest observations were made with the 32-inch (81-cm) refractor of the Meudon Observatory near Paris. Antoniadi originated what is termed the *Antoniadi scale*, a measure of the quality of the observing conditions, or *seeing*, recorded in numbers from I (perfect) to V (atrocious).

apastron
The point in the mutual orbits of a double star at which the two components are farthest apart, or in the orbit of a planet at which it is farthest from its parent star.

aperture synthesis
Technique in radio astronomy which uses a number of small radio dishes to build up the same view of the sky as would be seen by one enormous dish. It overcomes the impossibility of building fully steerable radio telescopes with diameters much greater than a few hundred yards or meters. In aperture synthesis, a line of interconnected radio dishes tracks a source for 12 hours as it crosses the sky. During this period, and as a result of the Earth's rotation, the line traces out a ring in space. As seen from the pole this ring is a perfect circle; but in practice, because radio sources are usually far from the pole, the line of dishes will be foreshortened, giving an elliptical aperture. Some of the dishes may then be moved and the observations repeated the next day. All the readings are stored in a computer and eventually combined to make a complete map. The largest aperture synthesis instrument is the 3-mile (5-km)-long radio telescope at Cambridge, England. It is able to see the radio sky in more detail than any other radio telescope.

aphelion
The point in an object's orbit that is farthest from the Sun. It is on the opposite side of the orbit from the *perihelion*, the nearest point to the Sun. For objects such as planets, with nearly circular orbits, the difference between aphelion and perihelion distances is relatively slight. But for comets, which have very elongated orbits, the difference can be considerable.

apochromat
Telescope object glass that has been highly corrected for color error (CHROMATIC ABERRATION). The typical ACHROMATIC lens brings two colors (usually red and blue) to a common focus by using two different types

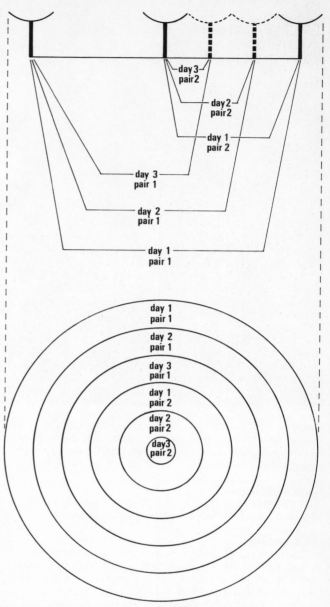

The dishes of an aperture synthesis telescope are connected together into pairs. Any number of dishes can be used. The greater the number, the quicker the observation is completed, but the greater the construction cost. By making one or more of the dishes movable, a series of pairs with different spacings can be built up on successive days. As the Earth spins, the pairs trace out the area of an imaginary single dish whose diameter equals the maximum length of the line.

of glass in combination. In an apochromat, three lenses are used, thus bringing a third color, usually violet, to the same focus. This greatly reduces the bluish halo of so-called secondary spectrum that a refracting telescope shows around bright stars. Such lenses are often called *photovisual*, because the violet rays to which early photographic emulsions were primarily sensitive focus near the same plane as the visual (yellow) rays. A near-approach to apochromatic quality can be obtained by using special varieties of glass in a two-lens objective.

apogee

The greatest distance from Earth of an orbiting object such as a satellite. The opposite point in its orbit is termed the *perigee*.

Apollo program

The space project that landed a total of 12 American astronauts on the Moon. On May 25, 1961, President John F. Kennedy set the goal of landing a man on the Moon and returning him safely to Earth before 1970. At that time there had only been one American manned spaceflight—the short suborbital mission of Alan Shepard. One early idea for reaching the Moon was to build an enormous rocket, to be called Nova, which would launch a single spacecraft to land on the lunar surface and then take off again; this method of reaching the Moon was called *direct ascent*. A rival idea was called *Earth-orbit rendezvous*, in which the spacecraft would be assembled from parts launched into Earth orbit by several smaller rockets. By the end of 1962 NASA had decided on a third method, called *lunar orbit rendezvous*. Only one rocket would be needed to launch both the Apollo spacecraft and a smaller vehicle called the lunar module, which would separate to make the eventual lunar landing. The idea had first been proposed by the Russian rocket theorist Yuri Vasilievich Kondratyuk (1897–1942) about 1916.

Apollo hardware. The three-man Apollo command module is conical in shape, with a maximum width of 12 feet 10 inches (3.9 m) and total height of 10 feet 7 inches (3.2 m). It has five windows: one in the hatch, two at the sides, and two facing forward for observing rendezvous operations. At the command module's apex is a tunnel and removable hatch, through which astronauts can crawl when docked with another craft. Around this tunnel are stored the parachutes for return to Earth and inflatable bags to right the spacecraft after splashdown.

The command module's heat-insulated walls vary in thickness from 0.7 inch (1.8 cm) at the apex to 2.7 inches (6.9 cm) at the base, which takes the full friction of reentry. The walls contain small gas jets for aligning the craft during reentry.

Inside the command module are three couches for the astronauts, an equipment bay for navigation, and stowage lockers for food and spacesuits (they wear the suits during launch, reentry, and other critical maneuvers such as docking). The couch armrests contain hand controls for guiding the spacecraft. Two astronauts can sleep in hammocks slung below the couches.

Above the spacecraft at launch is the launch escape tower, with small rockets powerful enough to pull the command module away to safety in the event of the launch vehicle's malfunctioning. The escape tower is jettisoned once the launcher's second stage has successfully ignited.

Behind the command module is the cylindrical service module, which supplies oxygen, water, and electric power to the command module. It houses the antennae used for high-quality communication with Earth. The service module is the same width as the command module, but over twice the length—24 feet 5 inches (7.4 m). Its main feature is the large service propulsion system (SPS) engine, used to make major speed changes. Around the walls of the service module are four sets of rockets, called the reaction control

system (RCS). These each contain four small thrusters for various maneuvers and for orienting the combined command and service modules (called CSM for short) in space. On the later Apollo missions, previously empty bays of the service module were fitted with equipment to photograph and study the Moon from orbit. Tape and film from these experiments had to be retrieved by spacewalk, because the service module is jettisoned before reentry.

The overall weight of the CSM is about 65,000 lb. (29,500 kg). When bound for the Moon the CSM was launched by the Saturn 5 rocket; but the CSM can be put into Earth orbit by the smaller Saturn 1B (see SATURN ROCKETS), as was done in the Apollo 7 flight. For lunar landings, the LUNAR MODULE was stored at the top of the Saturn 5's third stage, below the CSM, and was extracted by a docking maneuver once underway.

The missions. The spacecraft and its rockets were checked initially in experimental unmanned flights. The first manned flight of the CSM was to have been on February 21, 1967, with astronauts Virgil I. Grissom, Edward H. White, and Roger B. Chaffee. But during a simulated countdown on January 27 a fire in the capsule killed all three men. The Apollo program was delayed for 18 months while the accident was investigated and the Apollo craft redesigned to eliminate fire hazards and to introduce a quick-opening hatch; the rocket that was to have launched the ill-fated crew was instead used for the Apollo 5 test flight.

When it finally flew, the first manned Apollo, numbered Apollo 7, was described as "101 percent successful," accomplishing more tests in orbit than the original flight plan had envisaged. The next Apollo was the first manned launching by a Saturn 5; it put the first three men in orbit around the Moon. Apollo 9, which stayed in Earth orbit, was a trial flight of the spidery lunar module; and Apollo 10 was a complete dress rehearsal for the first lunar landing, including a trip in the lunar module to within 9 miles (14.4 km) of the lunar surface.

In the event, Apollo 11 met Kennedy's deadline with five months to spare; Apollo 12 repeated the achievement before the decade of the sixties was out. Although Apollo 11 had landed a few miles off target, Apollo 12 came down within a few hundred yards of the long-dead Surveyor 3 pathfinder probe, parts of which the astronauts brought back for examination on Earth. Apollo 13 was a near disaster; an explosion in an oxygen tank in the service module cut out nearly all the spacecraft's electrical supply, and the astronauts had to rely on the lunar module's electrical power and rockets to return them safely to Earth.

Apollo 14 marked a trend toward longer Moon walks and more scientific experiments—the astronauts set up explosive charges and operated a thumper device for seismic sounding of the Moon's outer layer. Apollo 15 saw the introduction of an electrically powered LUNAR ROVER vehicle, which allowed the astronauts to drive around and collect samples, while from orbit a mapping camera in the service module photographed the Moon in detail, and a small automatic satellite was released to send back data on the Moon's gravitational field.

The Apollo 16 mission made the only landing in the lunar highlands, to collect material for comparison with the lowland plains samples gathered by the other

Apollo flights. All records were broken by the final lunar mission, Apollo 17, which included the longest Moon walk and the greatest amount of Moon samples ever returned. One discovery was orange soil, at first thought to be signs of volcanism but later identified as glass beads produced by melting from the heat of a meteorite impact.

The Apollo program ceased with Apollo 17, at least in part because of cuts in the NASA budget and the lack of challenge from the Soviet Union, which had run into difficulties with its own manned space activities. Probably no further Americans will visit the Moon until the 1980s at the soonest. However, some remaining Apollo hardware has been used in exploits such as SKYLAB and the joint APOLLO–SOYUZ TEST PROJECT.

An important feature of the Apollo program was the continuous TV coverage. On the later flights, the color camera, remote controlled from Earth, was called the 'third astronaut' as it televised the lunar activities. This view shows the camera mounted on the Lunar Rover and the parabolic antenna which transmitted the picture to Earth. Apollo 17 astronaut Harrison Schmitt works beside a large lunar boulder in the background.

Unmanned Apollo test flights

Mission	Launch date	Results
Apollo 1	February 26, 1966	Suborbital test launch with Saturn 1B; CSM not sent into orbit
Apollo 2	July 5, 1966	Orbital test of Saturn 1B; second stage sent into orbit, but no spacecraft carried
Apollo 3	August 25, 1966	Suborbital test with Saturn 1B; CSM survived high-speed reentry
Apollo 4	November 9, 1967	First launch of a Saturn 5 rocket, and first test of Apollo CSM in orbit. The unmanned command module was blasted back into the atmosphere at the same speed as reentry from the Moon, to test its heat shield
Apollo 5	January 22, 1968	Unmanned test flight of the lunar module on its own; launched into Earth orbit by Saturn 1B
Apollo 6	April 4, 1968	Second test flight of Saturn 5, launching CSM into Earth orbit

Manned Apollo missions

Mission/Crew	Launch date/ Splashdown	Results
Apollo 7 *Walter M. Schirra Donn F. Eisele R. Walter Cunningham*	October 11/ October 22, 1968	Earth-orbital test flight of three-man CSM; launch by Saturn 1B
Apollo 8 *Frank Borman James A. Lovell William A. Anders*	December 21/ December 27, 1968	First manned Saturn 5 launch; 10 orbits of Moon in Apollo CSM
Apollo 9 *James A. McDivitt David R. Scott Russell L. Schweickart*	March 3/ March 13, 1969	Earth orbital test of CSM and lunar module; launched by Saturn 5
Apollo 10 *Thomas P. Stafford John W. Young Eugene A. Cernan*	May 18/ May 26, 1969	Full dress rehearsal of Moon landing, in lunar orbit; 2½ days spent orbiting Moon
Apollo 11 *Neil A. Armstrong Michael Collins Edwin E. Aldrin*	July 16/ July 24, 1969	Armstrong and Aldrin make first manned lunar landing, on July 20 in Sea of Tranquillity
Apollo 12 *Charles Conrad Richard F. Gordon Alan L. Bean*	November 14/ November 24, 1969	Conrad and Bean land on November 19 in Ocean of Storms
Apollo 13 *James A. Lovell John L. Swigert Fred W. Haise*	April 11/ April 17, 1970	Landing attempt canceled after explosion in oxygen tank damages spacecraft
Apollo 14 *Alan B. Shepard Stuart A. Roosa Edgar D. Mitchell*	January 31/ February 9, 1971	Shepard and Mitchell land on February 5 in Frau Mauro region of Moon
Apollo 15 *David R. Scott Alfred M. Worden James B. Irwin*	July 26/ August 7, 1971	Scott and Irwin land on July 30 at Hadley rill. First use of lunar roving vehicle
Apollo 16 *John W. Young Thomas K. Mattingly Charles M. Duke*	April 16/ April 27, 1972	Young and Duke land in Descartes highlands on April 21
Apollo 17 *Eugene A. Cernan Ronald E. Evans Harrison H. Schmitt*	December 7/ December 19, 1972	Cernan and Schmitt land on December 11 at the edge of the Sea of Serenity, near the crater Littrow

Apollonius of Perga (c.262–c.190 B.C.)
Greek mathematician credited with introducing the idea of EPICYCLES and eccentrics (off-center circles) to help explain the erratic motion of the planets in the sky. Such geometrical devices were subsequently prominent in the cosmological system of PTOLEMY. In his work on geometry, Apollonius studied and named

the ellipse, the parabola, and the hyperbola. Although he did not realize it, the ellipse is in fact the shape that describes the actual orbits of the planets.

Apollo–Soyuz Test Project (ASTP)
Joint mission in Earth orbit made by an American Apollo craft and a Soviet Soyuz spaceship in July 1975. The two were launched into similar orbits, and the Apollo caught up and docked with Soyuz. Apollo carried a special docking tunnel to serve as an airlock, enabling astronauts to move between the two spacecraft. One purpose of the mission was to test the docking tunnel for possible rescue of stranded spacemen, although the main importance of ASTP was political.

The Apollo-Soyuz flight in July 1975 offered the opportunity for the U.S. and Soviet Union to make their spacecraft mutually compatible. Here, the 6.6 ton Soviet Soyuz craft, with solar panels extended, is seen from the Apollo Command Module. Soyuz is 24 feet (7 m) in length.

apparent magnitude
The brightness with which a star appears as seen from Earth. It differs from intrinsic brightness, or ABSOLUTE MAGNITUDE.

apsides
The points in an orbit at which the bodies are either at their closest or at their farthest, such as perihelion and aphelion for objects in the solar system, or periastron and apastron for objects in orbit around another star. The line joining the two points is called the *line of apsides*.

Apus (the bird of paradise)
A faint constellation in the south polar region of the sky. It was named by Johann Bayer in 1603.

Aquarius (the water carrier)
One of the constellations of the zodiac, lying in the equatorial region of the sky. The Sun passes through Aquarius in the last half of February and first half of March. Aquarius is prominent in the evening sky during the northern hemisphere autumn. Although there are no particularly brilliant stars in Aquarius, it contains the globular cluster M2 as well as other objects of interest to astronomers, including the planetary nebula NGC 7009, called the Saturn nebula because of its resemblance to that planet.

Aquila (the eagle)
A prominent constellation lying on the celestial equator in part of the Milky Way. It is visible in the evening sky during the northern hemisphere summer. The brightest star in Aquila is ALTAIR. With the stars Vega in Lyra and Deneb in Cygnus, Altair forms the so-called summer triangle. There are several interesting variable stars in Aquila, and a number of novae have occurred within its boundaries.

Ara (the altar)
A southern hemisphere constellation that lies in the Milky Way south of Scorpius. It once formed part of the zodiac. There are no particularly bright stars.

arc minute, arc second
Small units of angular measure, 1/60 and 1/360 of a degree, respectively, used to record angular diameter or angular separation. They are not to be confused with minutes and seconds of time.

Arcturus
Red giant star in the constellation Boötes, the herdsman. Arcturus, also called Alpha Boötis, is the fourth-brightest star in the sky, of magnitude − 0.06. Its radius is 23 times that of the Sun, and its luminosity is equal to about 100 Suns. Arcturus is 36 light-years away.

Aerial view of the Arecibo 1,000-foot fixed radio telescope. The receiver is mounted in the box under the curved horizontal truss, 870 feet above the wire mesh surface. It is reached from the control room, **foreground**, by a walkway stretching from the base of the nearest tower.

Arecibo Observatory
Site of the world's largest single radio astronomy dish, 1,000 feet (305 m) in diameter. The dish is slung like a hammock in a natural hollow between hills in northern Puerto Rico near the city of Arecibo. The instrument, completed in 1963 and resurfaced in 1974, is used for ionospheric studies and radar mapping of the Moon and planets as well as for radio astronomy. The telescope cannot be steered, but the sky from 43° north to 6° south can be covered by moving feed aerials, which are supported above the dish from three towers around its rim.

Argelander, Friedrich Wilhelm August (1799–1875)
German astronomer responsible for the compilation of the BONNER DURCHMUSTERUNG, one of the world's greatest star catalogs, which lists positions and magnitudes for stars down to magnitude 9.5. The catalog sacrificed accuracy for quantity, however, and in 1867 Argelander proposed a new and greater project to the Astronomische Gesellschaft (German Astronomical Society). This resulted in the highly accurate AGK series of catalogs. In 1837 Argelander had published a study of the observed motions of stars (their PROPER MOTIONS), which confirmed the 1783 suggestion of William HERSCHEL that the Sun is moving through space toward a point in the constellation Hercules; this point is termed the SOLAR APEX. Argelander also introduced the so-called *step method* of visually estimating the brightness of a star by comparing it with stars of known magnitude. This technique in effect founded the serious study of variable stars, only a few of which were known at the

time. The step method is still used by amateur variable-star observers today.

Ariel
A satellite of Uranus, discovered in 1851 by the English amateur astronomer William Lassell. Ariel orbits Uranus once every $2\frac{1}{2}$ days at a distance of 119,000 miles (192,000 km). Its diameter is about 900 miles (1,500 km).

Ariel satellites
Series of British scientific satellites, launched by the United States. Ariel 1 was the first international satellite; like Ariel 2 it was built in the U.S., with only the experiment packages supplied by Britain. From Ariel 3 onward the entire spacecraft has been British-built.

Satellite	Launch date	Remarks
Ariel 1	April 26, 1962	Studies of the ionosphere and solar radiation
Ariel 2	March 27, 1964	Atmospheric and radio-astronomy studies
Ariel 3	May 5, 1967	Ionospheric and radio-astronomy studies
Ariel 4	December 11, 1971	Radio-astronomy experiments
Ariel 5	October 15, 1974	X-ray studies; mapped the sky at X-ray wavelengths and examined specific sources in detail

Aries (the ram)
A constellation of the zodiac, visible in the night sky during the northern autumn, and lying between Taurus and Pisces. Aries is not a particularly bright group; its importance has been in its location. About 2,000 years ago the Sun was in Aries as it passed from south to north of the celestial equator, an event called the vernal, or spring, EQUINOX. This point became known as the first point of Aries. However, through a slight wobble of the Earth's axis called PRECESSION, the location of the vernal equinox no longer lies in Aries; nevertheless, the term "first point of Aries" is still occasionally used.

Aristarchus of Samos (flourished 280–264 B.C.)
Greek astronomer who calculated the relative sizes of the Earth, Moon, and Sun. He attempted first to measure the relative distances from Earth of the Moon and Sun. With his results he compared the sizes of the Sun and Moon with that of the Earth. (The actual size of the Earth was later measured by ERATOSTHENES.) Although Aristarchus' value for the Sun's size was too small, he was still able to show it is far larger than the Earth. He therefore proposed that the objects in the sky should revolve around the Sun, rather than the Earth as was generally supposed. His contemporaries did not agree with his logic, and the concept of a heliocentric Universe was rejected until the time of COPERNICUS.

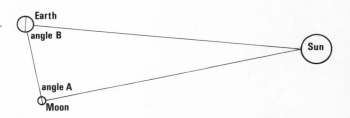

1   How Aristarchus sought to measure the distance of the Sun and Moon. At exactly half Moon, the angle between the Sun, Moon and Earth (angle A) is a right angle (90°). To complete the triangle, Aristarchus measured the angle between the Sun, Earth and Moon (angle B) at the same instant; he could then calculate the relative lengths of the sides Earth–Moon and Earth–Sun. He found that the Sun is about 19 times farther away than the Moon. The true figure is actually about 400 times; the error arose because angle B is difficult to measure with accuracy.

2   Both the Sun and Moon appear the same size in the sky; but since the Sun is more distant than the Moon, it must be correspondingly larger in size. The Sun and Moon can then be put on a scale drawing, as shown here, using Aristarchus' value for their relative distances. Observations of lunar eclipses showed the Greeks that the size of the Earth's shadow at the distance of the Moon was $2\frac{2}{3}$ the Moon's diameter. Putting this shadow onto the drawing allowed Aristarchus to add the Earth to scale; he then read off the sizes and distances of the Sun and Moon in terms of the Earth's size. He made the Moon's diameter $\frac{1}{3}$ that of the Earth, which is close to the true value of about $\frac{1}{4}$. He thought the Sun was seven times the Earth's diameter, which is too small because of the error in his estimate of the Sun's distance.

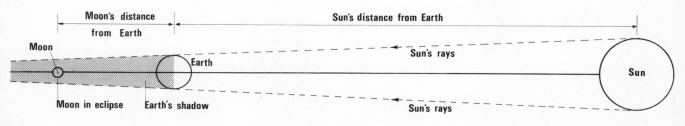

Aristotle (384–322 B.C.)

One of the major Greek philosophers, whose authority was cited by astronomers and physicists for almost 2,000 years. Aristotle took the system of heavenly spheres that EUDOXUS had originated to explain the motions of celestial bodies, and developed it by adding another 22 spheres, whose purpose was to counteract the effect of one planet's spheres on another's, so that neighboring planets did not affect each other's motion. Although Eudoxus probably regarded his spheres as only a mathematical device, Aristotle believed that they were real, solid orbs of crystal. Though Aristotle's scheme of the heavens was superseded by that of PTOLEMY, the notion of solid crystalline spheres persisted until the time of Tycho BRAHE.

Aristotle rejected the idea that the Earth spins on its axis or that it orbits the Sun, and his authority was used to dismiss all such notions until the time of GALILEO. Yet Aristotle's astronomical views were not all misleading: he argued convincingly for the spherical nature of the Earth, pointing out the way in which stars rose above or dipped below the horizon as one traveled north or south, and also noting the curved edge of the Earth's shadow at lunar eclipses. His extensive writings contained both his own investigations and summaries of existing Greek thought; thus we find from Aristotle that the Greeks knew the Moon was spherical, and also understood the true reason for the Moon's phases.

This armillary sphere, made in 1554 by Volpaja, enabled the Earth to be set to any position beneath the celestial sphere. The wide band is the Zodiac, which runs between the rings marking the two tropics.

armillary sphere

An obsolete instrument once used to convert star positions directly from the observed AZIMUTH and ALTITUDE to the celestial equivalents (RIGHT ASCENSION and DECLINATION). It consisted of a number of interlocking circles representing the celestial equator, the ecliptic, and other GREAT CIRCLES such as the horizon, so that a star, once sighted, could have its celestial coordinates read off on the scales. Early armillaries, as used by Hipparchus, Ptolemy, and Arab astronomers, worked in a vertical position; they achieved their highest development when Tycho BRAHE installed several, up to 9½ feet (3 m) in diameter, at his observatory (1576–1601). They were mounted on polar axes and in some cases were graduated in units of 10 arc seconds. Armillary spheres were never, however, as accurate as quadrants, which obtained direct readings of altitude and azimuth and could be constructed more rigidly, and they were little used after Tycho's time.

Armstrong, Neil Alden (b. 1930)

The first man to walk on the Moon, he was commander of the historic Apollo 11 mission of July 1969. Armstrong, a test pilot, was selected as an astronaut in 1962. He was command pilot of the Gemini 8 mission in March 1966, which performed the world's first space docking. However, a jammed thruster caused the spacecraft to spin dangerously, and Armstrong and copilot David Scott had to undock and make an emergency landing.

As commander of Apollo 11, Armstrong piloted the lunar module *Eagle* to a safe landing on the Moon's Sea of Tranquillity. As he stepped on to the Moon's surface on July 21, 1969, Armstrong said: "That's one small step for a man, one giant leap for mankind." Armstrong spent a total of 2¼ hours on the lunar surface. He collected the first lunar rock, and helped Edwin Aldrin fill boxes with a total of 50 lb. (22 kg) of samples. He also put out scientific experiments and took photographs of the surroundings. After the Apollo 11 mission Armstrong left the astronaut corps to become professor of aerospace engineering at the University of Cincinnati.

ascending node
See NODES.

asteroid

A small rocky body moving in an elliptical orbit around the Sun; asteroids are also called *minor planets*. Thousands exist, ranging from about 600 miles (1,000 km) in diameter down to small boulders. They are too small to have atmospheres and their Sun-scorched and meteorite-pounded dusty surfaces would be hostile to all life-forms. Asteroids frequently collide among themselves, producing debris that occasionally falls to Earth as METEORITES.

Discovery of asteroids. Most asteroids orbit the Sun in the gap between Mars and Jupiter, where the German astronomer Johann BODE predicted in 1772 that an undiscovered planet would lie. A group calling themselves the "celestial police" were organized by the German astronomer Baron Franz Xaver von Zach (1754–1832) to search for this new body. On January 1, 1801, the Italian astronomer Giuseppe PIAZZI, not one of the group, discovered the first asteroid, Ceres. But the celestial police soon discovered other bodies moving in similar orbits. The German astronomer Wilhelm Olbers found the asteroids Pallas and Vesta in 1802 and 1807; Juno was discovered by the German astronomer Karl Ludwig Harding (1765–1834) in 1804. These four asteroids are considerably brighter than the rest of the swarm. A fifth, Astraea, was not discovered until 1845; it was found by the German Karl Ludwig Hencke (1793–1866), who two years later found Hebe. Discoveries then followed rapidly, with 100 asteroids found by 1868, 200 by 1879, and 300 by 1890. This early hunting involved laborious visual searching; it was revolutionized in December 1891 when Max Wolf discovered an asteroid by the use of photography. The large numbers of faint asteroids discovered with this technique presented serious problems of identification until modern computers allowed all known asteroids to be readily followed. Minor-planet centers have now been established at Cincinnati and at Leningrad.

About 2,500 asteroids now have precisely known orbits, of which about 1,850 have been officially numbered. Some 2,000 more were sighted once but were lost through lack of further observations. The full number of minor planets is immense. Surveys suggest that there are about 50,000 asteroids which become brighter than, or equal to, magnitude 21. The total mass of the asteroid belt is roughly 0.0004 the Earth's mass, or about $2\frac{1}{2}$ million million million tons; this is twice the mass of the largest aseroid, Ceres.

Asteroid groups. The asteroids can be conveniently classed in three groups according to their orbits: the main belt, the Apollo and Amor group, and the Trojan asteroids. The main belt comprises over 95 percent of known asteroids, moving in slightly elliptical paths between the orbits of Mars and Jupiter. Those on the well-defined inner edge of the belt lie firmly in the plane of the solar system, but moving outward the orbits become more scattered. The Apollo and Amor group have highly elliptical, more sharply inclined orbits; at their farthest points (aphelion) they are in the main belt, but their closest points to the Sun (perihelion) lie among the inner planets. Thus, bodies such as 433 Eros, 1566 Icarus, 1620 Geographos, and 1685 Toro, can come quite close to the Earth. The Trojan asteroids move along in the same orbit as the planet Jupiter.

The asteroids of the main belt orbit the Sun with periods between about 2 and 6 years, compared with 11.86 years for Jupiter. The distribution is not smooth, and there are distinct groups and gaps (the so-called Kirkwood gaps). The gaps occur at precise fractions of Jupiter's period, where orbits would be repeatedly perturbed by Jupiter's gravitational pull. The principal breaks occur at periods of 4.0, 4.8, and 5.9 years, corresponding to ratios of 3:1, 5:2, and 2:1 of Jupiter's period, and there are many minor gaps.

In the 1920s, the Japanese astronomer Hirayama Seiji (1874–1943) found certain orbital groupings among minor planets. About 25 percent of asteroids can be placed in 10 such Hirayama families, each caused when two major asteroids collided in the past. Some families are compact, while others are very dispersed, suggesting that collisions have occurred at widely different times in solar system history.

Nature of asteroids. The name *asteroid* (star-like) was coined 150 years ago by William Herschel; even with today's giant telescopes few asteroids appear

Characteristics of Representative Asteroids

Asteroid	Period years	Orbit Perihelion distance a.u.	inclination degrees	diameter km	rotation period hours	Albedo %	Surface Type	Comment
1566 Icarus	1.12	0.187	23.0	1.4	2.3	17.8	Silicaceous	Apollo group
1620 Geographos	1.40	0.83	13.0	0.8 × 1.0 × 4.0	5.2	20.9	Silicaceous	Apollo group
433 Eros	1.76	1.13	10.8	7 × 16 × 35	5.3	14.2	Silicaceous	Amor group
8 Flora	3.27	1.86	5.9	150	13.6	16.8	Silicaceous	main belt
18 Melpomene	3.48	1.80	10.2	141	14.0	14.0	Silicaceous	main belt
4 Vesta	3.63	2.55	7.1	503	10.7	26.4	Basaltic?	main belt
192 Nausikaa	3.72	1.82	6.9	92	—	20.0	Silicaceous	main belt
887 Alinda	4.00	1.16	9.0	4.4	—	14.6	Silicaceous	Amor group
5 Astraea	4.13	2.10	5.3	116	16.8	17.7	Silicaceous	main belt
15 Eunomia	4.30	2.15	11.8	270	6.1	15.5	Silicaceous	main belt
3 Juno	4.36	1.99	13.0	226	7.2	19.0	Silicaceous	main belt
324 Bamberga	4.39	1.78	11.3	230	8.0	3.6	Carbonaceous	main belt
1 Ceres	4.60	2.55	10.6	955	9.1	7.2	Carbonaceous	main belt
2 Pallas	4.61	2.11	34.8	558	10.5	8.7	Carbonaceous	main belt
624 Hektor	11.55	5.02	18.0	50 × 50 × 210	6.9	2.8	Unusual—not identified	Trojan group
944 Hidalgo	13.70	1.98	43.1	—	—	—	—	Unique body

as anything other than star-like points, making diameter measurement very difficult. In 1971, however, two new methods of determining asteroid diameters were developed. The radiometric method involves the careful measurement of a minor planet's brightness at visual and infrared wavelengths; these figures indicate whether an object of given brightness is small and reflective, or larger but darker, from which can be calculated both visual ALBEDO (reflectivity) and asteroid diameter. The polarimetric method involves a study of the polarization of sunlight reflected by an asteroid to yield the asteroid's albedo which, with the visual brightness, reveals the diameter. This new work gives diameters which are considerably larger than early estimates, and which indicate that asteroids are very dark bodies. Albedos vary from the exceptionally low 2.8 percent for 624 Hektor and 747 Winchester, to as much as 26.4 percent for 4 Vesta, with the majority in the range 7 to 18 percent. (In comparison, carbon black has an albedo of about 3 percent, the Moon—a very dark body—6.7 percent, and the planet Venus 76 percent.)

The amount of light reflected by asteroids generally varies regularly by up to 1.5 magnitudes. This suggests they are rotating, irregular-shaped bodies, probably fragments from asteroid collisions. A typical example is the Martian moon PHOBOS which, like its companion DEIMOS, is believed to be a captured asteroid. In some cases, the deduced shapes are extreme: 433 Eros, for instance, is a spindle-shaped body, while the Trojan asteroid 624 Hektor is either a tumbling cylinder made of iron—a stone cylinder would disintegrate—or a twin system with two close stony bodies revolving around their common center of gravity. The three largest asteroids and a handful of smaller ones show little light variation and are probably near-spherical in shape.

Light reflected from asteroids at different wavelengths can be compared with powdered minerals in the laboratory to give broad indications of asteroid surface compositions. About 90 percent of asteroids so far studied have surfaces resembling either carbonaceous meteorites or normal stony or stony-iron meteorites (see METEORITE). Most stony asteroids have diameters of about 60 to 125 miles (100–200 km), while carbonaceous bodies are common in larger and smaller sizes and tend to predominate in the outer parts of the belt. 16 Psyche and a group of smaller asteroids have featureless spectra suggesting nickel-iron bodies, and 4 Vesta is unique both in its high albedo and in a spectral curve which resembles that of ACHONDRITE meteorites.

Origin of the asteroids. Wilhelm Olbers first suggested that the asteroids might be the fragments of a disrupted planet. But the general opinion now is that, at the origin of the solar system, a cloud between two and four times the Earth's distance from the Sun formed an original family of perhaps 15 to 30 asteroids with diameters from 60 to 600 miles (100–1,000 km). Collisions among these led to the highly fragmented system now visible, with only Ceres, Pallas, Vesta, and possibly a handful of others of the original bodies surviving intact. Asteroids of the Apollo and Amor group have such eccentric and highly tilted orbits that they may not have been formed in the same fashion at all, and may instead be the remains of extinct short-period comets.

astigmatism

The image fault produced by a lens or mirror in which different diameters focus rays into different planes. *Axial astigmatism* (the formation of an astigmatic image when the instrument is pointed directly at the object) is caused by a deformed optical component, and the image of a star at the best focus appears as a cross or lozenge. Off-axis or *abaxial astigmatism* arises when rays strike an uncorrected optical system at an angle, so limiting the diameter of the usable field. Off-axis astigmatism increases with decreasing focal ratio of the mirror or object glass. It is rarely important visually because of the restricted field of view, but is particularly objectionable in wide-field photographic instruments. The SCHMIDT TELESCOPE was designed primarily to solve the problem.

astrolabe

An ancient astronomical instrument, used for observing the altitudes of stars. In its simplest form it consisted of a gradual disk hanging vertically, with a sight or *alidade* which was turned to point to the object. The Arabs adopted the instrument, graduating the *tablet* or face to show the altitudes of various bright stars at different times, so that the local time could be determined. By substituting different tablets, the instrument could be used in different latitudes. In this form it was widely employed as a navigational aid until the more accurate sextant supplanted it in the 18th century.

A Persian astrolabe dating from AD 1221, without the alidade or sighting rule. Gears are used on this version to operate a calendar movement on the reverse side.

Its modern form, the *prismatic astrolabe,* is a very accurate instrument designed to measure the instant at which a given star comes to an altitude of 60°. This enables the latitude of the observing site to be determined. The instrument works by comparing the direct position of the star in the sky with that observed reflected from a mercury bath. Both images are fed, via a 60° prism, into a horizontal telescope. The observer then sees two stars apparently moving toward each other. When they merge, or come alongside, the star's altitude is exactly 60°. If the latitude of the site is already known, the prismatic astrolabe can be used to determine local time.

astrometry

The measurement of positions of objects in the sky; it is thus also termed *positional astronomy.* Astrometry is the basic work of national institutions such as the U.S. Naval Observatory and the Royal Greenwich Observatory. The accurate star positions they provide are needed by navigators and surveyors as well as by astronomers. The observatories also monitor the Earth's rate of spin, by reference to the passage of stars across the sky, to provide a basis for our time system. Although regular time signals are now derived from atomic clocks (see ATOMIC TIME), star checks are still needed to ensure that the signals do not get out of step with the Earth's slightly varying rate of rotation.

Foundations. Modern astrometry grew from the foundations laid by star catalogers such as HIPPARCHUS, Tycho BRAHE, James BRADLEY, Friedrich BESSEL, Friedrich ARGELANDER, and Sir George AIRY. Their work is distinct from that of observational astronomers such as Galileo and his successors, who were concerned with the visible features of celestial objects, or that of astrophysicists, who study the actual nature of celestial objects.

The first astronomical observations were entirely astrometrical; they led to the establishment of the first crude calendars, and eventually to a true understanding of the movement of the planets around the Sun. Since the invention of the telescope, astrometric observations have become more accurate; they have revealed that stars are not truly still in the sky as the ancient astronomers had believed. Therefore, even if all star positions were accurately tabulated, they would need to be resurveyed after a decade or so to take account of changes. Actually, the resurveying is continuous, and it is marked by an ever-increasing standard of accuracy.

Star movements. The celestial coordinates of an object drift continually because of the wobbles in the Earth's motion called PRECESSION and NUTATION. In addition, stars orbit the Galaxy at various speeds; they thus show varying shifts, known as their PROPER MOTION, the study of which aids astronomers in understanding the Galaxy's structure.

Nearby stars show a shift in position, called PARALLAX, as we view them from different places in the Earth's orbit. The parallax shift is greater the nearer the star is to us, and provides an important direct measurement of star distances; the study of parallax is therefore a major concern of astrometrists. Eventually, far better parallax measures will be possible by placing two telescopes wide apart in the solar system, giving a far longer baseline for observations

than the diameter of the Earth's orbit to which we are currently restricted.

Our most reliable information on the masses of stars is deduced from the orbits of visual binary stars, which are another important astrometric target. Astrometrists can also detect companion objects, too faint to be seen directly, from the slight wobble they produce in the proper motions of some stars. This technique has revealed the existence of a planetary system round at least one nearby star (see BARNARD'S STAR).

New developments. Modern radio astronomy instruments such as APERTURE SYNTHESIS telescopes and giant INTERFEROMETERS produce positional results far more precise than optical observations. To correlate the two systems, astronomers must now carefully measure the positions of known optical counterparts of radio sources, such as quasars, pulsars, and radio galaxies. In the 1980s, astrometry will be vastly improved by telescopes in orbit, such as the LARGE SPACE TELESCOPE. From space, the blurring effect of the Earth's turbulent atmosphere is avoided, which will allow positional measurements far more accurate than any obtainable from the ground.

Astronomer Royal

Honorary post in British astronomy, originally associated with the directorship of the Royal Greenwich Observatory. When King Charles II founded the Royal Observatory in 1675 he appointed John Flamsteed his first "astronomical observator." In 1971 the post of astronomer royal was separated from the Royal Observatory's directorship; the first man to be appointed astronomer royal under this new arrangement was the Cambridge radio astronomer Professor Sir Martin Ryle.

A separate post of Astronomer Royal for Scotland, created in 1834, is attached to the directorship of the Royal Observatory, Edinburgh. From 1791 to 1921 there was a post of Royal Astronomer for Ireland.

astronomical unit (a.u.)

The average distance between the Earth and Sun, equivalent to 92,955,832 miles (149,597,910 km). Distances within the solar system are frequently expressed in astronomical units. There are several ways of establishing its precise value, the modern method being to measure by radar the distance of another body in the solar system, usually the planet Venus. The time taken for a radar echo to return from Venus reveals its distance from Earth with great accuracy. The relative distances of the planets from the Sun are known from their orbital periods (see KEPLER'S LAWS), as if on a scale map. Measuring the separation between two planets reveals the scale, and allows the actual distances of the planets from the Sun to be worked out. The distance between the Earth and a space probe in orbit around the Sun, as revealed by tracking data, can provide a useful check on the scale of the solar system. An earlier means of calculating the distance between two objects in the solar system involved measuring the parallax of minor planets that pass close to the Earth. This method was used in 1931 when the minor planet Eros passed about 0.15 a.u. (14 million miles, 22½ million km) from Earth. But the new radar techniques give far better results, accurate to within a few miles.

astronomy, history of

Astronomy is probably man's oldest natural science. The passing year was marked by the progress of the seasons, and the Sun's altitude above the horizon indicated the time of day. From these simple calculations of time and position astronomy grew.

The earliest written astronomical records come from the fertile zone at low northerly latitudes that includes the flood-plains of the Nile in Egypt, the Tigris and Euphrates in Mesopotamia, and similar regions in India and China, and later Central America. In these areas, large-scale agriculture and trade developed as early as 4000 B.C.

Time-keeping was essential, in order to determine the correct season for planting and harvesting, and the appropriate days for religious festivals. The construction of a good working calendar was thus the major task facing the earliest astronomers. The length of the Moon's cycle of phases and the length of the year bear no simple relationship to one another: the Moon's cycle is just over $29\frac{1}{2}$ days, while the solar year lasts approximately 365 days (about 11 days more than 12 lunar months). The easiest solution seems to have been adopted—an extra month when the calendar grew too far out of step with the crops. From more detailed observations of the Moon, the Babylonians discovered by about 300 B.C. that almost a complete number of lunar months fitted into both 8 and 19 solar years; this latter relationship was rediscovered by the Greeks, and is known as the METONIC CYCLE.

Practical requirements fostered the development of astronomy. More accurate observations revealed that both the Sun and Moon moved across specific constellations during a year, so that the stars visible at a particular time of night are characteristic of the season. This resulted in the regular observation of stars and star groups and enabled a further check to be made on the calendar. For example, the Egyptians noted the correlation between the Nile floods and the *heliacal rising* of the star Sirius at Memphis (the star's first appearance in the morning sky before the Sun). Though the early astronomers gained considerable knowledge of the movements of celestial bodies, they gave little thought to the physical explanation of these motions. Not until the age of the Greek philosophers were theories sought to explain the phenomena.

Greek astronomy. While the Babylonians and Egyptians laid the foundations, it was left to the Greeks to consolidate early astronomy. Greek writers claimed to have learned from the Babylonians, but the difference between Greek and Babylonian astronomy was that while Babylonia was a single school of thought the Greeks lived in smaller city states, each producing its own scholars. Greek writings reveal a wide variety of notions, not all of them eventually accepted. The greatest names are those of ARISTOTLE, HIPPARCHUS, and PTOLEMY, and it was their views which eventually held sway.

Aristotle taught that the Earth is a sphere—indeed its diameter was later measured by ERATOSTHENES. His ideas about the structure of the Universe were less accurate, however: he fixed the Sun, Moon, and stars to crystal spheres which rotated about the Earth. In order to account for the RETROGRADE MOTIONS of some of the planets (an apparent backward movement

actually caused by the Earth overtaking them), a total of 55 spheres were necessary.

The greatest contributions to ancient astronomy came from one man—Hipparchus. His accurate measurements of the planets and his stellar catalog were unrivaled. Some three centuries later, Ptolemy took Hipparchus' work as the basis of his "Greater Collection," which eventually became known as the *Almagest*. Although Ptolemy's own observations have been questioned, it was through his work that astronomy was carried through to the Middle Ages. His geocentric theory of the Universe, with its numerous EPICYCLES and DEFERENTS, held sway for over 1,000 years.

After the decline of Greco-Roman civilization, the world's scientific knowledge was preserved in the hands of the Arabs. To them we owe many of our present star names. They refined the science of measurement, but few real theoretical advances can be attributed to them.

The revolution. When COPERNICUS published in 1543 details of his new *heliocentric* theory of the solar system, with the Sun, rather than the Earth, at the center of the planets' orbits, it did not immediately revolutionize astronomy. The classical Greek notions were too well established for it to be regarded as anything other than an aid to the calculation of planetary positions—and it seemed only slightly better than the accepted Ptolemaic *geocentric* system, in which the planets moved around the Earth. For 50 years, the situation simmered. But slowly the new ideas of Copernicus began to gain popularity, and the Church eventually had to take steps to suppress theories contrary to the classical view of the Universe.

To establish the new view of the Universe, it took the courage of GALILEO to report on the evidence of his own eyes that Aristotle was wrong, and the genius of KEPLER to refine the crude Copernican theory. The change began with the work of the Danish astronomer Tycho BRAHE, who realized the importance of good observations if any theory was to be confirmed. His observations of the Sun, Moon, and planets over many years were the most accurate that could be made without a telescope. It was fortunate that Tycho chose Kepler as his assistant, for on Tycho's death, Kepler brought his mathematical skill to bear on the new data. Eventually, he concluded that the Earth and planets *must* move around the Sun, in ellipses rather than in circles; and he went on to derive his famous laws of planetary motion.

With the invention of the telescope, Galileo in 1609 attacked Aristotle's and Ptolemy's ideas from another direction. He saw spots on the Sun, innumerable stars in the Milky Way—and moons moving around Jupiter, showing that there were bodies in the solar system of which Aristotle had not dreamed. His stand against Aristotelian dogma made him a natural champion of the Copernican theory.

While in Italy the Church was trying to suppress Galileo and his followers, the scientists of Protestant northern Europe were able to say and teach more or less what they thought. Thus it was men such as Christiaan HUYGENS, Edmond HALLEY, and above all Isaac NEWTON, who were eventually to break through the barriers which had existed for 2,000 years.

Newton and telescopes. Newton was not an astronomer. But in his theories of gravitation and of motion, and in his observations of the nature of light, he did more for astronomy than many an observer. Newton's laws, combined with those of Kepler, gave a virtually complete description of the movements of the planets, a description still valid today.

For a hundred years after Newton's invention of the reflecting telescope, little use was made of its advantages. Great observers such as CASSINI, FLAMSTEED, and Halley, continued to use simple refracting telescopes for their charting and measuring of heavenly bodies. In 1758 John DOLLOND introduced the achromatic lens, which gave improved, color-free images, and telescopic astronomy could begin in earnest. Yet the man who more than anyone else begun the study of the stars themselves, William HERSCHEL, used reflecting telescopes.

His discovery of Uranus and his subsequent studies of stars and nebulae were entirely due to the excellent telescopes which he constructed himself. Other astronomers still preferred to use refractors, but the high quality of objective lenses by men such as FRAUNHOFER could not really offset the light-gathering power of large reflectors.

During the 19th century, ASTROPHYSICS began to emerge in its own right. The studies of the spectrum began by Fraunhofer mark the first attempts at an understanding of the nature of the stars, work which was refined and extended by such men as William HUGGINS and Norman LOCKYER.

The many advances that have taken place during this century are due not so much to the work of individuals as to the improvement in observational techniques, such as the combination of large reflectors and photography, and to the introduction of powerful computers. The trend is toward larger and more advanced instrumentation used by teams of scientists rather than by individuals.

astrophotography

The use of photography has almost completely supplanted visual observation in modern astronomical research. Early emulsions (c.1840) were so slow that several minutes' exposure was required to record the Moon, a feat first performed by John W. DRAPER. The first star to be photographed was Vega, by W. C. BOND in 1850, with an exposure of 100 seconds using the 15-inch (38-cm) refractor of Harvard College Observatory. The British amateur astronomer Warren de la Rue (1815–1889) invented the spectroheliograph in 1858 and began a daily series of photographs of the Sun. Since the time of these pioneers, photographic sensitivity has increased by about 50,000 times, and practically all new telescopes are designed as photographic instruments.

Techniques. Except for small wide-angle patrol cameras, pure refracting systems are rarely used in astrophotography, because no combination of lenses can bring all colors to exactly the same focus. Mirrors are almost universally employed, often in combination with weak correcting lenses to give a flat focal plane or one of wider angle. Most new large telescopes are CASSEGRAIN TELESCOPES, modified to a form known as Ritchey-Chrétien, which is free from COMA, an aberration that reduces the useful field of view. The SCHMIDT TELESCOPE is used where fields of several degrees have to be covered on one plate.

Research has produced greatly improved photographic materials. Silver bromide, the main light-sensitive agent, is affected most strongly by violet light, but treating the emulsion with suitable dyes has extended its sensitivity across the visible spectrum and into the infrared. Special sensitizing has produced emulsions suitable for the long-exposure photographs needed for faint astronomical objects. Glass plates are preferred to film because of their permanence and suitability for accurate measurement.

Applications. Astrophotography can supply many kinds of information. The recording of faint stars and galaxies requires large aperture and long exposure times (up to perhaps 5 hours). Positional work for stars, minor planets, or comets, needs a large plate scale and a short exposure time. The determination of star brightnesses (PHOTOMETRY) is done on plates covering a wide field of view, to provide comparison stars of known brightness. Multicolor photometry requires filters of different color.

A number of specialized instruments have been developed for solar photography. In general, fixed telescopes of very long focal length fed by a moving mirror system (see COELOSTAT) are used, giving a direct solar image several inches across; whole-disk records now date back over a century. The invention of the CORONAGRAPH makes it possible to photograph the brighter regions of the solar atmosphere in the absence of an eclipse.

Photography has had little, or only very recent impact, in the observation of close double stars and the resolution of fine planetary detail. Both are highly dependent upon atmospheric steadiness, and the best planetary photographs have been obtained from high-altitude observatories such as the PIC DU MIDI OBSERVATORY in France.

The application of photography to astronomy has increased the size of the usefully-observable Universe a hundredfold, and the vital field of spectrum analysis would be practically impossible without it because of the faintness of the images. In general, the photographic plate can record stars up to three magnitudes fainter than the eye could detect using the same instrument; without it we should know little about the nature of our Galaxy, stellar evolution, or the construction of the Universe.

astrophysics

The application of physics to the study of the Universe, as distinct from the measurement of accurate celestial positions (ASTROMETRY). The Universe acts as a giant natural laboratory in which we can observe matter under conditions far more extreme than exist on Earth. Astrophysics developed from the new science of SPECTROSCOPY—the analysis of starlight—which 100 years ago showed that the stars are balls of incandescent gas, of similar physical nature to our Sun. Under the guidance of pioneers such as George Ellery HALE, astrophysics rapidly became a major part of observational astronomy. More recently, the study of matter outside the Earth has been increasingly important to our general understanding of physics, especially processes involving strong gravitational fields and nuclear energy.

Modern solar studies. Astrophysicists are still baffled by many mysteries involving our own Sun. The broad outline of a typical star's structure was

explained 50 years ago by Sir Arthur EDDINGTON, and in 1939 Hans BETHE outlined the nuclear processes that make stars glow. But details of the Sun's behavior do not fit the pattern. One problem is that, although theories predict that particles called neutrinos should be produced in great quantities by the nuclear reactions that power the Sun, no neutrinos have yet been detected. Either the theories are wrong, or our Sun is not a typical star.

Even the origin and behavior of major features such as the Sun's magnetic field and SUNSPOTS are poorly understood. Other phenomena, such as the approximately 11-year variation in the Sun's activity (the SOLAR CYCLE) have yet to be explained by astrophysics. Research into these and other problems of solar structure is now aided by observations from satellites and space probes that study the Sun at wavelengths blocked by the Earth's atmosphere from astronomers on the ground, and which measure the so-called SOLAR WIND of particles streaming from the Sun as it gusts with variations in solar activity.

Other stars. The great wealth of detail we can observe on the Sun is not visible on other stars because of their vast distances. They are therefore treated en masse rather than individually, by combining measurements from many different stars to give average values of characteristics such as brightness, temperature, and size. Astrophysicists study how stars are born, evolve, and die; their composition and its variation between different generations (see POPULATIONS, STELLAR); why they occur in galaxies; and why galaxies themselves cluster together.

Spectroscopy not only reveals details of the temperature and composition of stars; sometimes,

The Southern Cross rises to the left of the moonlit dome of the Anglo-Australian Telescope in this time-exposure showing star trails circling the South Celestial Pole— the simplest type of astrophotography.

Spectra are essential in interpreting the light from stars. A thin objective prism mounted in front of a telescope elongates all the star images into spectra, and in this star field in Cygnus normal stars (showing vertical dark absorption lines on a bright background) can be readily distinguished from the very hot Wolf-Rayet stars (with bright emission lines).

where direct observation shows one star, the spectroscope reveals light from two stars (see SPECTROSCOPIC BINARY). The motions of such twin stars in orbit around each other produce changes in the spectral lines (caused by the DOPPLER EFFECT), from which astronomers can deduce the period of the orbits and the masses of the stars.

Information on the masses, diameters, and temperatures of stars provides the raw data for creating mathematical models of stars in computers. The computer calculates how temperature, pressure, and density vary from the center of a given star to its surface. The model can even be allowed to evolve in the computer, revealing how stars age. The predictions of the star's surface temperature and composition can then be checked with spectroscopic observations of real stars.

Between the stars. Astrophysics is now probing the tenuous matter between the stars. Radio and infrared astronomy reveal many complex molecules in the cool dust and gas clouds of our Galaxy (see INTERSTELLAR MOLECULES). These include so-called organic molecules which are based on carbon, the backbone of life. If such complicated molecules exist in space, they must have been present on Earth from the time it formed, providing building blocks for the assembly of the first life forms (see LIFE IN THE UNIVERSE). Also between the stars are heavy, high-speed particles called COSMIC RAYS, ejected by the violent processes that astronomers observe in the Universe.

Deep space. Astrophysics provides the basic observations to help determine the origin of the Universe. In the 1920s, the American astronomer Edwin HUBBLE proved that there are galaxies outside our own, and that their motions show the Universe is expanding. This led to the idea that the Universe began in a BIG-BANG explosion and has been evolving ever since. Astrophysicists looking deep into space (and therefore far back into time) have found confirming evidence that the Universe looked different in the past: long ago there were objects called QUASARS, emitting vast amounts of radiation at all wavelengths. The advent of RADIO ASTRONOMY has given astrophysicists a new window on the Universe, revealing what appear to be vast explosions in distant galaxies (see RADIO GALAXY). Such powerful events strain the limits of known physical processes. Radio observations have also revealed the so-called

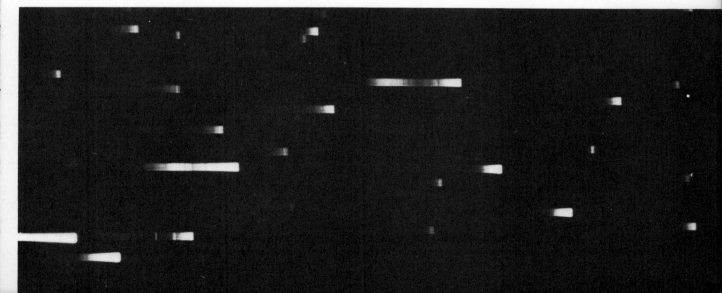

Atlas rocket

BACKGROUND RADIATION, believed to be a relic of the big-bang fireball in which the Universe formed (see also COSMOLOGY).

The new astrophysics. Artificial satellites are opening up new windows on the Universe, particularly for ultraviolet, X-ray, and gamma-ray emissions, which come from intensely hot, energetic objects—quasars in deepest space and small, compact stars in our own Galaxy, possibly including black holes. The objects involved have high gravitational fields; and this provides new information about the way gravity can be converted into energy. The observations help test Einstein's general theory of RELATIVITY, which is concerned with the nature of gravity and provides the best overall theory for describing the Universe. The theory also predicts that gravity waves should ripple through the Universe; one aim of astrophysics in the next decade will be to detect these waves, using sensitive laboratory detectors which oscillate as the wave passes, and observations from space platforms monitored by laser beams.

Back to earth. Astrophysics embraces both old knowledge and new discoveries. The laws of CELESTIAL MECHANICS, which describe the motions of planets in gravitational fields, are today also applied to the motion of spacecraft and the orbits of double stars that include X-ray sources (see X-RAY ASTRONOMY).

But to most people, developments in understanding the Earth and its immediate environment are likely to be the outstanding product of astrophysics. Space-probe studies of other planets in our solar system, and of the relations between the Sun and Earth, are building a new and better picture of our planet. The discovery of the Earth's MAGNETOSPHERE (the extension of the magnetic field into space), and how it interacts with the SOLAR WIND, may explain changes in the atmosphere that affect weather patterns and climate. Studies of the whole solar system's movement through space, and its interaction with interstellar material, could explain such dramatic changes in the Earth's history as the ice ages.

An Atlas rocket with Centaur second stage, launching the Surveyor 4 Moon probe in 1967. Atlas by itself could only be used for orbital launches; with Centaur or Agena second stages, payloads could be sent to the Moon.

Atlas rocket

Major American space launcher, modified from an intercontinental military missile which was first successfully test flown on December 17, 1957. The Atlas was used to launch America's first men into orbit in the MERCURY PROJECT. With upper stages such as AGENA and CENTAUR, Atlas rockets are still used to launch U.S. satellites and space probes. The Atlas has three engines which ignite at lift-off. The two outer engines, each developing 185,000-lb. (84,000-kg) thrust, fall away after $2\frac{1}{2}$ minutes, while the central sustainer engine of 60,000-lb. (27,300-kg) thrust continues to burn; because of this arrangement, the Atlas is often termed a $1\frac{1}{2}$-stage vehicle. The main body of Atlas has a diameter of 10 feet (3 m). Its height and weight vary according to mission, but average around $82\frac{1}{2}$ feet (25 m) and 300,000 lb. (140,000 kg).

atomic time

A means of timekeeping based on oscillations of an atom of cesium, a soft, silver-white metal. The cesium atom oscillates over 9 billion times a second. Counting these oscillations gives the basis of an extremely accurate method of timekeeping, which has now been adopted worldwide. The first such clock was built in 1955; modern versions are accurate to one second in many tens of thousands of years, and accuracies of one second in a million years may eventually be possible. The rotation of the Earth was the previous basis of time signals. But since the start of 1972, the world's legal time system has been based on the readings of an atomic clock. Atomic clocks give far greater accuracy than any previous method of timekeeping, and have revealed that the rotation of the Earth is not constant. In addition to occasional fluctuations, there is an overall tendency for the Earth's rotation to slow down. Extra seconds, called leap seconds, must occasionally be introduced into atomic time to keep it roughly in step with the Earth's rotation. One or two leap seconds are introduced each year.

ATS satellites

Series of satellites testing satellite applications of new scientific and technical developments in communications, observation, weather and

Earth-resource monitoring, and navigation. The initials ATS stand for Applications Technology Satellite. The original ATS program grew out of experience with the SYNCOM communications satellites, and the emphasis has been on communication, including the relaying of television pictures, and experiments in educational broadcasts and medical consultation.

A new generation of ATS satellites started with ATS-6. Using a 30-foot (9.1-m) antenna and far more powerful transmitters than any previous communications satellite, it relays education and health instruction broadcasts to small ground stations; it was first located over the United States to provide links with remote areas in the Rockies, Appalachians, and Alaska. ATS-6 was later moved to a point over Kenya, where it broadcast to several thousand stations in India. ATS-6 also acts as a link between other satellites and the Earth, as for example in relaying data and pictures from the Apollo-Soyuz joint flight in 1975; this is possible because of the much higher orbit of ATS.

a.u.
Abbreviation for ASTRONOMICAL UNIT, a yardstick for expressing distances within the solar system.

Auriga (the charioteer)
Prominent constellation in the northern hemisphere of the sky. It is visible in the night sky during the northern winter, lying on the edge of the Milky Way between Gemini and Taurus. The brightest star in Auriga is the yellow-colored CAPELLA. Auriga contains two star clusters visible in binoculars, and several double stars of interest to astronomers. One of these is the yellow supergiant Epsilon Aurigae, 3,300 light-years distant, which is orbited by a dark companion that eclipses it, causing the star to vary from magnitude 3.4 to 4.5 every 27 years. This companion may be a black hole, or it may be surrounded by a cloud of dark dust or gas—possibly a forming planetary system. A spectacular double star in Auriga, Zeta Aurigae, consists of a red giant orbited by a bright blue star.

aurora
An atmospheric phenomenon caused by the impact on the upper atmosphere of atomic particles from the Sun that become trapped in the Earth's magnetic field. Aurorae are most commonly seen near the magnetic poles. In the northern hemisphere they are called the *aurora borealis,* in the southern hemisphere the *aurora australis.* Their popular name is the northern lights or southern lights.

Aurorae appear as diffuse light areas in the sky, often red, yellow, or green in color. Auroral displays can be shaped like arches or folds of drapery, and aurorae often shimmer and move to produce a truly astounding spectacle.

Aurorae occur at heights above 60 miles (100 km). They are more frequent during maximum periods of the Sun's 11-year cycle of activity, when particles are sprayed out from solar flares. Norwegian scientists were foremost in studying aurorae in the early part of this century. They found that the colors of aurorae were mostly produced by molecules of oxygen and nitrogen in the upper atmosphere as they were bombarded by particles from the Sun, chiefly electrons. One recent discovery has been that aurorae in the north and south hemispheres are linked, with atomic particles oscillating back and forth between the hemispheres along the Earth's magnetic lines of force. Numerous satellite and rocket experiments have been made to learn more about the aurorae, but not all the processes occurring are fully understood yet.

azimuth
The bearing of an object around the observer's horizon; it is the coordinate at right angles to the object's ALTITUDE. Azimuth is measured in degrees from north to a point on the horizon directly beneath the object.

B

Baade, Walter (1893–1960)
German-born American astronomer who discovered there are two essentially different populations of stars, and whose determination of a new distance for the Andromeda galaxy led to a doubling in the scale of intergalactic distances. Using the 100-inch (254-cm) Mount Wilson telescope during the early 1940s, Baade made the first photographs of individual stars in the central region of the Andromeda galaxy. He found that they were reddish in color, in contrast to the blue-white stars found by Edwin HUBBLE in the spiral arms. Baade termed these population II and I respectively (see POPULATIONS, STELLAR). After World War II, Baade used the new 200-inch (508-cm) telescope to study CEPHEID VARIABLE stars in both populations, and in 1952 found that the population I (spiral-arm) stars were brighter—and, thus, farther away—than had previously been supposed. Technically, each population of Cepheids thus had its own PERIOD-LUMINOSITY RELATION. The distance to the Andromeda galaxy had been calculated from observations of population I stars, but using a period-luminosity law that was in fact applicable only to population II stars. The calculated distance was therefore wrong. Baade's new value showed that the Andromeda galaxy was roughly 2 million light-years away, over twice as far as had been thought. Baade's discovery of stellar populations advanced the study of stellar evolution, and his recalibration of intergalactic distances substantially altered concepts about the size and age of the Universe.

Babcock, Harold Delos (1882–1968)
American astrophysicist who, with his son Horace Welcome Babcock (b. 1912), in 1951 developed an instrument for recording magnetic fields on the Sun. This instrument, the solar magnetograph, operates by measuring the effect of magnetic fields on spectral lines (the ZEEMAN EFFECT). The solar magnetograph has revealed regular reversals of the Sun's magnetic polarity. In 1964, Horace Babcock was appointed director of the Mount Wilson and Palomar Observatories (now Hale Observatories). Horace Babcock is also noted for his discovery in 1946 and subsequent study of MAGNETIC STARS.

background radiation
Radiation coming from space at radio and far

infrared wavelengths, believed to be energy left over from the BIG-BANG origin of the Universe. Such radiation was independently predicted by the American physicists George GAMOW in 1948 and Robert DICKE in 1964, and first measured by the American physicists Arno Penzias (b. 1933) and Robert Woodrow Wilson (b. 1936) in 1965. Measurements at various radio and infrared wavelengths show that the radiation is consistent with that emitted by a source at a temperature of 2.7°K (−454.8°F), the actual temperature of the Universe. Theory predicts that the Universe would have reached its present temperature as it expanded after the initial big-bang explosion. The radiation was actually emitted, at a temperature of several thousand degrees, but cooled as the density of matter in space thinned out and could not absorb all the energy from the big-bang explosion. This probably happened when the Universe was between 100,000 and 1 million years old (one hundred-thousandth to one ten-thousandth its present age). The exact figure is uncertain because the present density of the Universe is unknown.

Baikonur

Official name for the Soviet space launching site at TYURATAM, near the Aral Sea about 1,300 miles (2,100 km) southeast of Moscow.

Baily, Francis (1774–1844)

English astronomer, best known for describing the phenomena known as BAILY'S BEADS, which appear at a solar eclipse. Baily was a founder of the Royal Astronomical Society. He became widely known for his painstaking analysis and editing of many classic star catalogs, notably those of Ptolemy, Tycho Brahe, Hevelius, and Flamsteed, and for improvements in the

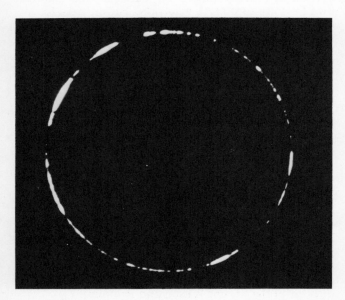

A complete ring of Baily's beads, photographed during the near-total solar eclipse of May 1966. These "beads" are seen during eclipses which occur when the apparent size of the Moon is very slightly less than that of the Sun, allowing the edge of the brilliant solar disk to shine through irregularities in the Moon's rugged limb.

Nautical Almanac. In 1842 he produced a figure for the density of the Earth—5.66 times that of water—which is within 3 percent of the modern value.

Baily's beads

Beads of light at a solar eclipse caused by the Sun's light shining between mountain peaks along the edge, or limb, of the Moon. As the Moon moves across the Sun's face just before totality, the last sliver of sunlight is broken into shining drops by the jagged lunar mountains. As the eclipse ends the same effect occurs at the opposite edge of the Moon. The phenomenon lasts for only a few seconds. Often one bead shines much brighter than others to give what is termed the "diamond ring" effect. At an annular eclipse, the Moon's disk does not quite cover all the Sun, and a complete ring of Baily's beads is seen. The phenomenon is named after the English astronomer Francis BAILY who described it at an annular eclipse in 1836.

Barnard, Edward Emerson (1857–1923)

American astronomer, the leading visual observer of his day and a pioneer of astrophotography. His most famous discovery was of a rapidly moving star close to the Sun, now called BARNARD'S STAR. In 1892 he discovered Jupiter's fifth moon, the first to be found since Galileo's time. In 1889 he began photographing the Milky Way in detail, and showed that its dark regions are not true gaps but are caused by areas of obscuring gas; his *Photographic Atlas of Selected Areas of the Milky Way* was published posthumously in 1927. Barnard also discovered 16 comets. Observing with the Lick 36-inch (91-cm) refractor in 1892–1893, he discovered craters on Mars, though his observations were not published at the time.

Barnard's star

The second-closest star to the Sun, 5.9 light-years away. Barnard's star, lying in the constellation Ophiuchus, is a faint red dwarf of visual magnitude 9.5. The star is named for E. E. BARNARD, who discovered in 1916 that it has the fastest PROPER MOTION of any star across the sky, amounting to 10.3 arc seconds a year. In about 180 years, therefore, it moves across an area of sky equal to the apparent diameter of the Moon. According to Peter VAN DE KAMP, Barnard's star is orbited by at least two planets about the size of Jupiter. Barnard's star is approaching us at 67 miles (108 km) per second. In about 10,000 years' time it will be the closest star of all, 3.8 light-years distant, of magnitude 8.6 and with a proper motion of over 25 arc seconds a year.

Bayer, Johann (1572–1625)

German astronomer who in 1603 published *Uranometria,* the first atlas covering the entire sky. It charts over 2,000 stars, 1,000 of them taken from the observations of Tycho Brahe. Observations of the southern heavens came from the work of Dutch navigators. Bayer created 12 new constellations—Apus, Chamaeleon, Dorado, Grus, Hydrus, Indus, Musca, Pavo, Phoenix, Triangulum Australe, Tucana, and Volans—to accommodate many of the southern stars. He assigned Greek and Latin letters to stars in approximate order of their brightness, a system still used today.

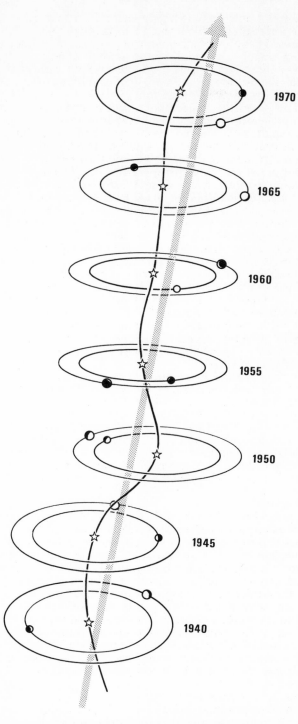

The orbits of two planets around Barnard's star are deduced from the wobble of the star's observed motion. After Peter van de Kamp.

B.D.
Abbreviation for the BONNER DURCHMUSTERUNG star catalog.

Bean, Alan Lavern (b. 1932)
American astronaut who landed on the Moon with the Apollo 12 mission, and commanded the second crew to visit the Skylab space station. Bean made his first space flight in November 1969 as lunar module pilot on Apollo 12. With his commander Charles Conrad, Bean made two Moon walks, spending a total of seven hours on the lunar surface. In 1973 he led the second three-man Skylab crew on what was then a record-breaking stay in space of 59 days.

Belyaev, Pavel Ivanovich (1925–1970)
Soviet cosmonaut, commander of the Voskhod 2 mission in March 1965 during which Alexei Leonov crawled from the craft to perform the first walk in space. Belyaev had to control the spacecraft's return to Earth manually when the automatic landing system failed. Belyaev died from peritonitis following an operation for stomach ulcers.

Bessel, Friedrich Wilhelm (1784–1846)
German astronomer who made the first measurement of a star's distance. Bessel became director of Prussia's Königsberg Observatory in 1813. There he measured the exact positions of over 50,000 stars. The PROPER MOTIONS of a number of these stars indicated they were relatively close to the Sun. Bessel chose a star with large proper motion, 61 Cygni, and from careful observation found in 1838 that it showed a slight yearly shift in position as observed from different parts of the Earth's orbit (see PARALLAX). The amount of shift revealed the star's distance, which Bessel calculated for 61 Cygni at 10.3 light-years; the modern value is 11.2. In 1844 Bessel announced the presence of unseen companions to Sirius and Procyon, based on an analysis of their irregular proper motions. The companion stars later were discovered visually and are now known to be white dwarfs. Bessel's method is still used to reveal faint companions of stars.

Betelgeuse
A red giant star which marks the right shoulder of Orion; it is also known as Alpha Orionis. Betelgeuse is so large that the orbit of Mars would fit inside it. Because of its huge size it is slightly unstable, and varies irregularly in size and light output. Its magnitude changes from about 0.4 to 1.3 and its size from 300 to 420 times that of the Sun. Betelgeuse is roughly 650 light-years from the Sun.

Bethe, Hans Albrecht (b. 1906)
German-born American physicist who showed in 1938 how stars could generate energy in their interiors by the fusion of hydrogen atoms into helium, through what is called the CARBON-NITROGEN-OXYGEN CYCLE. With the American physicist Charles Louis Critchfield (b. 1910), Bethe also showed that a direct reaction between hydrogen nuclei to produce helium, now called the PROTON-PROTON CHAIN, could account for energy generation in cooler stars. The carbon-nitrogen-oxygen cycle is now believed to operate inside stars hotter than the Sun, while the proton-proton chain predominates in the Sun and cooler stars. Bethe's was the first detailed explanation of stellar energy processes, and it earned him the 1967 Nobel Prize in physics.

Big bang
The supposed event which started the Universe's expansion. Measurements of the current expansion of

the Universe (see HUBBLE'S CONSTANT) suggest that the big bang occurred about 18 billion years ago. At the time of the big bang, all matter in the Universe must have been squeezed into a hot, dense, primeval fireball. Conditions in the fireball were so extreme that not even individual atoms could have existed there. The cosmological expansion was much faster immediately after the big bang explosion than it is now. In this rapid expansion, many complicated processes occurred in a very short time. After about a millionth of a second from the beginning of the expansion, the fireball temperature was on the order of a trillion degrees—hotter than the interiors of the hottest stars, and not until after a few minutes was the temperature low enough for atomic nuclei to form. Calculations show that only helium would have formed in abundance; the rest of the matter remained as uncombined protons, later to form hydrogen atoms. Observations reveal that about 25 percent of the present Universe is helium, presumably mostly of primeval origin (see also ELEMENTS, ABUNDANCE OF).

Within a few hundred seconds the temperature had dropped too low for further nuclear processes to take place, but not until about a million years later did the fireball cool sufficiently for atoms to form. When this happened the material became transparent, allowing the fading glow of the cooling primeval fireball to remain visible ever since. It is now detected as a feeble BACKGROUND RADIATION at a mere 2.7° above ABSOLUTE ZERO.

Much interest attaches to the very earliest stages of the big bang. Before a millionth of a second, little is known of the physical condition of the fireball, but very general considerations of gravity, which governs the explosive motion of the early Universe, lead us to predict that the density of matter would have been limitless, forming a so-called *singularity* at the beginning of the expansion. If so, this moment represents the real "creation" of the Universe. In the future, physics may allow us to probe further back into the earliest stages of the expansion, even to the moment when the structure of space-time was breaking up; possibly this will change our present simple picture of the big bang.

Big Dipper
See URSA MAJOR.

binary star
See DOUBLE STAR.

black hole
A theoretical object whose gravitational pull is so strong that nothing can escape, not even light. Black holes were first suggested in 1795 by the French mathematician Pierre LAPLACE, on the basis of Newton's theory of gravity. In 1939 the American physicist J. Robert Oppenheimer (1904–1967) showed that similar objects are also predicted by Einstein's theory of gravity (the general theory of RELATIVITY).

If a star is greatly compressed—to a few miles or kilometers across in the case of our Sun—its surface gravity would then be so great that nothing could prevent it from shrinking without limit. Such an event may happen in a SUPERNOVA explosion, when the core of the exploding star is tightly compressed. Inside a critical radius, known as the SCHWARZSCHILD RADIUS,

space and time become highly distorted. The effect is to imprison anything which falls through this radius. Therefore black holes can grow in size as they sweep up more material, but no amount of energy can enable a body inside the Schwarzschild radius to reach the outside again. Even light cannot escape; the collapsing star appears totally black.

The fate of the compressed object strains our comprehension. Theory predicts that in a few thousandths of a second the collapsing matter collides in the center of the star at a single mathematical point. The density of the star would then rise to infinity; the very fabric of space-time itself would become so violently curved and distorted that it could no longer exist within the black hole. No one knows what actually happens at the center, except that the star (or any observer who has fallen in after it) cannot reemerge through the Schwarzschild radius into the same Universe again. For during the fleeting destruction of the star, an infinite span of time will have elapsed in the outside world. The star has traveled, quite literally, out of our Universe, leaving only a "hole" in space—a black hole. Recently, it has been discovered that even this black hole slowly disappears. Subtle effects cause the hole to glow faintly. As time goes on, the temperature slowly rises, until eventually the hole simply disappears completely in a flash of radiation.

Although black holes cannot be observed directly, they may be detectable in orbit around other stars. In such a situation, matter from the visible star would be pulled into the black hole, heating up strongly and emitting X-rays as it falls in. Recently, satellites have discovered a number of X-ray sources in the direction of double-star systems. The X-ray source in Cygnus called Cyg X-1 may be just such a system containing a black hole (see X-RAY ASTRONOMY).

blue giant
A large, hot, and bright star. A typical blue giant has a temperature of 20,000°K (36,000°F), a radius 10 times that of the Sun and a brightness of 20,000 Suns. Blue giants can arise in two ways. They may be massive young stars that burn out very quickly, some or all exploding as SUPERNOVAE. The brightest stars of the PLEIADES cluster are blue giants of this type. Or they may be old stars at a late phase of their evolution, passing between RED GIANT and WHITE DWARF stages. This type of blue giant is found in globular clusters. Both types lie at the top right of the MAIN SEQUENCE of stars in the HERTZSPRUNG-RUSSELL DIAGRAM, a plot of star brightnesses against their color.

Bode, Johann Elert (1747–1826)
German astronomer, popularizer of the so-called BODE'S LAW, a formula that roughly describes the distances of the planets from the Sun out to Uranus. Bode was director of the Berlin Observatory, where in 1801 he produced *Uranographia,* the first successful attempt to chart all stars visible to the naked eye, and in which he also included the first systematic delineation of constellation boundaries.

Bode's law
A number series that roughly coincides with the average distance of the planets from the Sun outward to Uranus. It is named for the German astronomer

Johann Bode, who pointed out the relationship in the 1772 edition of his book *Introduction to the Study of the Starry Sky*. According to Bode, if the distance from the Sun to Saturn (then the outermost planet known) is taken as 100, then Mercury is separated from the Sun by 4 such parts. The distance to Venus is 4 + 3 = 7, to Earth 4 + 6 = 10, and Mars 4 + 12 = 16. Then at 4 + 24 = 28 comes a gap. Jupiter is 4 + 48 = 52 parts away, and Saturn 4 + 96 = 100.

When the planet Uranus was discovered in 1784, it seemed to fit this scheme, though not with perfect accuracy. Bode's law led to a search for the missing planet between Mars and Jupiter, resulting in the discovery of the asteroids. The law was also used as a starting point by astronomers in search of planets beyond Uranus. However, it clearly breaks down toward the edge of the solar system. According to modern views, the planets would have come naturally to take up station in orbits where they least perturbed each other; such an arrangement would spontaneously produce some kind of Bode's law progression of distances in any planetary system.

Bode's "law" is thus not really a law at all. What is more, it was first put forward in 1772 by the German mathematician Johann Daniel Titius (1729–1796), and is now often referred to as the Titius-Bode law.

bolide
An exploding or fragmenting bright meteor (see FIREBALL).

bolometer
A device for measuring radiation emitted by an object at all wavelengths, invented in 1878 by the American astronomer Samuel Pierpont Langley (1834–1906). Radiation falling on the bolometer's detector causes a rise in temperature, which changes the electrical resistance of a circuit; the amount of resistance change reveals the intensity of the incident radiation. The so-called *bolometric magnitude* of a star is its total brightness over all wavelengths; *bolometric correction* is the difference between the visual and bolometric magnitudes of a star.

Bond, William Cranch (1789–1859)
American astronomer, pioneer of the use of photography in astronomy, who in 1850 took the first photograph of a star, Vega. Bond was the founder and first director of the Harvard Observatory, and was succeeded by his son George Phillips Bond (1825–1865). Together the two mapped the sky photographically, studied comets, and discovered the eighth satellite of Saturn, named Hyperion, and Saturn's transparent inner ring, called the crepe ring. George P. Bond explored the potential of photography for measuring star brightnesses and recording positions. In 1857, he took the first photograph of a double star, Mizar.

Bondi, Sir Hermann (b. 1919)
Austrian-born British mathematician, who with Thomas GOLD proposed the STEADY STATE theory of cosmology in 1948, maintaining that the appearance of the Universe has been constant throughout time. Bondi has done major work on the theory of gravitational radiation and other areas of astrophysics, notably the internal structure of normal stars like the Sun. From 1967 to 1971 he was director-general of the European Space Research Organization, ESRO; he then became chief scientific adviser to the British Ministry of Defence.

Bonner Durchmusterung
One of the world's major star catalogs, prepared by the German astronomer Friedrich ARGELANDER over a period of 25 years. The catalog was first published in 1859–62 and contained 324,198 stars down to about magnitude 9.5, from the north pole to $-2°$ in declination. The stars were plotted on charts published in 1863 the best of their time. Although not as accurate as later catalogs, the *Bonner Durchmusterung* retains its importance because of the large number of stars listed. Argelander's colleague Eduard Schönfeld (1828–1891) extended the catalog to $-23°$ in 1886, plotting 133,659 stars.

booster
Term often used to describe the first stage of a launch rocket, or sometimes the whole rocket. *Strap-on boosters* are subsidiary rockets that augment the thrust of the first stage at lift-off. They fall away as the rocket climbs. VOSTOK and TITAN III are examples of rockets that use strap-on boosters.

Boötes (the herdsman)
A large constellation in the northern hemisphere of the sky, best placed for observation in the spring. The brightest star in Boötes is the red giant ARCTURUS. Boötes is found at the end of the handle of the Plow (Ursa Major); it is shaped like a kite, with Arcturus at its tail. Boötes contains a number of interesting double stars, including Epsilon Boötis, also called Izar or Pulcherrima. Of visual magnitude 2.59, it is 230 light-years away; the main star is also a SPECTROSCOPIC BINARY.

Borman, Frank (b. 1928)
American astronaut, commander of the Apollo 8 mission, which made 10 orbits of the Moon on December 24–25, 1968, before returning to Earth. Borman, a qualified aeronautical engineer, became an astronaut in 1962. His first space flight was with copilot James Lovell in Gemini 7 during December 1965, when Gemini 6 joined them in orbit for the first rendezvous between two manned spacecraft. Borman and Lovell went on to break the duration record for space flight, becoming the first men to spend two weeks in space.

Boss, Lewis (1846–1912)
American astronomer, responsible for one of the best-known and most accurate modern star catalogs, the *Boss General Catalogue*. Lewis Boss died after publishing only a preliminary catalog containing 6,188 stars. But his son Benjamin Boss (1880–1970) completed the catalog, finally published in 1937, with 33,372 stars. From his work on star positions, Lewis Boss found in 1908 that the stars in the HYADES appear to be moving away from the Sun en masse. This common movement provides an important trigonometrical method of finding the distance of the cluster (see MOVING CLUSTER METHOD).

Bowen, Ira Sprague (1898–1973)

American astrophysicist, who made extensive and important studies of the spectra of nebulae, showing that mysterious green-colored emission lines which had been attributed to an unknown element called "nebulium" were actually caused by ionized oxygen and nitrogen atoms under the near-vacuum conditions of space (see FORBIDDEN LINES). Bowen became director of the Mount Wilson and Palomar Observatories (now Hale Observatories) in 1948. He was responsible for the installation of the 200-inch (508-cm) reflector, and designed many subsidiary pieces of equipment for use with large telescopes.

Bradley, James (1693–1762)

English astronomer, discoverer of the ABERRATION OF STARLIGHT and NUTATION. Bradley's uncle was James Pound (1669–1724), a clergyman and amateur astronomer who introduced him to the study of the stars. Bradley became professor of astronomy at Oxford in 1721 and devoted the rest of his life to science.

At the private observatory in Kew, near London, of Samuel Molyneux (1689–1728), a politician and amateur astronomer, Bradley began in 1725 to look for the effect of PARALLAX in star positions. Nearby stars should show a slight motion during the year because of the Earth's changing position in its orbit. Bradley concentrated on the star Gamma Draconis, which appeared nearly overhead. He soon found a shift, but it was too large and in the wrong direction to be parallax. Bradley realized the shift was due to the Earth's motion across the path of the incoming starlight, which slightly distorted the true position of the star. This unexpected effect, which he announced in 1728, is called the aberration of starlight. It was not what Bradley was looking for, but it was nonetheless the first direct observational proof of the Earth's motion around the Sun. From the amount of aberration Bradley calculated the speed of light at 183,000 miles (295,000 km) per second—very close to the modern value.

Bradley was made astronomer royal in 1742. At Greenwich his accurate observations confirmed another shift in the position of several stars, which could not be explained by aberration. In this case the entire Earth was nodding slightly in space; this effect is called nutation. Bradley believed it occurred because the Moon's orbit moves once around the Earth every 18.6 years. He extended his observations over two decades before he finally announced the effect and its cause in 1748.

Bradley continued to log accurate star positions throughout his time at Greenwich. His observations were later assembled into an important star catalog by the German astronomer Friedrich BESSEL.

Brahe, Tycho (1546–1601)

Danish astronomer, the greatest observer of the pretelescopic era. His painstaking records of planetary positions provided the German mathematician Johannes KEPLER with the raw data for his laws of planetary motion, which for the first time gave an accurate description of planetary orbits and finally established the modern view of the solar system.

Tycho, as he is usually called, was born into a noble family and had an adventurous upbringing. Although

The greatest observer of the pre-telescopic era, Tycho Brahe. His discoveries brought him wealth and fame during his lifetime, yet today we remember him for the observations which he bequeathed to his assistant— Johannes Kepler.

sent to study law, Tycho was so impressed by a solar eclipse in 1560 that his interest was turned to the study of the sky. All his time was spent reading about and observing the heavens, and he became dismayed when he realized from his observations that existing tables of planetary motion were seriously in error. He therefore resolved to prepare the most accurate observational records possible, and built instruments larger and more accurate than any before.

The quality of his work became clear in 1572, when an exploding star, or supernova, flared up in the sky. Tycho's observations showed that the supernova must lie far beyond the atmosphere and in the realm of the stars, which had always been considered perfect and unchanging. This was a major blow to astronomical tradition. Tycho's book on the star, *De nova stella,* made his reputation, which was confirmed in 1577 when he was able to show that a bright new comet also lay far beyond the Moon. Tycho proved that the comet of 1577 moved in an orbit that took it among the planets. This refuted the age-old view that the planets orbited the Earth on solid crystal spheres; the comet would have shattered those spheres had they existed. But Tycho could not bring himself to accept the alternative view of COPERNICUS that all objects orbited the Sun. Instead, he proposed his own system, in which the planets (and comets) orbited the Sun, which in turn went around the stationary Earth.

In 1576 the king of Denmark granted Tycho the island of Ven between Denmark and Sweden, where he set up a castle called Uraniborg and an observatory called Stjerneborg. This became the astronomical center of the world, over which Tycho reigned like a

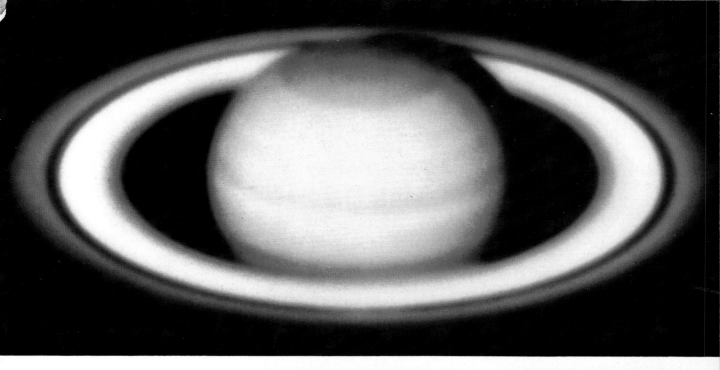

Above Saturn, photographed by Stephen Larson with the 61-inch (155-cm) Catalina telescope in Arizona. The different colors of the outer rings, A and B, show up well; ring C, inside them, is faint and can only be seen here where it hides the planet's surface.

Right Jupiter—the largest planet in the solar system, with its enigmatic clouds. The bright zones are the tops of clouds of ammonia gas, while the darker belts probably consist of gas descending to deeper regions of the planet. The exact nature of the Great Red Spot is still uncertain.

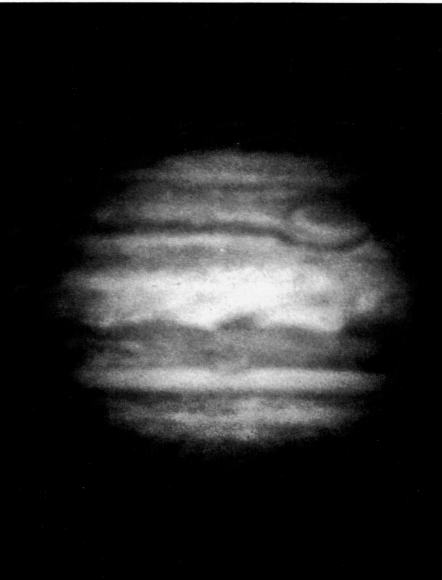

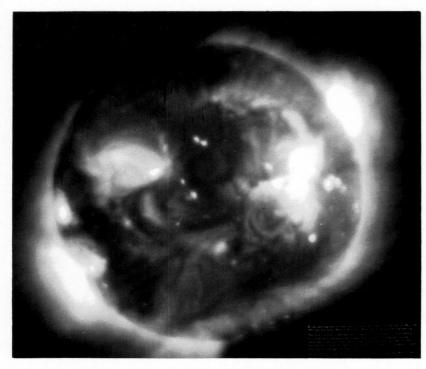

Left This X-ray view of the Sun taken by Skylab astronauts shows features previously unexpected. The X-ray corona consists of loops, probably following the lines of magnetic fields. Also shown here is a "coronal hole"—where the X-ray emission is restricted in some way.

Below These superimposed images of the Sun were taken by astronauts on a Skylab mission. The colors, produced in a laboratory reconstruction, are false: the brightness of the image at any point corresponds to its X-ray temperature. The spike at the top of the view extends for a million miles into space.

true king. At Ven he produced series of observations more accurate than any before, taking positions of stars and planets to the limit of naked-eye accuracy and measuring the length of the year to within one second. In Tycho's lifetime nearly every astronomical constant was revalued, and the calendar drastically reformed.

Tycho, however, was an exceptionally arrogant and argumentative man. In a student duel in 1565 he had lost the bridge of his nose, which he replaced with a bizarre metallic substitute. In 1588, the king of Denmark died; the new king was not prepared to tolerate him, and so after 20 years on Ven, Tycho left. Eventually, in 1599, he settled at Benatky Castle near Prague, where he had been appointed imperial mathematician to the Emperor Rudolf II. There he embarked on an analysis of the lunar and planetary observations he had collected at Ven. In 1600 he was joined by the young Johannes Kepler. The following year Tycho died, leaving his observations to Kepler in the hope that he would be able to prove the Tychonic theory of the heavens. Ironically, Kepler used Tycho's remarkable work to establish instead the Copernican theory of the solar system.

Brans-Dicke cosmology

A variant of Einstein's theory of gravity (general RELATIVITY) developed in the early 1960s by the American physicists Carl Henry Brans (b. 1935) and Robert H. DICKE, in which a new type of field produced by matter throughout the Universe plays an important role. The Brans-Dicke theory proposes that this field changes the geometry of space and time, thus affecting the motion of bodies in a gravitational field. One consequence is that the force of gravity would appear to grow progressively weaker as the Universe expands. Although the predicted weakening would be very slight (less than one part in ten billion each year), the accumulated effect over astronomical time scales would have a drastic effect on the solar system, making the Sun, for example, much brighter in the distant past. Observations of lunar and planetary motions have so far failed to provide convincing evidence for the weakening of gravity.

bremsstrahlung

Radiation emitted by high-speed particles, particularly electrons, as they are suddenly slowed down or scattered by atoms. The word *bremsstrahlung* is German for "braking radiation." As the particles are retarded, part of their energy is converted into electromagnetic radiation; the intensity and wavelength of the radiation depend on the rate of retardation. Bremsstrahlung is usually associated with the emission of radiation in stars and galaxies.

Bruno, Giordano (1548–1600)

Italian philosopher, a Dominican monk, who was an early supporter of COPERNICUS' theory that the Earth and other planets orbit the Sun. Bruno envisaged an infinite Universe, with endless suns each having their own retinue of planets. He believed that these planets might be populated, and pointed out that the inhabitants would think they were at the center of the Universe. He was thus among the first to show that our view of the Universe is purely relative. Bruno was burned at the stake by the Inquisition for his

opposition to the idea that any absolute truth could exist. Fear of the revival of similar heretical philosophies was also behind the Inquisition's persecution of GALILEO.

Burbidge, Geoffrey (b. 1925) and Eleanor Margaret (b. 1922)

British husband-and-wife team of astrophysicists, who with William Fowler and Fred HOYLE showed in 1956 that heavy elements are constantly being built up from light ones inside stars, and were therefore not all produced in the supposed big-bang explosion at the start of the Universe. Margaret Burbidge's observations of the rotation of galaxies led to the first accurate estimates of galactic masses. The Burbidges also discovered that QUASARS show several different red shifts in their spectral lines, indicating material is being ejected at high speed. From 1972 to 1973 Margaret Burbidge was director of the Royal Greenwich Observatory, the first woman to hold the post.

Bykovsky, Valery Fyodorovich (b. 1934)

Soviet cosmonaut who made the longest solo space flight in history. He spent almost five full days in the Vostok 5 spacecraft during June 1963. He was joined in orbit for three days by female cosmonaut Valentina Tereshkova in Vostok 6. Bykovsky was selected as a cosmonaut in 1960.

Byurakan Astrophysical Observatory

Major Soviet observatory, located near the city of Yerevan in Armenia. The observatory was founded in 1946 by Viktor AMBARTSUMIAN, its current director. Among other instruments are a Schmidt telescope with a 39-inch (100-cm) corrector lens and a 52-inch (132-cm) mirror. Its largest telescope is a 102-inch (260-cm) reflector.

C

Caelum (the chisel)

A small and faint constellation in the southern hemisphere of the sky near the foot of ERIDANUS. Its name was given by de Lacaille in the 1750s.

calendar

A timetable for reckoning days and months of the year. The first calendars go back to primitive times, when the Moon's $29\frac{1}{2}$-day cycle of phases was the guide. As man developed an agricultural way of life, he required a calendar linked to the seasons, which entailed using the Sun, not the Moon. The calendar now adopted throughout the world is solar. Some lunar calendars still exist for determining religious festivals, but since the year contains 365 days, and no $29\frac{1}{2}$-day period can be exactly divided into it, lunar and solar calendars remain separate. The basis of our present calendar is Roman, although the Romans originally used a lunar, and not a solar system of reckoning. In the eighth or seventh century B.C. a ten-month lunar calendar had been drawn up with months of 30 days and 31 days, giving a total of 304 days; the balance of 61 days was left as a gap. In due course extra days were added until the total reached 355. Yet this "Roman Republican Calendar" still did not keep in step with the seasons, and it was modified to a solar one by inserting an extra or "intercalary" month. This was added between February 23 and 24 once every two years; it contained either 27 or 28 days and, when it appeared, the remaining five days of February were omitted. The intercalary month thus gave an additional 22 or 23 days every two years which, added to the 355-day year, gave a solar calendar of 366¼ days.

The intercalary month was inserted on the order of the priesthood, but by the middle of the first century B.C. the calendar had reached a most unsatisfactory state. In consequence, Julius Caesar consulted the Greek astronomer Sosigenes (fl. 1st century B.C.), and decided to abandon a lunar calendar altogether. The year was taken as 365¼ days in length, a total of 90 days were intercalated, and the year beginning March 1, 45 B.C. became January 1 of the Julian calendar. To obtain the odd quarter day, one intercalary day was inserted every fourth year between February 23 and 24.

Various changes in the lengths of months were introduced into the Julian calendar; our present system arose after July was named for Julius Caesar, August after Augustus, and the arrangement of 30 and 31 days was adopted, with 28 days allotted for February. The Julian calendar contained no weeks, merely business and nonbusiness days. The seven-day week did not arrive until the fourth century A.D.

The adoption of 365.25 days for the length of the year is only an approximation; the precise figure is 365.242199 days. This small difference of only 11 minutes 14 seconds per annum mounts up to 1½ days in two centuries, and 7 days in 1,000 years. By 1545 the date of the vernal equinox was 10 days out, and the determination of Easter was affected. By 1572, the year of the accession of Pope Gregory XIII, a new and more correct calendar was prepared with the help of the Bavarian Jesuit astronomer Christopher Clavius (1537–1612). The year was taken as 365.2422 days; and it was agreed that 3 out of 4 century years should not be leap years; in practice a century year is not a leap year unless exactly divisible by 4. Thus 1700, 1800, and 1900 were not leap years, but 2000 will be. Most Roman Catholic countries adopted the Gregorian calendar in 1582; other countries followed much later, with Britain and America changing in 1752.

Callisto
Second-largest satellite of Jupiter, 3,100 miles (5,000 km) in diameter, discovered by Galileo in 1610. It orbits Jupiter every 16 days 16 hours 32 minutes at an average distance of 1,169,800 miles (1,882,600 km). Callisto's mass is about 1.5 times that of our Moon, but its density is only 1.65 times that of water; to account for this low density, one theoretical model suggests that the satellite has a muddy core surrounded by a deep, slushy mantle topped with a thin icy crust. Some dark markings have been charted on Callisto, so that at least part of its surface may be covered with dust.

Cambridge Radio Observatory
See MULLARD RADIO ASTRONOMY OBSERVATORY.

Camelopardalis (the giraffe)
A large but faint constellation near the north pole of the sky, introduced in 1624 by the German mathematician Jakob Bartsch (1600–1633), a son-in-law of Johannes Kepler. Its name is sometimes also written as *Camelopardus*.

Cancer (the crab)
A constellation of the zodiac, best visible in the northern winter. The Sun passes through Cancer from late July to mid-August. Cancer is the faintest of the zodiacal constellations, having no particularly bright stars. Its most famous feature is the star cluster called Praesepe, also known as the beehive, bearing the catalog numbers M44 and NGC 2632. Praesepe contains several hundred stars 572 light-years away.

Cancer, Tropic of
The farthest latitude north of the equator at which the Sun appears overhead during the year. The latitude of the Tropic of Cancer corresponds to the angle of the Earth's axial tilt—23° 26' 32" in 1976 (the angle changes slightly with time because of the perturbing effects of the Sun, Moon, and planets). The Sun is overhead at noon on the Tropic of Cancer at the summer SOLSTICE, around June 21 each year. About 2,000 years ago, the Sun lay in the constellation of Cancer at the summer solstice. But the drifting effect of the Earth's axis called PRECESSION has carried the summer solstice into Gemini, and before the end of the century it will have moved into Taurus. Its southern hemisphere equivalent is the Tropic of CAPRICORN.

Canes Venatici (the hunting dogs)
A constellation of the northern hemisphere of the sky lying between Boötes and Ursa Major, best seen in the northern hemisphere spring. It was introduced in 1536 by the German cartographer Petrus Apianus (Peter Bienewitz or Bennewitz; 1495–1552), but reached its present form on the star map of Johannes HEVELIUS. The brightest star, Alpha, was named Cor Caroli (Charles' heart) by Edmond Halley; it is a double star of total magnitude 2.8. Canes Venatici contains the famous Whirlpool galaxy, M51 (NGC 5194), the first galaxy in which spiral structure was recognized, by Lord ROSSE in 1845. The constellation also contains a bright globular cluster, M3 (NGC 5272).

Canis Major (the greater dog)
A prominent constellation in the southern hemisphere of the sky, containing the brightest star in the sky, SIRIUS. From the northern hemisphere, Canis Major is best seen during winter. The star Epsilon, also known as Adhara, is 18 times the Sun's diameter, of magnitude 1.50; it has a companion of magnitude 8.1.

Canis Minor (the lesser dog)
A small constellation lying below Gemini in the equatorial region of the sky, best seen during the northern hemisphere winter; with Canis Major it represents the dogs of Orion, near whose feet it lies. The major star of Canis Minor is PROCYON.

Cannon, Annie Jump (1863–1941)
American astronomer, responsible for the Harvard system of classifying star spectra, now universally used by astronomers. Miss Cannon's work at the

Harvard College Observatory showed that stars can be grouped into a small number of classes related to their color. These types are assigned the letters O, B, A, F, G, K, M—ranging from hottest to coolest; each letter is also subdivided from 0 to 9 (see SPECTRAL TYPE). Miss Cannon was the major author of the massive *Henry Draper Catalogue,* issued in nine volumes between 1918 and 1924, which contains the spectral classification of 225,300 stars; later supplements brought the total to over 350,000 (see also DRAPER, HENRY). Miss Cannon also published two volumes of variable stars, in 1903 and 1907, the second containing about 2,000.

Canopus
The second-brightest star in the sky, of magnitude −0.73, also called Alpha Carinae. Canopus is a yellow supergiant, 25 times the diameter of the Sun, and 110 light-years away. Canopus, named for the mythical Greek helmsman, is often used as a guide star for spacecraft navigation.

Cape Canaveral
A sandy promontory on the Atlantic coast of Florida, the main launching site for U.S. space missions. Cape Canaveral was first used as a missile test site in 1950, with the launch of modified V-2 rockets. Rows of permanent launchpads were later constructed close to the shoreline for missile tests and space launchings; each type of rocket requires its own launch facilities. The Cape is now shared by NASA's KENNEDY SPACE CENTER, which controls civilian space programs, and the Cape Canaveral Air Force Base, which handles military launches. NASA built the Vehicle Assembly Building (V.A.B.) farther inland on Merritt Island; in this enormous building the giant Saturn family of rockets for the Apollo lunar landing program and the Skylab space station missions were assembled before being driven to the launchpad. The V.A.B. has been modified to handle the SPACE SHUTTLE. Other rockets are assembled on the launchpad. Rockets launched from the Cape fly southeastward over the Atlantic, where they are tracked by the Air Force's Eastern Test Range facilities based on islands, ships, and aircraft. The Eastern Test Range was originally established for tracking missile tests and has its headquarters at Patrick Air Force Base, 15 miles (24 km) south of the Cape. Cape Canaveral is used for launches by the Army, Navy, and Air Force as well as by NASA. In 1963 it was renamed Cape Kennedy, but in 1973 reverted to its original name.

Cape Observatory
Familiar name for the Royal Observatory, Cape of Good Hope, which was founded in 1820. The Scottish astronomer Sir David Gill (1843–1914) was Royal Astronomer at the Cape from 1879 to 1907, where he took numerous photographs of the southern skies. These were analyzed in the Netherlands by Jacobus KAPTEYN to produce the *Cape Photographic Durchmusterung,* a listing of 454,875 stars down to magnitude 9.5 between declination −19° and the south pole, published between 1896 and 1900. At the start of 1972, the Cape Observatory merged with the Republic Observatory in Johannesburg to form the SOUTH AFRICAN ASTRONOMICAL OBSERVATORY, which has its headquarters at the Cape.

Capella
The seventh-brightest star in the sky, also known as Alpha Aurigae. It is a double-star system consisting of two yellow stars each about three times the Sun's mass, orbiting every 104 days. Their light combines to give Capella an apparent magnitude of 0.09; the system is 45 light-years away.

Capricorn, Tropic of
The southernmost latitude on Earth at which the Sun appears directly overhead at noon. This occurs on the winter SOLSTICE (about December 22). The latitude corresponds to the inclination of the Sun's path or ECLIPTIC to the celestial equator (23° 26′ 32″ in 1976). Although the Sun used to lie in the constellation of Capricorn at the solstice, the drifting or PRECESSION of the Earth's axis has now carried the position into Sagittarius (see also CANCER, TROPIC OF).

Capricornus (the sea goat)
An inconspicuous constellation of the zodiac, best seen during the northern hemisphere late summer; the Sun passes through the constellation from late January to mid-February. Its brightest star, Alpha, appears to be double; in fact, the two stars are unconnected, one being 116 light-years away and the other 1,100.

carbon-nitrogen-oxygen cycle (CNO cycle)
A chain of nuclear reactions by which energy is released in stars twice or more as massive as the Sun. The principal result of the reactions is to turn hydrogen into helium. The nucleus of a helium atom is a little

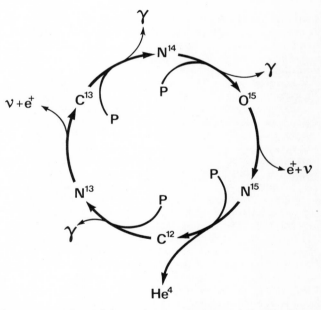

A representation of the carbon cycle (more fully, the carbon-nitrogen-oxygen cycle). Beginning at the bottom, a carbon 12 nucleus and a hydrogen nucleus (a proton) together form nitrogen plus a photon at gamma ray wavelengths (shown by γ). The nitrogen decays to carbon 13, giving off an electron and a neutrino (v) The reactions continue round the cycle with the emission of more photons, electrons and neutrinos until carbon 12 is reached again, with the overall consumption of four protons and the production of one helium nucleus.

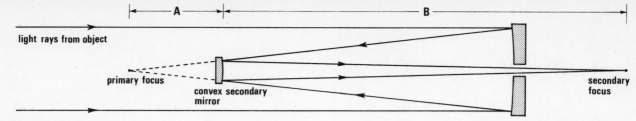

light rays from object

primary focus

convex secondary mirror

secondary focus

concave primary mirror

Cassegrain telescope

less heavy than the four hydrogen nuclei (protons) which fuse to make it; this small amount of missing mass is turned into energy, which is released to power the star. The reaction begins when protons are added one by one to a carbon nucleus, turning it into nitrogen and then oxygen nuclei, before it decays to its original form, ejecting a helium nucleus as it does so. The net result of the cycle is to convert hydrogen to helium, leaving the carbon unchanged to begin another cycle.

The excess mass is released as energy in the form of neutrinos, elusive subatomic particles, and gamma rays, energetic photons. The neutrinos speed directly out of the center of the star, where the nuclear reactions occur; but the gamma rays diffuse out only slowly, becoming degraded along the way to a much larger number of lower energy photons, which leave the star's surface in the form of visible light, and ultraviolet and infrared radiation. During the CNO cycle, three gamma rays and two neutrinos are created from the mass which is destroyed. In stars of about the Sun's mass or less, the CNO cycle still occurs, but it produces less of the star's energy than the PROTON-PROTON CHAIN.

carbonaceous chrondrite
A fragile type of stony meteorite, containing carbon. See CHONDRITE.

Carina (the keel)
A constellation in the southern hemisphere of the sky, formerly part of the ancient constellation of Argo Navis but made separate by Nicolas Louis de Lacaille. The main star in Carina is the brilliant CANOPUS, but the constellation's most famous feature is Eta Carinae, a spectacularly variable star about 6,800 light-years away, embedded in a bright, diffuse nebula called NGC 3372. After varying between fourth and second magnitude for two centuries, Eta Carinae surged up to become the second-brightest star in the sky in 1843. It has faded since to the limit of naked-eye visibility, at which it remains. The star is surrounded by an expanding gas shell that seems to have been thrown off in its outburst; some astronomers speculate that Eta Carinae may be a peculiar kind of supernova.

Carpenter, Malcolm Scott (b. 1925)
Second American astronaut to orbit the Earth. Carpenter, who had been backup to John Glenn for the first orbital flight, circled the Earth three times on May 24, 1962. He used up more fuel for maneuvering the spacecraft while in orbit than planned, and landed 250 miles (400 km) off target.

Cassegrain telescope
A type of reflecting telescope in which the image is observed through a hole in the main mirror. Light is collected, as in a normal reflecting telescope, by a

large concave primary mirror, and is reflected to a smaller convex mirror facing it. This secondary mirror, positioned inside the main mirror's focus, reflects the light back to a hole in the center of the main mirror. The secondary mirror thereby increases the FOCAL LENGTH of the telescope without lengthening the tube. The compactness of the Cassegrain design makes it favored for large reflecting telescopes. The primary mirror usually has a FOCAL RATIO between 3 and 5; the secondary mirror can increase this to an effective focal ratio of between 12 and 30.

The design was invented in 1672 by the French physicist N. Cassegrain. In the classical Cassegrain telescope, the primary mirror is a PARABOLOID and the secondary a HYPERBOLOID. Other combinations of curvatures may also be used. The Dall-Kirkham modification employs a spherical secondary mirror and a primary that is between a sphere and a paraboloid. The Ritchey-Chrétien type, named for the American G. W. RITCHEY and the French astonomer and optician Henri Chrétien (1879-1956), has hyperbolic curves on both mirrors and is free from COMA; it therefore has a wider usable field than other types. For this reason it is employed in most large telescopes.

Cassini, Giovanni Domenico (1625–1712)
Italian-French astronomer whose observations added significantly to knowledge of the solar system. As professor of astronomy at the University of Bologna, Cassini in 1665 measured the rotation period of Jupiter to within a few minutes, and the next year did the same for Mars. In 1668 he made tables of the motion of Jupiter's four bright satellites, which were later used by Olaus ROEMER in his measurement of the speed of light. In 1669 Cassini was invited to Paris, becoming director of the new Paris Observatory on its opening in 1671; he took French nationality in 1673 and changed his first names to Jean Dominique. At Paris, Cassini discovered four satellites of Saturn —Iapetus, Rhea, Tethys, and Dione—between 1671 and 1684, and in 1675 drew attention to the dark gap in the rings of Saturn now called Cassini's division. He also realized that the rings are not solid, but consist instead of countless tiny particles orbiting like little moons. In 1672, with the help of observations by Jean Richer (1630–1696), Cassini measured the distance of Mars, thereby producing by far the most accurate estimate of the scale of the solar system.

On his death, his son, Jacques Cassini (1677–1756), succeeded him as director of the Paris Observatory. Jacques compiled the first tables of the motion of Saturn's satellites, and determined the proper motion of the star Arcturus. César François Cassini de Thury (1714–1784), son of Jacques, became director of the Paris Observatory in 1756, and began a major map of France. This was completed by his own son, Jacques Dominique, comte de Cassini (1748–1845), who succeeded him as director.

Cassini's division

Gap about 1,700 miles (2,700 km) wide in Saturn's rings, separating ring A and ring B and named for its discoverer Giovanni Domenico CASSINI. It is not totally empty, but contains far fewer particles than do the adjacent rings. The division is caused by the gravitational effects of Saturn's satellites, notably Mimas. Particles in Cassini's division orbit Saturn with half the period of Mimas, and are thus strongly perturbed and tend to move into a different orbit. This explanation was advanced in 1867 by Daniel Kirkwood (see also KIRKWOOD GAPS).

Cassiopeia

A prominent constellation in the north polar region of the sky, lying on the edge of the Milky Way and named for a queen of Greek mythology. Its brightest stars form a shape like a letter W, the apex of which points toward the pole star. The apex star, Gamma, is an irregular variable star, which seems to throw off shells of material. Near the star Kappa was the supernova observed by Tycho Brahe in 1572; this is now a radio source, 11,400 light-years away. One of the most powerful radio sources in the sky, Cassiopeia A, lies near the star cluster labeled M52; this source is also the remains of a supernova.

Castor

Second-brightest star in the constellation Gemini, also known as Alpha Geminorum. It is a system of six different stars, with a total visual magnitude of 1.58. The two main stars, magnitudes 1.96 and 2.89, are in mutual orbit every 420 years; they are each SPECTROSCOPIC BINARIES, of periods 9.2 and 2.9 days respectively. The system is completed by a far-off pair of red dwarf stars, eclipsing every 19 hours 33 minutes and varying in magnitude between 9.1 and 9.6. Castor is 45 light-years away.

catadioptric system

An optical system that uses both lenses and mirrors to form an image. Combining reflection and refraction helps to improve telescopic performance (particularly in suppressing of COMA). One of the first such attempts was made by the Swiss optician Emile Schaer (1862–1931), who between 1913 and 1922 used a lens near the focus of a 39-inch (100-cm) reflecting telescope to improve its field of view for photography. Later, the American astronomer Frank E. Ross designed coma-correcting lenses for use with the large American reflectors. The first truly catadioptric system, however, was the wide-field telescope designed by Bernhard SCHMIDT. A similar idea, with a different type of lens, was employed in the MAKSUTOV telescope (1944). Another important catadioptric instrument is the extremely fast super-Schmidt telescope designed by the American optical engineer James Gilbert Baker (b. 1914). By introducing extra lenses into the system, catadioptric patterns allow the designer to correct simultaneously aberrations including coma, ASTIGMATISM, and field curvature.

Catalina Observatory

The observing station of the University of Arizona's Lunar and Planetary Laboratory, situated at 8,235 feet (2,510 m) in the Catalina Mountains. The LPL operates a 61-inch (155-cm) reflector for lunar and planetary photography, inaugurated in 1965. Nearby is the Mount Lemmon Infrared Observatory.

celestial mechanics

The study of the motion and gravitational interactions of bodies in space. Celestial mechanics is used to calculate the orbits of bodies, and to predict their positions for constructing tables of their motions. Similar techniques are now used to calculate the orbits of satellites and space probes; this branch is often termed *astrodynamics*. Celestial mechanics was born in 1687, when Isaac NEWTON set down his three laws of motion, and his discovery that all bodies attract each other by the force of gravity. The Swiss mathematician Leonhard Euler (1707–1783) then established classical methods for determining the motions of the Moon and comets. Careful measurements indicated that the long-range gravitational effects of the planets produced perturbations on solar system bodies. These effects were investigated by a series of brilliant French mathematicians in the mid-18th century. Alexis Claude Clairaut (1713–1765) computed the perturbations of HALLEY'S COMET by the major planets; Joseph Louis LAGRANGE discovered secular (time-dependent) effects in the solar system; and the marquis de LAPLACE proposed his nebular hypothesis for the formation of the solar system after considering the stability of rotating fluids. The greatest achievement of this period was the prediction of the planet Neptune from its perturbations on Uranus by John Couch ADAMS and independently by Urbain LEVERRIER. The start of the 20th century saw the introduction of further refinements by such workers as Simon NEWCOMB and the French mathematician Jules-Henri Poincaré (1854–1912). Shortly afterward, Albert EINSTEIN proposed his theory of RELATIVITY, which helped explain details in the motion of some bodies, particularly Mercury. The computational tradition of Newcomb and his colleagues was continued and extended by Dirk Brouwer (1902–1966), Gerald Maurice Clemence (1908–1974), and Wallace John Eckert (1902–1971); the latter introduced computer techniques which have revolutionized celestial mechanics. In 1951, the three published the monumental *Coordinates of the Five Outer Planets, 1653-2060,* which serves as the basis of all research involving the motions of the planets from Jupiter to Pluto.

celestial sphere

The imaginary sphere of the heavens, with the Earth at its center, which appears to rotate once every day. All astronomical objects appear to lie on the surface of this sphere. In ancient times the stars were believed to be points on a real sphere revolving around the Earth; the concept is retained as a useful device in establishing a coordinate system to specify the position of an object in the sky.

The Earth's daily west-to-east rotation makes the celestial sphere appear to rotate from east to west every 23 hours 56 minutes 4 seconds (a SIDEREAL day). The celestial sphere rotates about the north and south celestial poles, which are in line with the Earth's own axis; an observer at one of the terrestrial poles would therefore find the corresponding celestial pole directly overhead, while an equatorial observer would have them at his north and south horizons. Near the north celestial pole lies the bright POLE STAR. The altitude of the celestial pole above the observer's horizon is equal to his latitude. The celestial equator is the projection of the Earth's equator on the celestial sphere. Its maximum altitude on the celestial sphere is equal to 90° minus the latitude.

The position of an object on the celestial sphere is defined by the coordinate called *declination* (equivalent to latitude on Earth) and *right ascension* (the equivalent of longitude). Declination is measured in degrees north (+) or south (−) of the equator; the poles lie at 90° declination. Right ascension is measured in hours, from 0 to 24, corresponding to the celestial sphere's daily rotation; the hour number increases from west to east. The zero point of right ascension lies where the Sun crosses the celestial equator on its way north at the beginning of northern spring, the VERNAL EQUINOX. The Sun's apparent yearly path around the celestial sphere is termed the ECLIPTIC; it is actually a projection of the plane of the Earth's orbit. The ecliptic cuts the celestial equator at about $23\frac{1}{2}°$, the angle of the Earth's axial tilt.

Because the reference points on the celestial sphere are reflections of the Earth's own rotation, the wobble of the Earth's axis, called PRECESSION, causes the celestial poles to move, so that the lines of right ascension and declination are slowly changing. The annual drift averages about 3 seconds of right ascension and 15 arc seconds of declination (varying considerably in different parts of the sky), and for precise positional work the EPOCH (reference date) of observation must be stated.

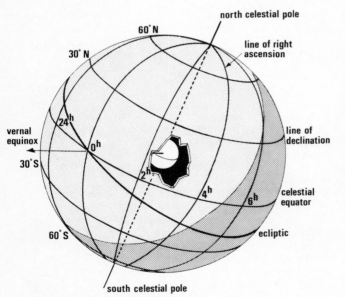

Centaur
American upper-stage rocket burning liquid hydrogen. It is 30 feet (9.1 m) long and 10 feet (3 m) wide, and has two engines that together produce 30,000 lb. (13,600 kg) of thrust, almost twice that of the AGENA second-stage rocket. The Centaur's first mission was on May 30, 1966, when, with an ATLAS first-stage booster, it launched Surveyor 1 to the Moon. The 117-foot (35.7-m) high Atlas-Centaur combination has since launched numerous planetary probes. The Centaur now serves as the top stage on a modified Titan III booster, to launch still heavier probes, such as the Vikings sent to Mars.

Centaurus (the centaur)
A prominent constellation in the southern hemisphere of the sky, lying in a rich part of the Milky Way. Its brightest star, ALPHA CENTAURI, is actually a triple system whose three components are the closest stars to the Sun. At the heart of the constellation is the globular cluster Omega Centauri, cataloged as NGC 5139. This is one of the brightest and richest of all globular clusters, containing hundreds of thousands of stars, 17,000 light-years away. The galaxy NGC 5128 is a famous radio source, Centaurus A; it seems to have ejected clouds of gas in explosions.

cepheid variable
A type of star which swells and contracts in size like a beating heart, varying in brightness as it does so. The time taken for a cepheid variable to complete one cycle of light changes is directly related to its average brightness: the longer the period, the brighter the star. Cepheid variables go through their pulsation cycle over periods of between 2 and 40 days, varying in brightness by up to about one magnitude. They are named for the prototype, Delta Cephei, discovered in 1784 by the English amateur astronomer John Goodricke. The cepheids are yellow supergiant stars, similar in color to the Sun but with masses 5 to 10 times greater. They are also much brighter, which makes them visible over long distances. Since their period of pulsation and their average brightness are linked, (the so-called PERIOD-LUMINOSITY RELATION), by measuring the period astronomers can calculate the cepheid's intrinsic brightness, its ABSOLUTE MAGNITUDE. Comparing this with how bright it appears in the sky its APPARENT MAGNITUDE reveals its distance. Cepheid variables therefore act as a standard for measuring distances, and have been of great importance to astronomers. In our Galaxy, cepheids lie along the spiral arms and occur in GALACTIC CLUSTERS; they are young and belong to the so-called population I of stars. To distinguish them from *Virginis stars,* a similar but older and

The celestial sphere, as seen from the outside. The Earth rotates counterclockwise as seen from above the north pole, so if one imagines the Earth to be fixed in space, the celestial sphere rotates clockwise—in this illustration, from left to right—once a day. The celestial equator, poles and lines of declination (latitude) are above their Earth counterparts. Where the ecliptic (the Sun's track) crosses the equator is the vernal equinox, where the Sun is located on March 21. This is the 0^h point of right ascension, and as the sphere apparently rotates all other points in the sky follow it after a given time interval. This gives rise to the lines of right ascension.

fainter class of stars with which they were formerly confused, population I cepheids are sometimes called *classical cepheids* (see VARIABLE STARS). In the 1971 *General Catalogue of Variable Stars,* 696 cepheid variables were listed.

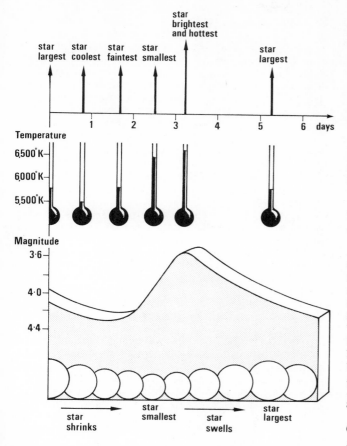

As a cepheid star pulsates in size, so its temperature and surface area change. The maximum brightness does not occur when the star is at its largest and comparatively cool, but shortly after it passes through its smallest stage.

Cepheus

A constellation near the north pole of the sky named for a king in Greek mythology. Its most famous feature is Delta Cephei, the prototype of the class of stars known as CEPHEID VARIABLES. Near Delta Cephei lies the famous double star Krüger 60, composed of 10th- and 11th-magnitude red dwarfs which orbit each other every 44.6 years; they are 12.8 light-years away. The star is named for the German astronomer Adalbert Krüger (1832–1896), who mistook two stars in the same line of sight for a real double. In 1890 the American astronomer S. W. Burnham discovered that one of the stars seen by Krüger did in fact have a genuine companion.

Ceres

Largest of the asteroids, and the first to be discovered, by Giuseppe PIAZZI in 1801. Its diameter is about 593 miles (955 km), and it orbits the Sun every 1,682 days (4.6 years) at an average distance of

257,120,000 miles (413,800,000 km). At its brightest, Ceres is just visible to the naked eye, and is easily followed in binoculars. The surface temperature of Ceres has been measured as around 160° K. Its mass is about 6.7×10^{-10} that of the Sun, and it has an average density about that of Mars or the Moon.

Cernan, Eugene Andrew (b. 1934)

American astronaut, commander of the final Apollo mission to the Moon. Cernan, a qualified aeronautical engineer, was selected as an astronaut in 1963. On his first space flight, Gemini 9 in June 1966, he performed a space walk lasting over two hours. In May 1969 Cernan flew on the Apollo 10 rehearsal for the first Moon landing, descending in the lunar module with Tom Stafford to within 10 miles (16 km) of the Moon. Cernan landed on the Moon on December 11, 1972, in the Apollo 17 mission. With geologist Harrison Hagan ("Jack") Schmitt (b. 1935) he roamed the lunar surface for a record total of 22 hours during a three-day stay.

Cerro-Tololo Inter-American Observatory

Astronomical observatory on Cerro Tololo Mountain in the foothills of the Chilean Andes, run by the Association of Universities for Research in Astronomy (AURA). The observatory, at an altitude of 7,100 feet (2,160 m), has headquarters 35 miles (55 km) northwest in La Serena on the Pacific coast. The observatory was officially opened in November 1967 and is used by astronomers from throughout the world. Its largest telescope is a 158-inch (400-cm) reflector, a twin of the reflector at KITT PEAK NATIONAL OBSERVATORY and the largest telescope in the southern hemisphere; it was installed in late 1974. Also at the observatory are a 60-inch (152-cm), a 36-inch (91-cm), and two 16-inch (40.6-cm) reflectors, and a 24-inch (61-cm) Schmidt telescope.

Cetus (the whale)

A large and straggling constellation of the equatorial zone of the sky, best seen during the northern hemisphere autumn. Its most famous object is the star Omicron, also called MIRA. This is the prototype of a class of stars which vary erratically in brightness over long periods of time. The star Tau Ceti is very much like our Sun; it is 11 light-years away, and thus one of the closest stars. Nearby in the sky is UV Ceti, a pair of twin red dwarfs among the smallest stars known; one of them is a special kind of FLARE STAR, which has given its name to a small group known as UV CETI STARS.

Chamaeleon (the chameleon)

A faint and insignificant constellation lying near the south pole of the sky. It was named by Johann Bayer in 1603.

Chandrasekhar, Subrahmanyan (b. 1910)

Indian-born American astrophysicist, responsible for important studies of stellar structure and evolution, including the nature of WHITE DWARF stars. During the early 1930s Chandrasekhar calculated that a white dwarf could not have a mass more than 1.44 times the Sun's mass; above this, it would compress itself into something denser still. This upper limit for a white dwarf's mass is know known as the

Chandrasekhar limit. Stars that are more massive must lose matter before they can become white dwarfs; this is believed to happen at the end of a star's life when its outer layers expand to form a PLANETARY NEBULA. However, if the star's mass is still above the Chandrasekhar limit, it will collapse into a BLACK HOLE. Chandrasekhar has also studied how stars transfer energy by radiation in their atmospheres; his classic book on the subject is *Radiative Transfer* (1950).

chondrite

The commonest stony meteorite, containing tiny round inclusions called *chondrules,* measuring about 1 millimeter across. Stony meteorites make up more than 92 percent of meteorite falls, and of these over 91 percent are chondrites. Chondrites have a mean density 3.6 times that of water, and consist of tiny fragments of various minerals and flecks of nickel-iron, with the chondrules packed among them. They are quite different in nature from terrestrial rocks, which suggests they have never been part of any large planetary body. The chondrules are normally the minerals olivine or pyroxene, melted and recooled. These inclusions are thought to represent primordial material, particles which condensed from the gas and dust cloud around the Sun.

Carbonaceous chondrites. An exceptionally fragile group of rare meteorites with mean densities about twice that of water. They contain up to 5 percent by weight of a black, tarlike, carbon-rich material. They also have considerable amounts of water, and so can never have been heated to too great a temperature. Many asteroids seem to have surfaces similar in composition to carbonaceous chondrites. Carbonaceous chondrites are probably the commonest type of meteorite in space, but they seldom reach Earth intact because of their fragility.

The material of which meteorites are made formed within a short period of about 100 million years some 4.6 billion years ago. The majority of chondrites originated in about six asteroids with diameters from 120 to 400 miles (200–650 km), orbiting 1.9 to 1.8 a.u. from the Sun. The chondrites were apparently broken off from these during collisions some 10 million years ago. The carbonaceous meteorites came from the surfaces of these asteroids, while the progressively more reheated chondrites originated at greater depths in the parent bodies. Some carbonaceous meteorites may also have originated in periodic comet nuclei.

Christie, Sir William Henry Mahoney (1845–1922)

Eighth astronomer royal. He joined the Royal Greenwich Observatory in 1870 as chief assistant to Sir George AIRY. With Edward Walter Maunder (1851–1928) Christie began the famous Greenwich series of solar photographs which provide valuable records of daily solar activity. Christie succeeded Airy as astronomer royal in 1881. His main contribution was in reequipping the observatory for astrophotography and spectroscopic studies, which continue today; he introduced some of the most famous Greenwich telescopes, such as a 28-inch (71-cm) refractor, and the 26-inch (66-cm) photographic refractor and 30-inch (76-cm) reflector jointly named for their donor, the surgeon and amateur astronomer Sir Henry Thompson (1820–1904).

chromatic aberration

The failure of a lens to bring light of all wavelengths to the same focal point, thereby producing color fringes around an image. Chromatic aberration occurs because each lens tends to act like a prism, splitting white light into a spectrum of colors. Glass refracts blue light more sharply than red, with the result that the "blue" image is formed slightly closer to the lens than the "red" image; the other colors of the spectrum are distributed in between. A single lens therefore produces a short line of colored images of the original object. The amount of chromatic aberration depends on the dispersive (color-spreading) power of the glass. Flint glass has a higher dispersion than crown glass, and a lens made of flint glass thus suffers from more serious chromatic aberration than a similar one made of crown glass. By combining two lenses of different dispersive power, the effects of chromatic aberration can be largely canceled out to produce a near color-free image (see OBJECT GLASS).

chromosphere

The layer of gas about 10,000 miles (16,000 km) thick above the Sun's visible surface (the PHOTOSPHERE). The chromosphere is less than one-thousandth the density of the photosphere; consequently it emits only a relatively weak light which is usually lost in the photosphere's brilliance. The chromosphere is visible only at eclipses or through special instruments such as the CORONAGRAPH. The name, meaning "sphere of color," arises because of the layer's distinct pinkish-red tone, caused by light emitted from hydrogen atoms at a specific wavelength termed Hα. The chromosphere is homogeneous only to heights of about 1,000 to 2,000 miles (1,600–3,200 km). Its upper region looks like a flaming forest because of the jets of hot gas, called SPICULES, which surge up from the photosphere to heights of as much as 10,000 miles (16,000 km), injecting material into the Sun's thin outer atmosphere, the CORONA. The temperature of the chromosphere rises from about 5,000° K at the top of the photosphere to over 1 million degrees where it gives way to the corona.

Circinus (the compass)

A faint and insignificant constellation in the southern hemisphere of the sky, adjacent to Centaurus. It was given its name by Nicolas Louis de Lacaille in the 1750s.

circumpolar

Term describing a celestial object that does not set when seen from a given latitude on Earth; as the term implies, the object appears to circle around the pole. For a star to be circumpolar, its angular distance from the pole must be less than the observer's latitude. From latitude 45° only objects within 45° of the pole are circumpolar; all other objects rise and set during the night. At the pole (90°) all celestial objects visible are circumpolar. But at the equator none are circumpolar; all objects seem to rise and set.

Clark, Alvan (1804–1887)

American instrument-maker whose company five times set the world record for manufacturing the largest telescope lens. His sons George Bassett Clark (1827–1891) and Alvan Graham Clark (1832–1897)

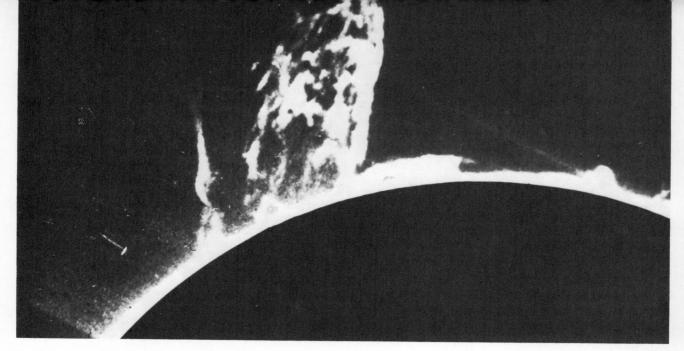

A shining arc of the Sun's chromosphere, seen here in a picture taken with a coronagraph, in which the brilliant photosphere is artificially eclipsed with a diaphragm. The great surge of gases, **left center**, is an eruptive prominence, some 190,000 miles (300,000 km) high, comprising chromospheric material suspended in the corona.

helped make the firm at Cambridge, Massachusetts, the world's leader in astronomical optics. The company's reputation was established in 1862, when Alvan Graham Clark, testing a newly completed 18½-inch (47-cm) refractor, discovered the white dwarf companion to the star SIRIUS. The Clark firm made the 26-inch (66-cm) refractor of the U.S. Naval Observatory in 1871, a 30-inch (76-cm) refractor for Pulkovo Observatory in 1884, and the 36-inch (91-cm) refractor of Lick Observatory in 1888. They completed the 40-inch (101-cm) telescope for Yerkes Observatory in 1897; this remains the largest refracting telescope ever made.

coalsack

A large dark cloud of dust and gas, lying in the Milky Way near the Southern Cross. The minute dust particles of the cloud scatter light from the stars behind it, giving the appearance of a jagged hole in the bright background of stars. The coalsack is the most prominent of all the dark NEBULAE, and lies 400 light-years from us. It is 40 light-years across and probably over 100 times more massive than the Sun. It is no different in composition from the bright nebulae, like that in Orion; it is dark only because no stars are embedded within to illuminate it. Such nebulae may be the birthplaces of stars; the coalsack is apparently starting to fragment into smaller, darker globules which will eventually form a star cluster. In time, the nebula will begin to glow from the light of protostars within. Similar clouds can be seen in the northern Milky Way. The so-called *northern coalsack* in Cygnus is almost as notable as its southern counterpart.

coelostat

A mirror system used to reflect the light from a celestial body into a fixed telescope; the name coelostat means "stationary sky." The method is frequently employed with long-focus solar tower telescopes. A plane mirror is mounted facing the celestial equator on an axis pointing to the celestial pole; if it is driven around this axis as the Earth rotates, the reflected beam not only remains stationary but does not alter its orientation during the object's passage across the sky. Normally, a second plane mirror is used to reflect the light from the coelostat into the telescope. For general astronomical use the simpler *siderostat* ("stationary star"), which gives a rotating field, is used. The siderostat consists of a single adjustable mirror, turning equatorially once in 24 hours. It reflects the beam toward the north or south celestial pole and into the objective of a suitably inclined telescope; such instruments are often known as *polar telescopes* since the tube is parallel to the Earth's axis.

Collins, Michael (b. 1930)

Command-module pilot on the Apollo 11 flight to the Moon in July 1969. Collins remained in orbit around the Moon while astronauts Neil Armstrong and Edwin Aldrin made the first manned lunar landing. Collins was selected as an astronaut in 1963. He first flew in space in July 1966 on the Gemini 10 mission, during which he made three space walks, on one of them moving across to a rocket the Gemini spacecraft had docked with. After Apollo 11, Collins left the astronaut corps to become director of the Smithsonian Air and Space Museum in Washington.

color index

A measure of the apparent color of a star, and hence of its temperature. Cooler stars emit comparatively more light at longer wavelengths and thus appear redder than hot stars. The color index is expressed as the difference in a star's brightness when measured at two selected wavelengths. The *international color index,* defined by E. C. Pickering about 1890, is the difference between the Photographic magnitude (blue light) and the Photovisual magnitude (yellow light). It is zero for white stars (Spectral type A0), positive for red stars, and negative for blue stars. Magnitudes are now seldom measured photographically, and the use of color filters with Photoelectric cells allows the color index between any two wavelengths to be found. The widely used UBV system utilizes the ultraviolet, blue, and yellow (visual) magnitudes.

color-magnitude diagram

A plot of the apparent brightness of stars against their Color index. It is similar to the Hertzsprung-russell diagram, which plots absolute magnitude against color. The color-magnitude diagram is often used for star clusters, since all members of the cluster are at the same distance, and the correction between apparent and absolute magnitude is constant for each star. In fact, the distance of the cluster can be accurately found from the amount of adjustment needed to bring its Main sequence, based on apparent magnitudes, into line with the main sequence of the Hertzsprung-Russell diagram, which is based on absolute magnitudes. The difference between the apparent and absolute magnitudes is the Distance modulus for the cluster.

Columba (the dove)

An insignificant constellation in the southern hemisphere of the sky near Canis Major; it was introduced during the 17th century.

coma

An optical defect in which a star appears distorted into a comet- or pear-like shape toward the edge of the field of view. It is caused by zones toward the edge of a lens or mirror which focus rays arriving obliquely into a short line of images of progressively larger diameter. Coma is one of the two principal off-axis aberrations (the other is Astigmatism) limiting the field of view of a camera or telescope. Its seriousness increases toward the edge of the field, but it is eventually swamped by astigmatism. Coma is more awkward than astigmatism in measuring star positions because comatic star images are not symmetrical. Achromatic objective lenses for refracting telescopes can be designed free from coma (if they are also free from Spherical aberration they are said to be *aplanatic*), but the Newtonian or classical Cassegrain reflecting telescope needs special correcting lenses.

Coma Berenices

A faint constellation between Leo and Boötes in the northern hemisphere of the sky, representing the hair of the Egyptian queen Berenice. Its main feature is an enormous cluster of many thousands of galaxies, several hundred million light-years away. Much nearer is the spiral galaxy M67, known as the black-eye galaxy because it contains a large area of dark dust. The north galactic pole lies in Coma Berenices, between the stars Beta and Gamma.

comet

A small icy body embedded in a cloud of gas and dust moving in a highly elliptical orbit around the Sun.

Comets spend most of their lives in frozen reaches far from the Sun, but periodically their orbits bring them close enough to be heated up and to release gas and dust clouds to form a hazy head, developing tails which always point away from the Sun. These have given comets their name: from the Greek meaning "long-haired one."

Comets were thought to be atmospheric phenomena until Tycho Brahe demonstrated that the comet he discovered in 1577 showed no shift in position through parallax, proving it was far beyond the Moon. Isaac Newton demonstrated that comet Kirch, discovered in 1680, moved in an orbit around the Sun in accordance with his theory of gravity. Edmond Halley, and later Wilhelm Olbers, greatly improved methods of determining the orbits of comets. The study of their physical structure and behavior began in the mid-18th century and is now an increasingly important field of research.

Discovery and naming. Comets are named for their discoverer or discoverers, with up to three names permitted. Lower-case letters are added as preliminary designations to record the order of discovery, the first comet of 1976 being designated 1976 a, the second 1976 b, and so on. Permanent designations are eventually allotted in the order of perihelion passage (closest approach to the Sun), the first comet to pass in 1976 being designated 1976 I, the second 1976 II, and so on. Amateur astronomers systematically scan the sky with telescopes and large binoculars in search of new comets; professional astronomers find faint comets accidentally on their photographic plates. The annual discovery rate has slowly increased; in 1750 one comet was found, in 1800 two, 1850 three, 1900 five, 1950 eight, and in 1975 eleven.

Orbits. By 1975, astronomers had observed 964 comets well enough to compute their orbits with accuracy. This total consists of 523 so-called *new* comets, moving in very long-period near-parabolic orbits, and 441 appearances of some 102 objects known as *periodic* comets, moving in short-period elliptical orbits (periods of less than 200 years). New comets move in orbits so long they will not return for thousands of years. Thus comet Bennett 1970 II, with an aphelion at 288 a.u., will not return for more than 1,000 years, while the famous comet Kohoutek 1973 f will take 75,000 years to complete one orbit. All comets are clearly members of the solar system moving in closed orbits. There is not one known case of a comet approaching from deep space. However, some comet orbits have been so perturbed by the planets' gravity they have left the solar system forever.

Comet perihelion distances vary from the record of comet van den Bergh 1974 g, 75 million miles (120 million km) beyond the orbit of Jupiter, to only 0.0048 a.u. for comet Thome 1887 I, within 14,000 miles (22,500 km) of the Sun's surface. But the majority cluster around 1 a.u., the Earth's own distance from the Sun. New comets have near-parabolic orbits which can be inclined at any angle to the planetary system. By contrast, short-period comets move in ellipses close to the plane of the planets. The short-period comets have their aphelions (farthest points from Sun) close to the paths of the giant planets, principally Jupiter. The so-called *comet families* were probably caused by the gravitational attractions of the planets.

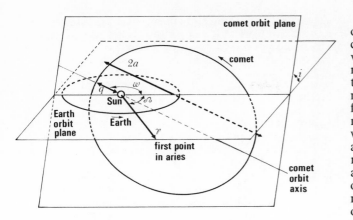

comet orbit plane

2a

comet

ω

q

Sun

Earth orbit plane

Earth

r

first point in aries

i

comet orbit axis

Ω

This diagram shows how a comet's orbit is defined in space. The longitude of its ascending node (Ω) from the First Point of Aries is measured in the plane of the ecliptic, while the longitude of perihelion (ω) is measured in the comet's plane. The inclination (i) is the angle between the two planes, while the perihelion distance (q) and the eccentricity define the size of the orbit. In any particular case, the moment of perihelion (T) must be given.

Anatomy of a comet. Despite the popular picture of a comet as a brilliant object with a fine tail, most comets are actually faint, diffuse, tail-less objects visible only with large telescopes. Comets consist of a nucleus, a head or *coma,* and a tail. The nucleus is the only solid body, with a diameter of from 1 to 30 miles (1.6–48 km), a low overall density, and a mass from about 10 billion to 100 trillion tons. A comet's nucleus is made of dust particles loosely compacted with water ice, together with frozen carbon monoxide and methane. The Sun's heat melts the nucleus, releasing huge volumes of gas which carry away dust and ice particles. This gas produces the head or *coma* of the comet consisting of water vapor, carbon monoxide, and OH (hydroxyl), with minor amounts of other molecules containing carbon, hydrogen, and nitrogen. These molecules are then broken up into smaller fragments to produce a PLASMA, or ionized gas. The coma is normally about 10,000 to 100,000 miles (16,000–160,000 km) across; the record is 1.1 million miles (1.8 million km) for comet Flaugergues 1811 I. Hydrogen, carbon, and oxygen atoms form an invisible enveloping cloud some 10 to 20 million miles (16 to 32 million km) across. If a comet becomes very active, the SOLAR WIND carries away dust and gas to form its tail. Prominent tails develop only in large

Characteristics of some Periodic Comets and their Orbits

Comet	Designation	Period (years)	Perihelion (astronomical units)	Aphelion	Orbital inclination (degrees)	Number of observed returns	Nucleus diameter (miles)	Current absolute magnitude*
Encke	1786 I	3.30	0.339	4.10	12.0	52	2	11.4
Tempel (2)	1873 II	5.26	1.364	4.68	12.5	15	1.5	10.2
Finlay	1886 VII	6.90	1.080	6.17	3.6	9	0.5	13.5
Faye	1843 III	7.41	1.616	5.98	9.1	17	1.0	11.5
Kearns-Kwee	1963 VIII	9.01	2.229	6.43	9.0	2	8	9.5
Tempel-Tuttle	1866 I	32.91	0.982	19.56	162.7	4	10	9.0
Halley	1682 I	76.09	0.587	35.33	162.2	27	25	4.6

*Absolute magnitude is the brightness of the comet when 1 a.u. from both Sun and Earth

Characteristics of some New Comets and their Orbits

Comet	Designation	Perihelion (astronomical units)	Orbital inclination (degrees)	Orbital eccentricity	Current absolute magnitude
Ikeya-Seki	1965 VIII	0.0078	141.86	0.99915	6.0
Seki-Lines	1962 III	0.0314	65.01	1.00000	6.5
Daido-Fujikawa	1970 I	0.0657	100.17	1.00000	10.5
Kohoutek	1973 f	0.1424	14.30	1.00000	6.5
Arend-Roland	1957 III	0.3160	119.94	1.00021	4.5
Bennett	1970 II	0.5376	90.05	0.99627	5.0
Suzuki-Saigusa-Mori	1975 k	0.8375	118.51	1.00000	9.5
Kohoutek	1970 III	1.7189	86.31	0.99911	10.0
Humason	1962 VIII	2.1334	153.28	0.98957	1.5
Wild	1968 III	2.6091	135.17	1.00000	8.5
van den Bergh	1974 g	6.0197	60.77	1.00000	9.0

active comets close to the Sun, and always point away from the Sun. The plasma in the coma consists mainly of ionized molecules of carbon monoxide, carbon dioxide, methylidine, and nitrogen, which are accelerated away from the head at speeds up to 400 miles (640 km) per second, forming a straight-rayed gas tail pointing directly away from the Sun. The inner coma is rich in dust grains about one micron (millionth of a meter) across. Being larger than the ions of the gas tail, these move more slowly away from the head, producing a fan-like dust-tail. These are always curved, but can appear straight when seen as the Earth passes through the plane of the comet's orbit. At such times, the dust particles may appear to stretch in front of the comet to produce an anomalous spike, or *anti-tail*. Comet Arend-Roland 1957 III was very dusty, releasing 70 tons per second of fine particles about 1 micron across, while the famous comet Kohoutek 1973 f released over 1,000 tons of mainly large particles every second. Comets also release small amounts of much heavier particles, of millimeter-, centimeter-, and even meter-size. These spread out slowly ahead of and behind the comet in its orbit to form a meteor stream. The tails of bright comets can be extremely long, usually about 5 to 25 million miles (8–40 million km), the record being 200 million miles (320 million km) for the great comet 1843 I.

Brightness. Comets brighten rapidly as they move toward the Sun, because sunlight excites their gases to fluorescence. New comets are dust-rich and do not brighten as much as periodic comets, which are gas-rich. Comets also show unexpected brightenings and fadings caused by changes in solar activity. They are very sensitive to the solar wind and provide excellent probes for the study of solar activity.

Periodic comets. Comets lose from 0.1 to 1 percent of their mass at each perihelion passage. Short-period comets have periods generally from five to seven years and thus come to perihelion frequently; they fade by 2 magnitudes or more per century as they decay. Periodic comets tend to be gas-rich because they have lost the dusty outer layers of their nucleus, revealing a more compacted center, which contains larger, meteor-sized particles. The force of gas escaping from a periodic comet's nucleus slightly changes its path over long periods of time. In some comets this effect increases as the comet ages, while in others the effect decreases. This suggests there are two types of nuclei: one a uniform icy-conglomerate which eventually dissipates completely, the other a nucleus with a central core of heavier compacted stony material which ultimately loses all its gas, leaving a rocky body like an asteroid. Some of the Apollo and Amor ASTEROIDS might well be such "dead" comets.

Numbers and origin. The total number of comets is enormous—about 10 million comets must have their perihelion points within the orbit of Neptune alone. Most new comets move sunward from a great distance, suggesting they originate in a vast cloud at a distance of 20,000 to 60,000 a.u. This is known as Oort's cloud, after the Dutch astronomer Jan OORT, and contains an estimated 100 billion comets. It is believed to have formed at the very edge of the solar system as the Sun and planets were being born. The comets in the cloud move in circular orbits inclined at all angles to the planetary system, until they are perturbed by passing stars into orbits that eventually carry them toward the inner solar system. Comets are clearly composed of material that has been deep-frozen for millions of years; and their investigation by spacecraft, currently being planned, will yield much valuable information on the early history of the solar system.

command module

The compartment in a manned spacecraft in which the crew sit during launch and landing. The command module contains equipment for communication, navigation, and life support, and controls for guiding the spacecraft. The idea of building spacecraft in sections, or modules, was introduced for the GEMINI series of manned flights; in Gemini, the crew compartment, called the reentry module, was attached to an adapter module, which contained bulky equipment such as fuel cells and oxygen tanks. It was in the APOLLO PROGRAM that the names command module and service module were introduced for these parts of the spacecraft. The single-man American MERCURY craft was entirely self-contained.

commensurability

An exact ratio between the orbital periods of two or more orbiting bodies. Orbital periods whose ratio is any whole number, or whole fraction, are said to be commensurable. If, for example, one body takes two or three times as long to complete an orbit as another, or a half or a third of the time, then the orbits of the two bodies are commensurable. The bodies will return to the same relative positions after two or three orbits; in consequence, they will have strong gravitational effects on each other. Some of the KIRKWOOD GAPS in the asteroid belt are believed to be due at least in part to commensurability with the orbit of Jupiter; and CASSINI'S DIVISION in Saturn's rings may be due to commensurability with Saturn's satellites.

communications satellites

Relays in space for sending telephone, radio, and television signals around the world. They provide many communications circuits far more cheaply than long-distance cables, and they can link any places in the world that have suitable ground equipment. Most countries are now in satellite communication via the INTELSAT network. Sets of three satellites are used, stationed at points above the Atlantic, Indian, and Pacific Oceans, because each satellite can only see about a third of the Earth below it.

Communications satellites use very short radio wavelengths called *microwaves,* which pass straight through the Earth's ionosphere without being reflected, unlike longer waves used for conventional radio transmission. Television transmission also uses very short wavelengths that cannot be bounced off the ionosphere. Without communications satellites it is difficult to send television pictures over long distances without using cables or chains of relay stations. Using communications satellites is much more reliable than bouncing radio waves off the ionosphere, where solar storms can cause sudden radio "blackouts."

A communications satellite ground station has an aerial like a radio telescope, usually 80 to 100 feet (25–30 m) in diameter. It sends radio signals out to

The large satellite is an Intelsat IVA communications satellite, workhorse of international 'phone and video circuits, with a capacity for 11,000 simultaneous two-way voice-quality channels. Compared with it is Early Bird, which in 1965 became the world's first synchronous orbit communications satellite.

Score and Courier were *active-repeater* satellites: they accepted messages and actively rebroadcast them. But at the same time, NASA was experimenting with the so-called *passive reflector* type of communications satellite. These were giant balloons that simply reflected signals back to Earth (see ECHO SATELLITES).

Telstar and Relay. Balloon satellites such as Echo have the advantage that anyone can bounce signals from them, at any frequency, and there are no electronic parts to go wrong. But the signal returned from a simple balloon reflector is as weak as a reflection from the Moon. Much better reception is obtained from satellites that amplify the signal before rebroadcasting it. This was spectacularly demonstrated in 1962 by TELSTAR, which carried the first transatlantic television signals and effectively opened the era of commercial satellite communications. Telstar was followed by a more powerful NASA satellite called RELAY.

Both Telstar and Relay were put into comparatively low orbits around the Earth. One possible communications system is to have a succession of such low-altitude satellites passing continuously across the sky; but rapidly moving satellites provide a tracking problem for ground stations, and satellites in low orbit can be damaged by radiation from the Earth's Van Allen belts.

Syncom and Intelsat. Satellites in higher orbits can cover a much greater area of the Earth. In the synchronous or geostationary type of orbit first envisaged by the English author Arthur Charles Clarke (b. 1917), they appear to hang stationary over one point on the equator. The first experimental satellite of this type was SYNCOM, introduced in 1963 by NASA. Geostationary satellites soon became standard for international communications. In 1964 an international organization named Intelsat was created to arrange a network of communications satellites for commercial use. The organization has continued to build, launch, and operate the Intelsat series of satellites.

Military communications. Because of their reliability, both the United States and the Soviet Union use satellites for military communications. In 1966 the first U.S. military satellites were launched in the Initial Defense Satellite Communication System (IDSCS); the system of 26 satellites was completed in 1968. Placed just below synchronous altitude, they drifted slowly around the Earth so that if one malfunctioned, another would soon appear to take its place. They were followed in 1971 by the first launches of the Defense Satellite Communication System (DSCS), which are larger and more powerful satellites in synchronous orbit, capable of communicating with small, mobile ground stations. Other tactical satellites being developed, such as the U.S. Army's Tacsat, allow communication with individual ships, aircraft, tanks, and jeeps.

Other satellites. Synchronous satellites cannot cover areas at latitudes higher than about 70°. The Soviet Union, because of its large land area in high northerly latitudes, instead uses a system of satellites inclined at 65° to the equator in elliptical 12-hour orbits. There is always a satellite in view of a ground station to ensure round-the-clock satellite coverage (see MOLNIYA SATELLITES). Countries with similar remote communities are turning to domestic satellites

the satellite, and receives others. Messages in each direction are carried on waves of different frequency; each of these so-called *carrier* waves can have hundreds of telephone calls superimposed on it. Depending on the number of carrier waves, the most powerful satellites can deal with several thousand telephone conversations at a time. Television pictures take up a great deal more space on the carrier, leaving room for fewer telephone messages.

History. The first attempt at communication using an artificial satellite was the U.S. government's Project SCORE (Signal Communication by Orbiting Relay Equipment), in December 1958. Score was simply an Atlas rocket in low orbit carrying a tape recorder and radio equipment. It broadcast a prerecorded Christmas message from President Eisenhower, and also recorded messages from ground stations and retransmitted them. Following this success, the U.S. Army Signal Corps in 1960 experimented with a satellite called Courier. This was a 500-lb. (227-kg), 52-inch (132-cm) diameter sphere studded with photocells to draw power from sunlight, that relayed teletype, voice, and facsimile data between military ground stations, before breaking down after 17 days.

for ease of communication, such as with the ANIK SATELLITES, WESTAR SATELLITES, and ATS SATELLITES.

Other nations with special communications requirements are now producing their own satellites. The Franco-German Symphonie satellite, launched on December 18, 1974, is placed in synchronous orbit at 11.5° west longitude for communications between Europe, Africa, and North and South America. During 1977 the European Space Agency plans to launch OTS (Orbital Test Satellite) for European communications.

Future satellites will need to work at higher frequencies, in order to carry more messages, and to have the information packed onto their carrier waves by more efficient techniques. Across the busy North Atlantic satellites will need special antennae to focus on the most important ground stations. Eventually, satellites may have on-board switching, so that they act like automatic telephone exchanges in the sky.

Comsat
The Communications Satellite Corporation, an American corporation set up in 1963 to own and operate commercial communications satellites. With other countries the Comsat Corporation operates the INTELSAT series; it also manages the Intelsat system on behalf of the other participants. The word comsat, without a capital letter, is often used as an abbreviation for COMMUNICATIONS SATELLITES.

Congreve, Sir William (1772–1828)
British rocket pioneer who developed a series of gunpowder-powered projectiles for military use; these were the first real improvements since the rocket's invention. Congreve's rockets carried explosive warheads and were used in several campaigns during the Napoleonic Wars; artillery companies and even

ships were equipped to fire Congreve rockets. The phrase "the rockets' red glare" in the "Star-Spangled Banner" is a reference to the bombardment of Fort McHenry by Congreve rockets during the War of 1812. Congreve rockets were later modified for firing life lines and signal flares.

conjunction
An alignment of astronomical bodies. A planet closer to the Sun than the Earth is said to be at *inferior conjunction* when it is between Earth and Sun, and at *superior conjunction* when it is on the far side of the Sun from Earth; the exact moment of conjunction is when the centers of the Sun and the planet have the same celestial longitude. A planet farther from the Sun than Earth is said to be in conjunction when it is behind the Sun as seen from Earth; when such a planet is opposite in the sky to the Sun, it is said to be at opposition. Two planets can be in conjunction with each other if their celestial longitudes are the same. By extension, the term is often used to mean any close approach between two planets.

Conrad, Charles (b. 1930)
American astronaut, commander of the second Apollo Moon-landing mission, and of the first Skylab crew. Conrad, a qualified aeronautical engineer, became an astronaut in 1962. His first flights were on Gemini 5 in August 1965, and Gemini 11 in September 1966. On November 19, 1969, he piloted the Apollo 12 lunar module to a precision landing near an old automatic Moon probe, Surveyor 3. During a 31½-hour stay on the Moon with Alan BEAN, he spent a total of 8 hours roaming the lunar surface. Conrad's fourth space mission was in May-June 1973, when he and his two-man crew spent a month in the Skylab space station. In early 1974 Conrad resigned from the astronaut corps to enter private business.

Constellations

Constellation	Genitive case	English name	Abbreviation	Approximate position		Area (square degrees)	Order of size
				α h	δ °		
Andromeda	Andromedae	Andromeda	And	1	+40	722	19
Antlia	Antliae	Air pump	Ant	10	−35	239	62
Apus	Apodis	Bird of Paradise	Aps	16	−75	206	67
Aquarius	Aquarii	Water carrier	Aqr	23	−15	980	10
Aquila	Aquilae	Eagle	Aql	20	+5	652	22
Ara	Arae	Altar	Ara	17	−55	237	63
Aries	Arietis	Ram	Ari	3	+20	441	39
Auriga	Aurigae	Charioteer	Aur	6	+40	657	21
Boötes	Boötis	Herdsman	Boo	15	+30	907	13
Caelum	Caeli	Chisel	Cae	5	−40	125	81
Camelopardalis	Camelopardalis	Giraffe	Cam	6	+70	757	18
Cancer	Cancri	Crab	Cnc	9	+20	506	31
Canes Venatici	Canum Venaticorum	Hunting dogs	CVn	13	+40	465	38
Canis Major	Canis Majoris	Greater dog	CMa	7	−20	380	43

Constellation	Genitive case	English name	Abbreviation	Approximate position α h	δ °	Area (square degrees)	Order of size
Canis Minor	Canis Minoris	Lesser dog	CMi	8	+5	183	71
Capricornus	Capricorni	Goat	Cap	21	−20	414	40
†Carina	Carinae	Keel	Car	9	−60	494	34
Cassiopeia	Cassiopeiae	Cassiopeia	Cas	1	+60	598	25
Centaurus	Centauri	Centaur	Cen	13	−50	1,060	9
Cepheus	Cephei	Cepheus	Cep	22	+70	588	27
Cetus	Ceti	Whale	Cet	2	−10	1,231	4
Chamaeleon	Chamaeleontis	Chameleon	Cha	11	−80	132	79
Circinus	Circini	Compasses	Cir	15	−60	93	85
Columba	Columbae	Dove	Col	6	−35	270	54
Coma Berenices	Comae Berenicis	Berenice's hair	Com	13	+20	386	42
Corona Australis	Coronae Australis	Southern crown	CrA	19	−40	128	80
Corona Borealis	Coronae Borealis	Northern crown	CrB	16	+30	179	73
Corvus	Corvi	Crow	Crv	12	−20	184	70
Crater	Crateris	Cup	Crt	11	−15	282	53
Crux	Crucis	Southern cross	Cru	12	−60	68	88
Cygnus	Cygni	Swan	Cyg	21	+40	804	16
Delphinus	Delphini	Dolphin	Del	21	+10	189	69
Dorado	Doradus	Swordfish	Dor	5	−65	179	72
Draco	Draconis	Dragon	Dra	17	+65	1,083	8
Equuleus	Equulei	Little horse	Equ	21	+10	72	87
Eridanus	Eridani	River	Eri	3	−20	1,138	6
Fornax	Fornacis	Furnace	For	3	−30	398	41
Gemini	Geminorum	Twins	Gem	7	+20	514	30
Grus	Gruis	Crane	Gru	22	−45	366	45
Hercules	Herculis	Hercules	Her	17	+30	1,225	5
Horologium	Horologii	Pendulum clock	Hor	3	−60	249	58
Hydra	Hydrae	Water snake	Hya	10	−20	1,303	1
Hydrus	Hydri	Lesser water snake	Hyi	2	−75	243	61
Indus	Indi	Indian	Ind	21	−55	294	49
Lacerta	Lacertae	Lizard	Lac	22	+45	201	68
Leo	Leonis	Lion	Leo	11	+15	947	12
Leo Minor	Leonis Minoris	Lesser lion	LMi	10	+35	232	64
Lepus	Leporis	Hare	Lep	6	−20	290	51
Libra	Librae	Scales	Lib	15	−15	538	29
Lupus	Lupi	Wolf	Lup	15	−45	334	46
Lynx	Lyncis	Lynx	Lyn	8	+45	545	28
Lyra	Lyrae	Lyre	Lyr	19	+40	286	52
Mensa	Mensae	Table Mountain	Men	5	−80	153	75

Constellations (continued from previous page)

Constellation	Genitive case	English name	Abbreviation	Approximate position α h	δ o	Area (square degrees)	Order of size
Microscopium	Microscopii	Microscope	Mic	21	−35	210	66
Monoceros	Monocerotis	Unicorn	Mon	7	−5	482	35
Musca	Muscae	Fly	Mus	12	−70	138	77
Norma	Normae	Level	Nor	16	−50	165	74
Octans	Octantis	Octant	Oct	22	−85	291	50
Ophiuchus	Ophiuchi	Serpent holder	Oph	17	0	948	11
Orion	Orionis	Orion	Ori	5	+5	594	26
Pavo	Pavonis	Peacock	Pav	20	65	378	44
Pegasus	Pegasi	Pegasus	Peg	22	+20	1,121	7
Perseus	Persei	Perseus	Per	3	+45	615	24
Phoenix	Phoenicis	Phoenix	Phe	1	−50	469	37
Pictor	Pictoris	Easel	Pic	6	−55	247	59
Pisces	Piscium	Fishes	Psc	1	+15	889	14
Piscis Austrinus	Piscis Austrini	Southern fish	PsA	22	−30	245	60
†Puppis	Puppis	Stern	Pup	8	−40	673	20
†Pyxis (= Malus)	Pyxidis	Compass	Pyx	9	−30	221	65
Reticulum	Reticuli	Net	Ret	4	−60	114	82
Sagitta	Sagittae	Arrow	Sge	20	+10	80	86
Sagittarius	Sagittarii	Archer	Sgr	19	−25	867	15
Scorpius	Scorpii	Scorpion	Sco	17	−40	497	33
Sculptor	Sculptoris	Sculptor	Scl	0	−30	475	36
Scutum	Scuti	Shield	Sct	19	−10	109	84
Serpens	Serpentis	Serpent	Ser	17	0	637	23
Sextans	Sextantis	Sextant	Sex	10	0	314	47
Taurus	Tauri	Bull	Tau	4	+15	797	17
Telescopium	Telescopii	Telescope	Tel	19	−50	252	57
Triangulum	Trianguli	Triangle	Tri	2	+30	132	78
Triangulum Australe	Trianguli Australis	Southern triangle	TrA	16	−65	110	83
Tucana	Tucanae	Toucan	Tuc	0	−65	295	48
Ursa Major	Ursae Majoris	Great bear	UMa	11	+50	1,280	3
Ursa Minor	Ursae Minoris	Little bear	UMi	15	+70	256	56
†Vela	Velorum	Sail	Vel	9	−50	500	32
Virgo	Virginis	Virgin	Vir	13	0	1,294	2
Volans	Volantis	Flying fish	Vol	8	−70	141	76
Vulpecula	Vulpeculae	Fox	Vul	20	+25	268	55

†The four constellations Carina, Puppis, Pyxis, and Vela originally formed the single constellation, Argo Navis, the Argonauts' Ship

constellations

Star patterns as seen from Earth, which provide a set of references for the recognition and identification of objects in the sky. Most ancient civilizations recognized similar patterns among the stars; from these our modern constellations have grown. The constellations in worldwide use today stem from those of the Greeks, listed by PTOLEMY in 150 A.D.; many of them probably originated with the Babylonians. Ptolemy listed 48 constellations. Twelve of them were the constellations of the ZODIAC, of much earlier origin; other constellations represented Greek mythological figures. New constellations were added in the 17th and 19th centuries, notably by star mappers such as Johann BAYER, Johannes HEVELIUS, and Nicolas Louis de LACAILLE; many of the more insignificant patterns eventually fell into disuse, leaving a total of 88 constellations. There were, however, no generally agreed constellation boundaries until 1930, when the International Astronomical Union decided upon regular boundaries following lines of right ascension and declination for 1875. These were drawn up in a two-volume atlas by the Belgian astronomer Eugene Joseph Delporte (1882–1955). The boundaries themselves are fixed with respect to the stars; however, because of PRECESSION they have now moved away somewhat from their original lines of right ascension and declination. Constellation names are in Latin, with Latin case endings. Thus the constellation name *Capricornus* becomes *Capricorni* when referring to a star in Capricornus (the genitive case): alpha Capricorni means "alpha of Capricornus." Modern astronomers retain the ancient constellations as convenient guides to the location of objects.

continuous creation

The continuous production of matter from nothing, postulated by the STEADY-STATE THEORY of the Universe. The steady-state theory starts from the assumption that the Universe looks the same at all times, as well as from all points in space. But since the Universe is observed to be expanding, the matter in it would become more spread out with time; and the proponents of the steady-state theory had to assume that matter is being continuously created to fill the space caused by the expansion.

The rate of creation was calculated at one hydrogen atom per liter of volume (about one quart) every 500 billion years, which is far too small to check in the laboratory. The matter created was thought to condense into galaxies in a continuous process. Fred HOYLE proposed that the created matter appeared from a *C-field,* a form of negative energy which he supposed to fill the Universe.

Strong doubt has now been cast on the steady-state theory by the discovery of the weak BACKGROUND RADIATION in space, which apparently originated in a completely different state of the Universe in the past. The steady-state theory now has few adherents, and the BIG-BANG cosmology currently in favor does not require the continuous creation of matter.

Cooke, Thomas (1807–1868)

One of the foremost telescope makers of the 19th century. He became well known in 1856, when Charles Piazzi Smyth (1819–1900), astronomer royal for Scotland from 1845 to 1888, took a 7-inch (18-cm) Cooke telescope to test seeing conditions on the island of Tenerife. The American firm of Alvan CLARK had just made an 18½-inch (47-cm) lens, the world's largest, and Cooke was commissioned in 1863 to build a 25-inch (64-cm) refractor by Robert Stirling Newall (1812–1889), a wealthy telegraph cable manufacturer and amateur astronomer. The lens was completed in 1868, but Cooke, exhausted, died soon after. The Newall telescope was given to the University of Cambridge in 1889, and in 1959 was installed at Mount Pendeli Observatory, 12 miles (19 km) northeast of Athens. In 1922 the Cook firm merged with an instrument company set up by Edward Troughton (1753–1835) and William Simms (1793–1860).

Cooper, Leroy Gordon (b. 1927)

Member of the first group of astronauts, selected in 1959, who made the sixth and last flight in the MERCURY series in May 1963. He orbited the Earth 22 times, more than all previous Mercury flights combined. In August 1965 he flew on the Gemini 5 mission, spending eight days in space and making 120 revolutions of the Earth, at that time a record. In mid-1970 he resigned from the astronaut corps to enter private business.

Copernicus, Nicolaus (1473–1543)

Polish astronomer whose heliocentric theory of the Universe demoted the Earth to the status of an ordinary planet, revolutionizing science and profoundly altering Man's conception of his world.

To Copernicus, a canon at a cathedral on the shores of the Baltic, astronomy was little more than a hobby, although his scientific knowledge was well known and he was consulted by the papacy on possible calendar reform. As a student of wide learning, Copernicus knew that certain Greek philosophers had suggested schemes of the Universe in which the Earth was not centrally placed. In the light of these alternative Greek ideas, he decided to take a fresh look at the contemporary view of the Universe, which he found unsatisfactory. According to the teaching he had received, Earth was stationary at the center of the Universe, surrounded by a sphere on which the stars were fixed. The movements of the planets were explained by a system of rotating circles known as deferents centered on the Earth and smaller ones, epicycles, whose centers lay on the circumference of the deferents. The movements of the circles around the central Earth should be regular; but to account properly for the observed motions, the deferents' centers had to be displaced from the center of the Earth, and thus from the center of the Universe. Copernicus believed he could overcome this inconsistency by placing the Sun at the center of the Universe.

Copernicus worked for many years to perfect his theory. A summary of his ideas was circulated, but he was reluctant to publish anything in detail in case so revolutionary an idea might meet with ridicule. In 1589 he was visited by the German mathematician Rheticus (Georg Joachim von Lauchen, 1514–1576). Rheticus stayed for two years and persuaded Copernicus to allow him to publish a small treatise on the theory and then to prepare a full text, which was published at Nuremberg in 1543. Called *De Revolutionibus Orbium Coelestium* (On the Revolution of the Celestial Spheres) it was marred by an unsigned preface by the German theologian Andreas Osiander (1498–1552), describing the theory as no more than a mathematical convenience. A copy reached Copernicus as he lay dying.

There is no doubt that Copernicus believed his

heliocentric theory to be a true description of the Universe as it actually was and, considering the storm of controversy it raised, so did others. It led eventually to new laws of physics, and a vastly bigger Universe than previously imagined; while the dethronement of Man as the center of all creation had the most profound moral and theological implications.

Coriolis force

A fictional force which appears to deflect the motion of bodies over a rotating surface, such as the surface of a spinning planet. On Earth, an object shot or thrown due north from the equator will land a little to the right (the east) of the target. This is because the eastward velocity of a point on the equator, which the projectile shares when it is fired, is greater than that of a point nearer the pole. Similarly, an object propelled toward the equator from the northern hemisphere would seem to veer to the right (in this case the west) of the intended course, as it dropped behind the rotational rate of the Earth nearer the equator. In the southern hemisphere, the projectile would seem to veer left.

This so-called *Coriolis effect*, named for the French physicist Gaspard Gustave de Coriolis (1792–1843), who drew attention to it in 1835, is responsible for the circular movements of weather features such as depressions and cyclones. It must be corrected for when plotting the path of a missile.

corona

The outermost layer of the Sun's atmosphere, beginning about 10,000 miles (16,000 km) above the visible surface. The name "corona" is the Latin for crown. The corona is observable at a total eclipse as a pearly light, comparable in brightness to the full Moon, streaming outward in fans and rays. It has no upper boundary but thins out gradually into interplanetary space. The outermost region of the corona is detected streaming past the Earth as the SOLAR WIND. The appearance of the corona varies through the SOLAR CYCLE. At solar minimum, when activity on the Sun's surface is concentrated toward the equator, the corona contains bright streamers that extend mostly from the equatorial region. At cycle maximum, there are streamers all around the Sun and the corona becomes more circular in appearance. The bright inner regions of the corona are caused by sunlight scattered from electrons; this is termed the K corona, from the German *kontinuum*. Farther out, the corona consists of sunlight scattered by dust particles; this is the F, or Fraunhofer, corona, which shows the dark FRAUNHOFER LINES of the Sun's spectrum. The smallest part of the corona's light is emitted from hot atoms. Mysterious emission lines in the corona's spectrum were once attributed to an unknown element named "coronium." In 1942 the Swedish physicist Bengt Edlén (b. 1906) showed that these lines were due to highly ionized atoms, which meant that the gases of the corona were exceptionally hot—modern investigations suggest a temperature of 2 million degrees K. The corona is probably heated by shock waves rising through the chromosphere from the photosphere.

The solar corona, during the total eclipse of February 25, 1952. Seen here at sunspot minimum, the corona is less dense than at other times and is concentrated toward the equator. Streamers of coronal gas, associated with active regions on the surface, surround lower latitudes, while radial plumes project from the poles.

Corona Australis (the southern crown)
A small but attractive constellation lying below
Sagittarius in the southern hemisphere, a counterpart
to the northern crown (CORONA BOREALIS). It lies in an
interesting region of the Milky Way, and contains
the globular cluster NGC 6541 about 14,000
light-years away.

Corona Borealis (the northern crown)
A constellation of the northern hemisphere of the sky
between Hercules and Boötes, best seen during spring.
It includes a famous irregular variable star,
R Coronae Borealis, which is the prototype of a class
of eruptive variables; its brightness changes from
5.8 to 12.5 erratically, and it can stay at maximum
for a year or two before it rapidly fades. Another
remarkable star, T Coronae Borealis, is a recurrent
nova; it erupted in 1866 and again in 1946.

coronagraph
A device invented by Bernard LYOT in 1930 for
observing the Sun's atmosphere in the absence of a
total eclipse. The inner brightest part of the corona
is about a million times fainter than the solar surface,
and is swamped by the glare of sunlight scattered in
the Earth's atmosphere. Lyot circumvented this
difficulty by establishing his coronagraph at the
high-altitude PIC DU MIDI OBSERVATORY, so reducing
atmospheric glare. To overcome the equally serious
glare caused by dust and diffraction in the telescope,
he designed the highly ingenious coronagraph. A
perfectly polished lens forms an image of the Sun
that is intercepted by an occulting disk, producing a
miniature eclipse, and then focused by a second lens
either for visual observation with an eyepiece, or on
a photographic plate. A series of diaphragms blocks
off the bright ring of diffracted light formed around
the objective. With these and other refinements,
observation of the corona in full daylight thus became
possible, and Lyot was able to observe PROMINENCES
visually and to photograph the corona out to about
7 arc minutes from the edge of the Sun. Other
coronagraphs have since been established at
high-altitude solar observatories throughout the
world.

Corvus (the crow)
A small and insignificant constellation in the
southern hemisphere of the sky, lying below Virgo.

cosmic microwave background
See BACKGROUND RADIATION.

cosmic rays
Nuclei of atoms, stripped of all their electrons,
shooting through space at speeds close to that of
light. Ninety percent of cosmic rays are hydrogen
nuclei (protons), and nine percent are helium nuclei
(alpha particles); the nuclei of all other elements
contribute only one percent of the total. High-speed
electrons, present in space in smaller numbers, are
also classified as cosmic rays. (The term cosmic "ray"
dates from early this century, when these fast particles
were thought to be high-energy X rays.)

The energy of a cosmic ray particle is a combined
measure of its mass and its speed. It is usually
expressed in electron volts (eV); 1 eV is the energy

gained by an electron when accelerated through an
electric potential of one volt. The energy of an air
molecule at room temperature is 0.1 eV, and the
energy of particles in an X-ray machine is about
10,000 eV. The fastest particles in the accelerators
used by nuclear physicists reach about 10^{11} (a
hundred billion) eV. Some cosmic rays have energies
as high as 10^{19} (ten million million million) eV.
Collisions of these very energetic particles with other
matter allow nuclear physicists to study reactions
impossible to achieve in man-made accelerators.

The so-called primary cosmic ray particles from
space do not reach the Earth's surface. Instead, they
collide with the molecules of the atmosphere,
smashing them into fragments, which in turn shatter
other nuclei in the atmosphere. This produces an *air
shower* of particles (secondary cosmic rays), which can
cover several thousand square yards or meters at
ground level.

The relatively rare cosmic rays with very high
energy (greater than 10^{14}, or 100 trillion, eV) can be
studied only by the air showers they produce.
Instruments scattered over several square miles on
the ground detect the shower, the size of which is
related to the energy of the particle which caused it.
The difference in arrival time at the various detectors
reveals from what direction the particle has come.
But lower-energy cosmic rays are abundant enough to
be studied directly, by instruments carried above the
atmosphere in balloons or satellites. Many of these
low-energy cosmic rays originate in the Sun during
solar flares and radio bursts, and their numbers
consequently fluctuate with the 11-year solar cycle.
True cosmic rays from deep space are thought to be
accelerated to their high speeds in exploding stars
(SUPERNOVAE) and their remnants (the fast-rotating
PULSARS). Some of the most energetic cosmic rays
may be produced in QUASARS.

Because cosmic ray particles are electrically
charged, their motion is influenced by magnetic
fields in space. The Earth's own magnetic field focuses
cosmic rays toward the poles, which makes it
difficult to determine their true direction or origin.
The degree to which a path is bent by a magnetic
field depends on the particle's energy. The most
energetic cosmic rays probably escape from the
magnetic fields of galaxies into space. Those that are
trapped in a galaxy produce electromagnetic
SYNCHROTRON RADIATION as they spiral in the
magnetic field. This process is most efficient for
electrons, which are several thousand times lighter
than atomic nuclei. Cosmic-ray electrons produce the
radio noise observed from our own and other
galaxies.

Cosmic rays can have important biological effects:
brain cells of men in space can be damaged by
primary cosmic rays, while on Earth, air-shower
particles are energetic enough to pass through the
body and alter the basic genetic material, DNA.
Mutations affecting the process of evolution may
therefore be caused by cosmic rays. Even major
changes may be possible once every several hundred
million years, when a supernova explodes within a
hundred light-years of the Sun. The resulting influx of
cosmic rays could wipe out entire species, and it has
even been suggested that the extinction of the
dinosaurs was due to such an event.

cosmogony

The study of the origin and evolution of individual objects in the Universe, such as stars and galaxies. Most of cosmogony is usually treated under ASTROPHYSICS and COSMOLOGY. The term cosmogony is now often restricted to the origin of the solar system (see PLANETS).

cosmology

The study of the origin and evolution of the Universe. Cosmology began as what we would now call physical geography, the study of the Earth, and became in Greek times the theory of planetary motions. After the invention of the telescope, cosmologists started to determine the structure of our Galaxy, the Milky Way, and then to study the motion of nearby galaxies. Present-day cosmology is largely concerned with the theory of space and time under the gravitational influence of matter.

Modern theories of cosmology. The light from galaxies shows a so-called RED SHIFT, in which all the SPECTRAL LINES appear at slightly longer wavelengths than normal. This is usually interpreted to be a result of the DOPPLER EFFECT, implying therefore that all galaxies are receding as seen from the Earth. The more distant the galaxy, the more it is red-shifted (HUBBLE'S LAW), meaning that the distances between all galaxies are increasing. An observer in another galaxy would thus see all other galaxies receding from him, and would conclude that he lives in an expanding Universe. Cosmologists generally believe that the Universe is expanding from an explosion, the so-called BIG BANG, which occurred some 18 billion years ago. The major competitor to the big-bang theory was the STEADY-STATE THEORY, which claimed that the CONTINUOUS CREATION of matter causes the Universe to expand. The steady-state theory has been under observational attack on two related fronts. First, the weak BACKGROUND RADIATION in space observed by radio telescopes is most easily explained as the left-over radiation from the big bang itself, much cooled by the expansion of the Universe. Second, distant galaxies appear to be packed together more closely than those nearer, showing that the Universe was more compact in the past; it thus cannot be unchanging. The steady-state idea has now lost favor, although it survives in mathematical variants by FRED HOYLE and Jayant Vishnu Narlikar (b. 1938).

The ultimate fate of the Universe may be either to expand for ever, or else to coast to a halt and collapse again, perhaps to re-explode. This is the theory of the OSCILLATING UNIVERSE, in which the big bang is sometimes called the *big bounce*. The Universe will collapse if it contains sufficient matter for gravitational attraction to overcome the outward momentum of the galaxies. To calculate the amount of matter in the Universe, we must know both the number of galaxies, which is relatively easy to calculate, and the average mass of a galaxy. Estimates of galaxy masses give answers differing by a factor of 100; moreover, there are no accurate estimates of the amount of matter (if any) between galaxies. Thus, the total mass of the Universe is not well enough known for its fate to be predicted by this test. An alternative approach is to examine the velocities of very distant galaxies. The light we see left these galaxies billions of years ago when, according to the oscillatory theory, the Universe would have been expanding faster than it is now. The most recent results show little slow down, and suggest that the Universe will expand indefinitely (see HUBBLE'S CONSTANT).

At the present stage, cosmological theory is also experimenting with various kinds of modifications to Einstein's General Theory of RELATIVITY. This theory sees the Universe as a skeleton of space and time on which the galaxies are studded, enabling us to see the space-time structure, just as the arrangement of atoms in a crystal enables us to see the crystal lattice. One such alternative is the BRANS-DICKE COSMOLOGY, which predicts a change in the force of gravity with time; the rate of change would affect the structure of the Earth and stars over their lifetimes. There is every possibility that astronomical observations will in the near future distinguish which space-time cosmology is correct (see also UNIVERSE).

Cosmos satellites

Continuing series of Russian Earth satellites, successors to the SPUTNIK series, introduced by Cosmos 1 on March 16, 1962. Among the stated intentions of the Cosmos program is the study of the Earth's upper atmosphere, radiation from space and from the Sun, radio propagation in the ionosphere, the Earth's magnetic field and radiation belts and meteoric matter, as well as the testing of new spacecraft; but the name Cosmos has also been used to conceal military satellites and probes that have failed to work.

The Russians introduced a new rocket, which they call the Cosmos launcher, for the start of the Cosmos program. It is a two-stage rocket, based on an intermediate-range ballistic missile code-named Sandal by NATO; the first stage has a thrust of 163,000 lb. (74,000 kg) and the second stage 24,250 lb. (11,000 kg). The first Cosmos satellite was launched from a former missile site called KAPUSTIN YAR, about 60 miles (100 km) southeast of Volgograd; all previous Soviet orbital launchings had been from TYURATAM. The Cosmos launcher, known in the West as C1, is capable of orbiting small satellites, up to about 1,200 lb. (550 kg). On August 18, 1964, the Russians introduced a more powerful launcher when they put three satellites, Cosmos 38, 39, and 40, into orbit together. This launcher, not named by the Russians but known in the West as B1, is believed to be based on the Skean intermediate-range missile, using a new restartable upper stage; it can orbit payloads of up to 7,500 lb. (3,400 kg). Heavier satellites are launched by the VOSTOK rocket also used for manned missions.

The launch of Cosmos 112 on March 17, 1966, saw the introduction of a third launch site, called PLESETSK, near Archangel in northern Russia. It can handle a wider range of rockets than Kapustin Yar, and has since become the major Cosmos launch site. Plesetsk is often used for military satellite launches; about half the Cosmos satellites are believed to have military purposes. Many of them are reconnaissance satellites, which eject a package of film for recovery after about two weeks in orbit, although some have been tests of a system for intercepting and destroying other satellites in orbit. Other military Cosmos are

communications and navigation systems.
The 500th Cosmos satellite was launched on July 10, 1972. Among the achievements of the series up until that date had been the automatic link-up in space between Cosmos 186 and 188 on October 30, 1967—this was the world's first unmanned rendezvous, and was a test flight of two prototype SOYUZ spacecraft. The Cosmos series was also used to develop an operational WEATHER SATELLITE system. Cosmos 122, launched on June 25, 1966, was an acknowledged forerunner of the system, carrying Earth sensors; on February 28, 1967, Cosmos 144 was launched as a prototype of the Meteor series of weather satellites.

A typical small Cosmos satellite is a cylinder about 5½ feet (1.8 m) long and 3½ feet (1.2 m) wide with hemispherical ends, weighing roughly 900 lb. (400 kg). However, the recoverable military reconnaissance satellites, and satellites used for biological studies, are modified Vostok capsules.

Cooperation with other Soviet bloc countries led to multinational experiments on board Cosmos 261, launched on December 20, 1968. This was followed by the Intercosmos series, the first of which, Intercosmos 1, was launched on October 14, 1969. By the end of 1976, the Soviet Union will have launched well over 800 satellites in the Cosmos series.

coudé focus

A focal point in a CASSEGRAIN TELESCOPE, produced by an arrangement of plane mirrors which diverts the light through the hollow polar axis to a stationary observing position. A coudé focus is particularly desirable for mounting heavy equipment such as a spectrograph, and the long light path gives a large-scale image valuable for detailed studies of spectra. Most large Cassegrain telescopes have a coudé focus. In the 200-inch (508-cm) telescope on Mount Palomar, for example, the coudé focus is formed in a constant-temperature room below ground level at an effective FOCAL LENGTH (mirror-image distance) of 500 feet (150 m). In refracting telescopes, plane mirrors may be placed behind the object glass to reflect the light along the polar axis to a coudé focus.

Crab Nebula

The expanding cloud of gas ejected by a star seen to explode in July 1054 A.D. This exploded star, or *supernova,* was observed by Chinese astronomers to reach magnitude —6 (brighter than Venus), remaining visible in daylight over 23 days. The resulting nebula was first noted telescopically in 1731 by the English amateur astronomer John Bevis (1693–1771), and later became the first object (M1) in the list of nebulae compiled by the French astronomer Charles MESSIER. It is easily visible in small telescopes as a hazy oval, lying in the constellation Taurus. Larger telescopes show that its total extent is 7 by 4 minutes of arc (12 by 7 light-years), although it is very faint toward the edges.

The Crab Nebula is about 6,300 light-years from the Sun. It emits powerful radiation at all wavelengths, from radio to X rays and gamma rays. This vast output, 25,000 times the luminosity of the Sun, is maintained by a rotating PULSAR at the nebula's center. The pulsar, only a few miles in diameter but as massive as the Sun, was originally the dense central core of the exploding star. It now rotates 30 times a second, flashing each time at radio, optical, and X-ray wavelengths. It is the fastest-rotating pulsar known, and the only one visible optically. Its radio pulses were discovered in 1968, and its optical flashes in 1969.

The pulsar is linked to the nebula by magnetic fields, which are braking its spin by one part in a million per day. The rotational energy lost is emitted as radiation by electrons spiraling in the magnetic field (SYNCHROTRON EMISSION). In this respect the optical emission from the Crab differs from that of all other known nebulae, which shine by the light emitted when electrons recombine with ionized atoms (see NEBULA). The Crab Nebula is also unique among

The Crab Nebula photographed in red light, which emphasizes the long gaseous filaments emitting the hydrogen red spectral line. The diffuse light from the center is synchrotron radition, caused by fast electrons from the pulsar, **lower of the two central stars**, moving through the nebula's magnetic field.

supernova remnants, which are generally ring-shaped and optically faint. This difference is probably again due to the pulsar. The expansion of the nebula was discovered in 1921 by the American astronomer John Charles Duncan (1882–1967), who compared photographs taken in 1909 and 1921; he took a confirming plate in 1938.

The Crab Nebula contains long thin filaments, whose light originates from recombining atoms. Each filament is surrounded by a magnetic field, which carries current of about a thousand million million amperes. These filaments were first observed by Lord ROSSE in 1844. His drawing vaguely resembled the pincers of a crab, which gave the nebula its name. The pulsar accelerates the filaments outward at about 700 miles (1,000 km) per second. The proportion of helium to hydrogen in the filaments is seven times greater than normal at the surfaces of stars. This additional helium was produced by fusion of hydrogen inside the star while it was shining steadily; the helium-rich interior became mixed with the surface layers after the explosion. The mass of luminous gas in the filaments is about equal to that of the Sun. There may also be several times more mass within the filaments which cannot be seen. The original star must thus have been considerably more massive than the Sun to have produced both the nebula and the central neutron star.

Crater (the cup)
A small and insignificant constellation in the southern hemisphere of the sky, lying between Leo and Hydra.

craters
Bowl-shaped depressions with raised rims formed by the fall of METEORITES. Shortly after the formation of the solar system, interplanetary space was dense with debris of all sizes; this has steadily been swept up by the planets, leaving visible effects on their surfaces. The unweathered faces of the Moon and Mercury abound with craters, while on Mars erosion has obliterated older features. On Earth, erosion is severe; even giant craters weather away rapidly, to leave only a few small recently formed craters clearly visible.

The Earth's atmosphere melts away the outer

The Arizona meteor crater—one of the best preserved on Earth. About 4,200 feet (1,280 m) across and 600 feet (180 m) deep, its exact dimensions are difficult to measure because it is not circular. The local rock is limestone, which has a natural grid of cracks. This has resulted in the crater being roughly square.

layers of any falling object from space and strongly decelerates bodies below 100 tons mass. The Earth's atmosphere cannot slow larger meteorites completely; their tremendous speed is converted into heat, causing a violent explosion. A 1,000-ton body would produce a 300-ton meteorite, striking the ground at 3 miles (5 km) per second. The ensuing explosion would shatter the meteorite, scattering fragments around a 500-foot (150-m) crater. Many such meteorite craters have been found. The Haviland crater in Kansas is 56 feet (17 m) across and was surrounded by more than 10 tons of pallasite meteorites. The Sikhote-Alin group of 24 craters in the U.S.S.R. was formed on February 12, 1947, by the fall of a giant fragmenting iron meteorite; the largest crater, 85 feet (26 m) across and 20 feet (6 m) deep, was caused by a 3-ton mass. The Henbury group of 13 craters in Australia, the largest 722 feet (220 m) across, are surrounded by many tons of iron meteorites.

An even larger example is the great meteor crater near Flagstaff, Arizona, one of the earliest meteorite impact features identified. It is a $\frac{3}{4}$-mile (1.2-km)-diameter, 600-foot(180-m)-deep bowl with a 150-foot (45-m) raised rim. A 250,000-ton, 240-foot(73-m)-diameter iron body impacting at 10 miles (16 km) per second produced the crater. Its rim is composed of folded and inverted layers of material pushed out from the center, while the rock under the floor of the crater is badly shattered and jumbled. Beneath this, however, the local sandstone is undisturbed. This impact explosion was so violent that it almost vaporized the meteorite; only the rear 20 feet (6 m) survived to scatter hundreds of tons of iron meteorites for 30 square miles (80 km^2) around the crater. The vaporized iron is now present as billions of tiny iron spherules in the surrounding desert soil.

Larger meteorites would be completely destroyed in the explosion. A $\frac{1}{2}$-mile(0.8-km)-diameter asteroid such as ICARUS would penetrate the surface rock for about 2 miles (3 km). The resulting explosion would have a peak temperature of 20,000° K at more than 5 million atmospheres pressure, vaporizing everything within a mile (1.6 km) of the impact and melting everything out to 1½ miles (2.4 km). The resultant crater would be 14 miles (22.5 km) across and 1½ miles (2.4 km) deep. The Ries basin in Germany is the eroded and weathered remains of just such a 14-mile crater. Even larger fossil craters (also known as *astroblemes*—literally, "star-wounds") have now been identified. The Vredefort ring in South Africa is the remains of a 25-mile (40-km) crater, an impact by a 1½-mile(2.4-km)-diameter asteroid, while the Sudbury structure in Canada is a similar crater which initiated widespread volcanic activity.

Several even larger craters are suspected, including the basin 217 miles (350 km) in diameter containing Lake Tengiz in Siberia, while the Nastapoka Island arc in Canada's Hudson Bay may be the only surviving portion of a 275-mile (440-km) crater. A steady stream of new probable fossil meteorite craters has been found in recent years, bringing the total number to about 200.

Crimean Astrophysical Observatory
Russian observatory near Simferopol in the Crimean Peninsula, at an altitude of 1,700 feet (560 m). It was originally sited at Simeis on the Black Sea shore, but was relocated after damage in World War II. Its main telescope is a 102-inch (260-cm) reflector completed in 1960; there is also a 49-inch (125-cm) reflector and other smaller telescopes. Located at the same site is the southern observing station of Moscow's Sternberg Astronomical Institute, which shares the observatory's facilities.

Crux (the southern cross)
The smallest, but one of the most famous of constellations. It lies in the southern hemisphere of the sky, next to Centaurus, of which it was a part until made a constellation of its own in 1673 by the French sailor Augustin Royer. Binoculars show that Alpha Crucis, also known as Acrux, is a double star, both components being brilliant white; their combined magnitude is 0.79. Beta Crucis, magnitude 1.24, is also a brilliant white star; next to it is the cluster NGC 4755, about 1,000 light-years away, centred on the star Kappa Crucis. Crux also contains the COALSACK nebula.

Culgoora Radioheliograph
A unique instrument at Culgoora, Australia, designed for radio observations of the Sun, particularly of the solar flares. It consists of 96 antennae, each 45 feet (13.7 m) in diameter, mounted in a circle 1.9 miles (3 km) in diameter; the aerials produce instantaneous radio pictures of the active areas around the Sun on cathode ray screens in the central control room. The radioheliograph, devised by the Australian radio astronomer John Paul Wild (b. 1923), began operation in 1967 at a frequency of 80 MHz; it has since been modified to work at double and half this frequency as well. The instrument forms part of the Culgoora Solar Observatory, which includes telescopes for optical observation of the Sun. The observatory is operated by the Radiophysics Laboratory of the Commonwealth Scientific and Industrial Research Organization (CSIRO).

culmination
The maximum altitude above the horizon reached by a celestial body on a given date. Unless the body moves significantly during the night, culmination is the moment that it crosses the *meridian,* the north–south line in the sky.

curvature of space
A distortion of space caused by the presence of matter. Such a distortion can be thought of as analogous to the distortion familiar on maps of the Earth using the Mercator projection. On the map the distortion occurs because the surface of the Earth is spherical, but the map's surface is flat. Polar regions such as Greenland consequently appear "stretched out." High-school geometry is "flat" (Euclidean) geometry, and is not correct on a spherical surface where, for example, initially "parallel" lines may meet. Until Einstein, it was believed that the geometry of space was Euclidean. However, Einstein's theory of RELATIVITY proposed that gravity "curves" space in a fashion analogous to the curvature of the Earth. Therefore the geometry of the Universe, or of the space surrounding any object, curves the path of matter and of light. Such a space distortion around the Sun was first confirmed from observations by Sir Arthur EDDINGTON during a solar eclipse in 1919. Radio telescopes have recently been used to measure the deflection by the Sun of radio waves from a distant QUASAR, and these very accurate measurements confirm the optical results.

Although difficult to visualize, the effects of space curvature are readily calculated. A curved Universe can be finite in size, yet have no boundary; it would thus be analogous to the Earth's surface, also finite but without any edge. If a person could travel in an apparently straight line far enough in any direction, he would find himself returning to the point from which he began.

Cygnus (the swan)
A prominent constellation of the northern hemisphere of the sky, best seen during the northern summer; because of its shape it is often called the Northern Cross. Cygnus lies in a dense part of the Milky Way and contains many objects of interest. Its major star is DENEB, one of the brightest in the sky. Beta Cygni, called Albireo, is one of the most beautiful double stars; its component stars are yellow and blue in color, of magnitudes 3.2 and 5.4 respectively. The first star other than the Sun to have its distance calculated was 61 Cygni; this was done by Friedrich BESSEL in 1838. One remarkable star in the constellation is P Cygni, a nova-like variable that was seen to flare up in the 17th century. Cygnus also contains the North American nebula (NGC 7000), an illuminated mass of gas so named because of its shape, and the Veil nebula (NGC 6992), which is part of the so-called Cygnus loop, the remnants of a supernova which exploded 60,000–70,000 years ago. One of the most powerful radio sources in the sky is Cygnus A, which lies near the star Gamma Cygni; it is thought to be an exploding galaxy (see RADIO GALAXY). X rays are detected coming from a double star near Eta Cygni; this source, called Cyg X-1, is believed to give the first direct observational evidence of a BLACK HOLE in space (see X-RAY ASTRONOMY).

D

δ

The symbol for the astronomical coordinate known as DECLINATION.

David Dunlap Observatory

Canadian observatory situated at an altitude of 800 feet (244 m) at Richmond Hill, Ontario, operated by the University of Toronto. Its main telescope is a 74-inch (188-cm) reflector, the largest optical telescope in Canada, which began operation in 1935. Since 1971 the University of Toronto has also operated a 24-inch (61-cm) reflector at the Las Campanas observatory in Chile.

Dawes, William Rutter (1799–1868)

English astronomer, originator of the famous *Dawes limit* formula for determining telescopic RESOLUTION. Dawes was a pioneer double-star observer, publishing a total of almost 2,800 measurements. From 1839 to 1844 he was in charge of the private observatory of George Bishop (1785–1861) in Regent's Park, London, where he made numerous observations. He was succeeded by John Russell Hind (1823–1895) who discovered 10 asteroids while there. Dawes calculated from his observations of double stars the so-called Dawes limit: the resolving power of a telescope measured in seconds of arc is 4.56 divided by the aperture measured in inches. This holds true assuming average seeing; under exceptional conditions the Dawes limit can be exceeded, though in many cases it will not be reached.

day

The time taken for the Earth to spin once on its axis; by extension, the rotation period of any planet. The rotation of the Earth can be measured relative to the stars (a SIDEREAL DAY) or to the Sun (a SOLAR DAY). Astronomers employ the sidereal day, but for other purposes the Earth's rotation with respect to the Sun is used. The two kinds of days differ slightly in length because the Sun's position against the star background changes as the Earth goes around its orbit. Actual solar time is, however, slightly irregular (see EQUATION OF TIME), and for normal timekeeping we use the *mean solar day,* which measures the Earth's rotation relative to an imaginary MEAN SUN, moving across the sky at a uniform rate. The mean solar day is divided into 24 hours of mean solar time. In terms of mean solar time, a sidereal day lasts 23 hours 56 minutes 4 seconds. The mean solar day is 24 hours 3 minutes 56.55 seconds of sidereal time.

declination

The celestial equivalent of latitude on Earth. The declination of an object north or south of the celestial equator is given in degrees and fractions of a degree, marked positive (+) for northerly declinations and negative (−) for southerly ones (see CELESTIAL SPHERE).

deferent

In early systems of planetary motion, the hypothetical circular path along which moved the center of a smaller circle, the EPICYCLE.

Deimos

Smaller and more distant of the two moons of Mars, discovered in 1877 by Asaph HALL. It orbits Mars every 30 hours 18 minutes at a distance of 14,590 miles (23,490 km) from the planet's center. Mariner 9 photographs show that Deimos is irregular in form and dotted with small craters. Its general shape resembles that of a squashed potato, with dimensions of 9 by 7½ by 7 miles (15 × 12 × 11 km).

Delphinus (the dolphin)

A small constellation in the northern hemisphere of the sky between Pegasus and Aquila. Four stars form a rectangle called Job's coffin. If the names of its two brightest stars, Sualocin and Rotanev, are spelled backwards, they give *Nicolaus Venator* (in Italian, Niccoló Cacciatore, 1780–1841), the assistant to Giuseppe PIAZZI, who cleverly named them.

Delta

A medium-sized rocket used to launch numerous scientific and communications satellites, notably the INTELSAT series. Delta is based on the Thor intermediate-range missile, with a liquid-fuel second stage and an optional solid-fuel third stage. There have been over 100 successful Delta launches since its introduction with Echo 1 in 1960. Improvements in design have steadily increased the size of the rocket and its payload capacity. In 1964, three small solid-fuel strap-on boosters were added to the first stage; this vehicle was named the Thrust Augmented Delta (TAD). In 1967 an improved version of the third stage produced the Thrust Augmented Improved Delta (TAID). The following year the Thor first stage was lengthened to give the Long-Tank Delta. The addition of strap-ons turns it into the Long-Tank Thrust Augmented Delta (LTTAD). Six or even nine strap-on solid rocket motors (SRMs) can now be added to the first stage; they are each 20½ feet (6.3 m) long and 31 inches (0.8 m) in diameter, and have a thrust of 52,000 lb. (23,500 kg). More advanced strap-ons each give a thrust of 84,000 lb. (38,100 kg). In 1974 a new variant was introduced called the Straight-Eight Delta because of its constant 8-foot (2.4-m) diameter. This has a first-stage thrust of 205,000 lb. (93,000 kg), second-stage thrust of 9,800 lb. (4,450 kg), and optional third stages of 10,000-lb. (4,540-kg) or 15,000 lb. (6,800-kg) thrust. The complete rocket, including a protective shroud covering the satellite, is 116 feet (35.4 m) high. Delta is capable of placing a satellite of 1,500 lb. (680 kg) into synchronous orbit, or one of 4,000 lb. (1,800 kg) into low Earth orbit.

Deneb

The brightest star in the constellation Cygnus, also called Alpha Cygni. It is a brilliant white star about 50,000 times more luminous that the Sun, of apparent magnitude 1.26, and about 1,500 light-years away. The name Deneb means "tail."

descending node

See NODES.

Dicke, Robert Henry (b. 1916)

American physicist who played an important part in the development of lasers, and later, in the 1960s, performed a delicate experiment which confirmed to one part in one hundred million that all bodies fall equally fast under gravity. This is called the principle of equivalence, and provides the physical basis of the general theory of RELATIVITY. Dicke developed a detailed extension of general relativity in which gravity grows progressively weaker with time (see BRANS-DICKE COSMOLOGY), and in 1964 suggested a search for the now-famous cosmic BACKGROUND RADIATION from the big bang.

diffraction

The apparent bending of light (or other radiation) around obstacles. It is because of diffraction that the edges of shadows are not sharp. A light beam passing the edge of an object is diffracted through a small angle, which varies with the wavelength: the longer the wavelength, the greater the bending. Scientists make use of this effect by means of a DIFFRACTION GRATING, which spreads white light into a spectrum of colors. Light from a star passing through a telescope is diffracted to form an image which is not a point but a disk, called the *Airy disk,* surrounded by successively fainter concentric circles or fringes. The smaller the telescope, the larger the Airy disk and hence the poorer its resolution, its ability to discriminate fine detail. In a reflecting telescope, the struts supporting the secondary mirror at the prime focus also cause diffraction, visible on photographs as the spikes on the images of bright stars. Diffraction was first noted by the Italian physicist Francesco Maria Grimaldi (1618–1663).

diffraction grating

A device for diffracting light into a spectrum, usually by employing a series of fine parallel lines ruled on a material such as glass. Light passing an obstacle is partly spread into a spectrum; by arranging a large number of narrowly spaced obstacles, light can be split into a series of spectra. Some of the light will pass straight through, but some will be partly diffracted into beams on either side of the undeviated light. These are called first-order beams; they will be dispersed into colors, with the longest wavelengths deviated the most. Farther away from the undeviated beam are fainter, more dispersed spectra; these are termed second-order, third-order, and so on. Thus, a grating can be used in a SPECTROSCOPE. Gratings are made by ruling machines that produce several thousand lines per centimeter. This is costly, and it is common to make transfer-like replicas using the original as a mold, in a material such as collodion or plastic. If the grating is in front of the telescope rather than at the focus, a coarser spacing is used, known as an *objective grating.* Reflection gratings are formed by ruling lines of an aluminized mirror. By varying the angle the grating can be *blazed* to throw most of its light into one order spectrum. The *Rowland grating,* invented in 1885 by the American physicist Henry Augustus Rowland (1848–1901), has lines ruled on a concave mirror, thus bringing the spectrum to a focus without the need for further optics.

Dione

Satellite of Saturn, fifth in order of distance from the planet, discovered by G. D. CASSINI in 1684. Its diameter is about 500 miles (800 km), and it is probably a rocky body, with a density of 3.2, similar to that of our Moon. Dione orbits Saturn every 2 days 17 hours 41 minutes at a distance of 234,600 miles (377,500 km).

direct motion

The movement of an object in its orbit from west to east. This is the normal direction of movement in the solar system; it is also called *prograde motion.* The opposite direction of movement, from east to west, is known as RETROGRADE MOTION.

Discoverer satellites

Series of U.S. Air Force satellites for scientific and military research. The Discoverer satellites were put into orbits over the Earth's poles. They were launched from the Western Test Range in California by Thor rockets, using an early version of the AGENA as an upper stage. In most Discoverers, an instrument package was ejected from the nose of the Agena after a specified number of orbits; it was either snared by an aircraft as it floated back to Earth under parachutes, or retrieved from the ocean after splashdown. The reentry practice gained with Discoverer was valuable for later manned programs; the capsule of Discoverer 13 was in fact the first object to be recovered from orbit. The first 15 Discoverers were engineering test satellites, but later capsules contained biological specimens to study radiation hazards in space and prototype equipment for military surveillance systems. After Discoverer 38, military surveillance and capsule recovery became classified subjects and subsequent satellites were not described.

Discoverer series

Serial number	Launch date	Results
1	February 28, 1959	No reentry capsule. World's first polar-orbiting satellite
2	April 13, 1959	Capsule ejected on orbit 17, but lost in Arctic
3	June 3, 1959	Failed to orbit
4	June 25, 1959	Failed to orbit
5	August 13, 1959	Capsule ejected, but into another orbit
6	August 19, 1959	Capsule ejected on orbit 17, but recovery failed
7	November 7, 1959	Capsule not ejected
8	November 20, 1959	Launched into wrong orbit; capsule overshot recovery area on orbit 15
9	February 4, 1960	Failed to orbit
10	February 19, 1960	Failed to orbit
11	April 15, 1960	Capsule ejected on orbit 17, but recovery failed
12	June 29, 1960	Failed to orbit
13	August 10, 1960	Capsule recovered from sea after 17 orbits; first recovery from orbit

(continued p. 58)

Discoverer **Series**

Serial number	Launch date	Results
14	August 18, 1960	First midair recovery, on orbit 17
15	September 13, 1960	Capsule ejected after 17 orbits, but lost in storm
16	October 26, 1960	Failed to orbit
17	November 12, 1960	Midair recovery after 31 orbits
18	December 7, 1960	Midair recovery after 48 orbits
19	December 20, 1960	No reentry capsule; carried infrared experiments
20	February 17, 1961	Capsule not ejected; equipment failure
21	February 18, 1961	No reentry capsule; infrared experiments
22	March 30, 1961	Failed to orbit
23	April 8, 1961	Capsule shot into different orbit; not recovered
24	June 8, 1961	Failed to orbit
25	June 16, 1961	Ocean recovery after 33 orbits
26	July 7, 1961	Midair recovery after 32 orbits
27	July 21, 1961	Failed to orbit
28	August 3, 1961	Failed to orbit
29	August 30, 1961	Ocean recovery after 33 orbits
30	September 12, 1961	Midair recovery after 33 orbits
31	September 17, 1961	Capsule failed to eject
32	October 13, 1961	Midair recovery after 18 orbits
33	October 23, 1961	Failed to orbit
34	November 5, 1961	Capsule not ejected
35	November 15, 1961	Midair recovery after 18 orbits
36	December 12, 1961	Ocean recovery after 64 orbits. Same launch rocket as Oscar 1 amateur radio satellite
37	January 13, 1962	Failed to orbit
38	February 27, 1962	Midair recovery after 65 orbits

dispersion

The spreading out of light into its constituent wavelengths to form a spectrum, occurring, for example, when light passes through a prism.

distance modulus

Technique for determining distances in astronomy by comparing an object's intrinsic brightness with its brightness as observed on Earth. An object's distance modulus therefore expresses the difference between the ABSOLUTE MAGNITUDE (actual light output) and the observed magnitude (apparent magnitude) of an object. One unit of magnitude corresponds to a difference in brightness of 2.512 times (see MAGNITUDE). Thus an object with a distance modulus of 15 magnitudes appears with only a millionth (2.512^{-15}) the brightness it would exhibit if it were at a distance of 10 parsecs (about 32.6 light-years), the standard distance used for calibrating absolute magnitude. Since the intensity of light diminishes according to the inverse square of distance, each magnitude of distance modulus corresponds to an increase in distance by a factor of $\sqrt{2.512}$, or 1.585. Thus a distance modulus of 15 magnitudes means a distance one thousand times (1.585^{15}) greater than the standard distance of 10 parsecs—in other words, 10,000 parsecs. In practice, brightness is also reduced by light-absorbing thin gas and dust in space, an effect which must be taken into account. The distances of objects such as galaxies are obtained by comparing the brightness of supergiant stars with similar stars of known distance in our own Galaxy. This produces an average distance for the entire galaxy.

Dollfus, Audouin Charles (b. 1924)

French astronomer and leading planetary expert, discoverer of Saturn's 10th moon, Janus. The son of an aviation pioneer, Dollfus made balloon ascents to obtain better views of the planets. He pioneered observations of the polarization of light from planetary surfaces, and concluded that the surface of Mars is covered by particles of an iron oxide, probably limonite (Fe_2O_3). His discovery of Janus was made on photographs taken at the Pic du Midi Observatory when the Earth passed through the plane of Saturn's rings in December 1966.

Dollond, John (1706–1761)

English lens maker who discovered by experiments in 1757 that an ACHROMATIC lens could be made by using two different types of glass, crown glass and flint glass. In 1752 John Dollond had joined the optical firm set up in 1750 by his son Peter Dollond (1730–1820); the lenses they made became world famous. In 1754 John Dollond made the first successful HELIOMETER. Peter Dollond in 1765 invented the triple achromatic lens, by placing two convex lenses of crown glass either side of the biconcave flint glass lens. In 1805 the firm was joined by Peter's nephew, George Dollond (1774–1852). John Dollond's daughter Sarah married Jesse Ramsden (1735–1800), who became a famous instrument maker in his own right, and in 1782 invented the eyepiece design that bears his name.

Dominion Astrophysical Observatory

Canadian observatory situated near Victoria, British Columbia, at an altitude of 750 feet (229 m). Its main telescope, a 73-inch (185-cm) reflector, began operation in 1918, and a 48-inch (122-cm) reflector was added in 1961. The associated Dominion Radio Astrophysical Observatory at Penticton, British Columbia, has an 84-foot (26-m) radio telescope and two large MILLS CROSS arrays. Both establishments are operated by the National Research Council of Canada.

Doppler effect

The change in frequency of waves emitted by an object as it moves toward or away from an observer. A familiar example, involving sound waves, is the change in pitch of sirens on emergency vehicles: the sound becomes higher pitched as the vehicle approaches but then lowers in pitch after the vehicle passes and moves away. The Austrian physicist Christian Johann Doppler (1803–1853) first drew attention to this effect in 1842. It holds true for any type of wave motion, including that of light.

Doppler used the analogy of a ship moving through equally spaced ocean waves (i.e., the distance from crest to crest is constant). If the ship is stationary, the waves will break on the bow, say, once every second. If, however, it is moving into the waves, they will break more frequently. To the observer on the ship, it would seem that the frequency of the waves was higher, or that the space between them, their wavelength, was shorter. The wavelength would grow longer, and the frequency drop, if the ship were moving away from the source of the waves. If, instead, the wave source is moving and the observer is stationary, the wavelength will appear longer as the source moves away, or shorter as the source approaches. For light, the longer wavelengths are redder; for sound, longer wavelengths are lower in pitch.

By measuring the shift in an object's spectral lines—toward the red end of the spectrum if it is receding, or toward the violet end if it is approaching—astronomers can establish the direction and speed of the object.

The RADIAL VELOCITIES of stars, the motions of DOUBLE STARS about each other, the rotation of stars, and the motions of galaxies (usually seen as a RED SHIFT) can all be determined in this way.

Dorado (the swordfish)
A constellation in the southern hemisphere of the sky, named by Johann Bayer in 1603. None of its stars is of particular interest, but Dorado contains the larger MAGELLANIC CLOUD. This has within it the so-called Tarantula nebula, NGC 2070, around the star 30 Doradus.

double star
Two stars linked by gravity; an alternative name for such a pair is a *binary star*. The first genuine double star to be discovered was MIZAR, observed by the Italian astronomer Giovanni Riccioli in 1650. This is a physical double star, in contrast to an apparent or optical double, a relatively rare occurrence in which two stars appear close together in the sky, but in reality are widely separated along the line of sight. It is estimated that only about 15 percent of all stars are single bodies like the Sun, while 46 percent are in double systems, in which one star is gravitationally attracted to the other and orbits it. The remaining 39 percent occur in multiple systems of three members or more. The separation between double stars ranges from the so-called *contact binaries* like W Ursae Majoris, in which the stars actually touch each other and share a common atmosphere, to stars so widely separated that they can be distinguished in a telescope. Astronomers suggest that the closest double stars—those with orbital periods less than about 100 years—were formed when a gas cloud condensing into a star split into two or more parts. Double stars with the longest periods may have been formed separately.

The closest binaries are revealed as double when their light is analyzed with a spectroscope (SPECTROSCOPIC BINARIES). Close binaries can also be detected if the Moon happens to pass in front of them, for their light will fade in two steps. The new technique of SPECKLE INTERFEROMETRY, in which a high-speed photograph of the star is de-blurred by illuminating it with a laser, can also produce images of two stars close together. The orbits of binary stars lie at completely random angles; they do not, for example, line up with the plane of the Galaxy. If the orbit is seen edge-on (an inclination angle of $90°$) the two stars may periodically pass in front of each other as an ECLIPSING BINARY.

Periods. In accordance with Kepler's third law (see KEPLER'S LAWS), the closer double stars tend to have shorter periods. The longest well-established period, 480 years, is of the binary Eta Cassiopeiae, easily seen to be double with binoculars. There are, however, stars whose periods are so long that they have nowhere near completed a single orbit since observations began—the period of PROXIMA CENTAURI around ALPHA CENTAURI is about a million years. The twin nature of some double stars is inferred simply because the stars have the same distance and share the same motion (PROPER MOTION) through space. The shortest periods are found among the binary stars having a small star such as a WHITE DWARF as one component. These can approach quite close, like AM Canum Venaticorum, with the shortest-known period—$17\frac{1}{2}$ minutes—of any ordinary double star. However, the X-ray novae having periods as short as 120 seconds discovered by the Ariel 5 satellite in 1975 may be binary stars in which both components are white dwarfs or NEUTRON STARS.

Evolution. The brighter star in a binary system is called the *primary,* the fainter the *secondary*. During the course of its evolution the primary expands, and its atmosphere may be gravitationally attracted to the secondary, the gas streaming from one star to the other. When this happens, the primary is said to fill its *Roche lobe,* and the binary star is said to be semidetached. So much gas may flow from the primary to the secondary that the secondary eventually becomes the new primary. Sirius, which has a companion that has evolved to a white dwarf, has now reached this state. The new primary also eventually expands, fills its Roche lobe, and the binary star again becomes semidetached, with the gas stream pouring from the new primary to the new secondary. Algol is such a star. The secondary may have become a white dwarf, neutron star, or black hole. When the gas stream falls onto such a small star it creates X rays; this is the cause of many of the X-ray stars discovered by the Uhuru (SAS-A) satellite in 1971. If so much material falls onto a white dwarf that its mass exceeds the maximum possible for such a star (the Chandrasekhar limit), it explodes as a NOVA.

Draco (the dragon)
A large and straggling constellation near the north pole of the sky. Its brightest star, Gamma Draconis, was observed by James Bradley in his discovery of the ABERRATION OF STARLIGHT. Thuban, Alpha Draconis,

was the north pole star about 2832 B.C.; the pole has since moved because of PRECESSION. Draco contains the planetary nebula NGC 6543, 1,700 light-years away.

Drake, Frank Donald (b. 1930)

American radio astronomer who made the first attempt to detect signals from other civilizations, called Project Ozma. At the National Radio Astronomy Observatory in West Virginia, Drake used an 85-foot (26-m) radio telescope to "listen" to two nearby stars resembling the Sun, Tau Ceti and Epsilon Eridani, for a total of 150 hours in 1960. Although no signals were heard, Project Ozma turned attention to the possibilities of interstellar contact by radio. In 1959, Drake's radio observations revealed the Van Allen radiation belts around Jupiter. Drake became director in 1971 of the National Astronomy and Ionosphere Center, which runs the 1,000-foot (305-m) Arecibo radio telescope. With this instrument Drake began to listen in 1975 for possible radio messages from super-civilizations in nearby galaxies (see LIFE IN THE UNIVERSE).

Draper, Henry (1837–1882)

American physician and amateur astronomer, a pioneer of astrophotography. His father, John William Draper (1811–1882), took the first photographs of the Moon in 1840, and of the Sun's spectrum in 1843. Henry Draper took the first successful photograph of a star's spectrum, that of Vega, in 1872; in 1880 he photographed the Orion nebula, and later also recorded its spectrum.

Henry Draper's widow endowed the Harvard College Observatory with funds to produce a classification of stellar spectra. A preliminary catalog containing 10,351 stars, the work of Williamina Paton Fleming (1857–1911), appeared in 1890. The main work, the *Henry Draper Catalogue,* containing 225,300 stars down to eighth magnitude, was published between 1918 and 1924. It was compiled by Annie J. CANNON, and it introduced the now-standard Harvard classification system (see SPECTRAL TYPE). She also compiled the *Henry Draper Extension* (1925–1936), listing 47,000 additional spectra of fainter stars. A further volume, published posthumously in 1949, contained another 86,000 stars. An entirely new edition of the catalog is now being prepared.

Dreyer, Johan Ludwig Emil (1852–1926)

Danish astronomer, compiler of the *New General Catalogue of Nebulae and Clusters of Stars,* containing 7,840 objects and published in 1888. Dreyer was appointed astronomer at the observatory of Lord ROSSE in 1874, and his interest in nebulae was stimulated by observations with Rosse's large telescope. Dreyer compiled his famous catalog, often simply called the NGC, while at Armagh Observatory where he was director from 1882 to 1916; it superseded John HERSCHEL's previous catalog. In 1895 and 1908 Dreyer published supplements, called the *Index Catalogues* (IC), of 1,529 and 3,857 newly found nebulae and clusters. Objects are often referred to by the NGC and IC numbers that Dreyer gave them. Dreyer was also a prominent historian of astronomy, editing works of Tycho Brahe (1908) and William Herschel (1912), and in 1906 publishing the standard *History of Astronomy from Thales to Kepler.*

dwarf stars

Name applied to ordinary stars like the Sun that are in the prime of their lives, shining by converting hydrogen into helium in their centers. The brighter, hotter dwarfs are as much as 65 times as massive as the Sun. Beyond this size, known as the *Eddington limit,* radiation pressure in their interiors blows them apart. The dimmer, cooler dwarfs (RED DWARFS) can be as small as 1 percent of the solar mass. Below this limit, the object would never become hot enough to make fusion occur and would be classified as a planet. In the HERTZSPRUNG-RUSSELL DIAGRAM the dwarfs lie on the MAIN SEQUENCE. Stars which are substantially larger and brighter than dwarfs are called GIANTS and SUPERGIANTS. There are a few stars lying just below the main sequence, called subdwarfs. These probably resemble ordinary dwarfs, but being older are of slightly different composition.

Dyson, Sir Frank Watson (1868–1939)

English astronomer royal from 1910 to 1933 (and astronomer royal for Scotland from 1906 to 1910). Dyson organized expeditions to observe the total eclipse of 1919; these showed that light from stars is bent by gravity as it passes close to the Sun, thus confirming a key prediction of Einstein's theory of general relativity. Dyson's own observations at numerous eclipses produced important information on the Sun's faint outer regions, the CHROMOSPHERE and CORONA. At Greenwich Dyson inaugurated the radio transmission of time signals in 1924, extending it to a worldwide service in 1927. In 1924 he introduced the free-pendulum type clock, designed by the engineer William Hamilton Shortt (1882–1971), which was used to maintain Greenwich time until the introduction of the QUARTZ-CRYSTAL CLOCK.

E

Early Bird

The first communications satellite launched by the INTELSAT organization. It was placed into synchronous orbit on April 6, 1965, over the Atlantic at longitude 35° west. Early Bird could carry 240 two-way voice circuits, or one television channel. It operated for $3\frac{1}{2}$ years, providing the first commercial satellite link between the United States and Europe; it was reactivated during July 1969 because of a temporary failure with an Intelsat 3 satellite. Early Bird was developed from the SYNCOM design.

Earth

The third planet from the Sun, and the only one known to support life. As seen from space, Earth would be termed the blue planet, because two-thirds of its surface is covered by seas. It is this predominance of water in liquid form that makes Earth hospitable to life as we know it. Often, however, much of the surface is obscured by white clouds caused by evaporating water vapor. By watching the Earth from space, another civilization would soon deduce the major features of the Earth's atmospheric circulation and of its weather patterns. We get such a view by the use of WEATHER SATELLITES. The Earth spins once on its axis every DAY and completes one orbit around the Sun roughly every $365\frac{1}{4}$ days (the YEAR). The Earth is roughly spherical in shape, with an average diameter of 7,918 miles (12,742 km); strictly, it has a slight bulge at the equator, giving an equatorial diameter 26.6 miles (42.8 km) greater than that from pole to pole. Its mass is 5.976×10^{21} tons, and its average density is 5.52 times that of water.

The Earth is only one of nine major planets in our SOLAR SYSTEM, orbiting the Sun. The Sun itself is merely one of about one hundred billion stars in the MILKY WAY Galaxy, many of which probably have planets of their own. Current astronomical opinion is that planet Earth is far from unique.

Formation of the Earth. The Earth and the rest of the solar system formed together with the Sun about 4.6 billion years ago when a cloud of cool gas and dust in space began to condense under the effect of its own gravity (see PLANETS). In the large central portion of the collapsing cloud, heat generated by the condensation process was sufficient to trigger nuclear reactions, forming a star—our Sun. But in the outer regions of the cloud, smaller, cooler objects formed— the planets, of which the Earth is a small, rocky example. Heat generated by the decay of radioactive elements in its interior, together with heating by the young Sun, would have driven away the lightest gases, leaving only the heaviest materials, such as the metal and rock that make up most of the Earth today. In its molten state, the elements making up young Earth would have become separated, with the heaviest (such as iron) sinking to the center to make a core, while the lighter rocks floated above, forming a mantle and crust. The continents have been subsequently built up by volcanic activity, and modified by erosion. Continuing radioactive decay still heats the Earth's interior, so that parts of it remain liquid. Slow convection in the warm mantle is believed to be the force that moves parts of the crust in the phenomenon known as continental drift.

Present structure of the interior. Our knowledge of the interior comes from studies of seismic waves from earthquakes and, in recent years, from nuclear explosions underground. Vibrations travel through the different layers of the Earth at different speeds, and are reflected and refracted like light waves traveling through glasses. Analysis of seismic vibrations recorded at many stations around the world allows seismologists to determine the nature and density of the underlying rocks.

The solid crust is only about 3 miles (5 km) thick under the oceans, but under the continents has an average thickness of 20 miles (32 km). Beneath this thin skin is a solid area of the mantle, extending down to about 60 miles (100 km), to which the crust is attached. This crust-mantle slab floats on a slush of upper mantle that extends down to about 150 miles (250 km). Under this slush lies the mantle proper, divided into two layers of different density and extending to 1,800 miles (2,900 km) depth. The outer, liquid core extends down from the bottom of the mantle to a transition layer at about 3,000 miles (5,000 km). The inner 800 miles (1,300 km) of our planet is made up of the iron-rich solid core.

The boundary between the crust and mantle is named the Mohorovičić discontinuity, or Moho, after the Croatian geophysicist Andrija Mohorovičić (1857–1936), who postulated it in 1909. The density of the material inside the Earth ranges from about 13.5 times the density of water at the center, down to 10 at the edge of the outer core, from 5.5 to 3.5 in the

The interior of the Earth, deduced from studies of the way earthquake waves are reflected and refracted as they pass through the Earth.

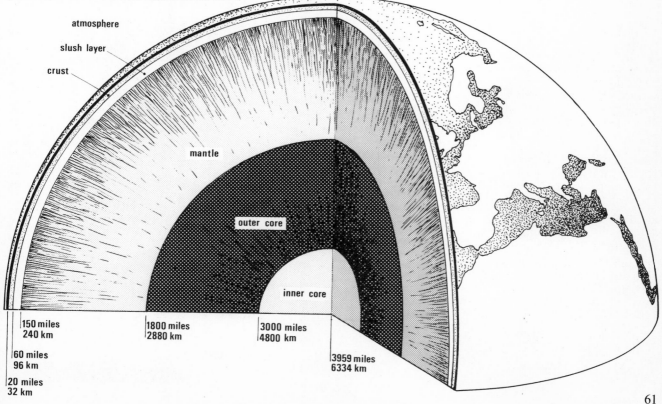

atmosphere

slush layer

crust

mantle

outer core

inner core

| 150 miles | 1800 miles | 3000 miles |
| 240 km | 2880 km | 4800 km |

60 miles
96 km

3959 miles
6334 km

20 miles
32 km

This view of Earth, taken by the Soviet spacecraft Zond 5 shows Africa and parts of Europe and Asia. It was taken on 21st September 1968, and thus at an equinox—the terminator line lies exactly along a line of longitude and passes through the poles. A belt of cloud lies along the equator, over the tropical rain forests, while a low pressure system covers Europe.

A profile of the Earth's atmosphere, showing the altitudes at which various phenomena appear. Even at the top of the mountains, men would find it difficult to breathe without additional oxygen.

mantle, and from 3.3 to 2.8 in the crust. The composition of the core is probably much the same as that of iron-nickel meteorites; the upper mantle is thought to be chiefly a silicate rock rich in the mineral olivine.

Surface features. The Earth's outer crust-and-mantle laminate is cracked like an eggshell into segments, termed plates, which are moved by the convection currents in the mantle below. Where two plates are separating, as along the center of the Atlantic Ocean, material wells up from inside the Earth to create new seafloor. Where two plates meet, old seafloor dips down into the Earth's interior, forming a deep trench, as at the Pacific edge of South America. Continents can be split apart as new oceans form, or they may collide to form mountain ranges. There is no evidence that similar processes occur on the other rocky planets of the solar system.

The Earth's surface is constantly being changed by erosion, caused by wind, ice, and water. Because of this, it is difficult to pick out on the surface of the Earth features corresponding to the craters of the Moon, Mars, and Mercury. A few large craters are known, and it seems very likely that all the inner planets of the solar system were subjected to the same intense bombardment by meteorites at some early stage in the evolution of the solar system (see CRATERS).

EARTH'S ATMOSPHERE

EXOSPHERE
250 mi
400 km

200 mi
320 km

IONOSPHERE

150 mi
240 km

90 mi
144 km

CHEMOSPHERE

45 mi
72 km

STRATOSPHERE

TROPOSPHERE

The atmosphere. The Earth's present atmosphere is the creation of volcanic activity, which released new gases, including large amounts of steam, to form a new atmosphere after the original mantle of light gases was lost to space. Most of the steam condensed to water, filling the oceans. The remainder of this early atmosphere seems to have been chiefly carbon dioxide, carbon monoxide, and nitrogen. The breakdown of these gases by sunlight and early forms of plant life produced an atmosphere that now contains about 78 percent nitrogen, 21 percent oxygen, a little less than 1 percent argon, together with traces of carbon dioxide, neon, helium, methane, krypton, and still rarer elements, as well as a variable amount of water vapor.

Like the Earth's interior, the atmosphere can be divided into several layers. The *troposphere* extends from the surface to an altitude of about 10 miles (16 km); this is the layer in which weather occurs, and at the top of the troposphere (the tropopause) the temperature is about $-60°$C. Above the troposphere is the *stratosphere,* a region in which temperature increases with altitude to a maximum of about $0°$C at 30 miles (50 km). The increase in temperature is caused by absorption of ultraviolet radiation from the Sun as oxygen (O_2) is converted into ozone (O_3). The ozone layer protects life on Earth by absorbing harmful ultraviolet radiation. Because the stratosphere becomes warmer with increasing altitude it is a very stable layer, preventing upward convection and acting as a "lid" on the weather systems of the troposphere below.

Above the stratosphere, temperatures decrease again in the *mesosphere,* down to a minimum of about $-100°$C at an altitude of 55 miles (90 km), and then increase again in the *thermosphere* until reaching equilibrium with space conditions. In this layer heating occurs because solar radiation splits oxygen molecules (O_2) into individual atoms (O). Speaking of the "temperature" of the tenuous outer layers is somewhat misleading, because although the molecules in the upper atmosphere may move at speeds corresponding to particular temperatures at sea level, the air is too thin to contain very much heat. It is in this region that the electrically-charged layers known collectively as the IONOSPHERE lie. AURORAE also occur in the thermosphere. Above about 300 miles (500 km) the thermosphere is also called the *exosphere,* because it is from this region that molecules from the atmosphere can leak away into space. Very high in the atmosphere charged particles interact with the Earth's magnetic field producing the radiation zones known as the VAN ALLEN BELTS, and the region known as the MAGNETOSPHERE, which for practical purposes marks the boundary of our home in space.

eccentricity
A measure of how elliptical an orbit is, or how far it deviates from being a true circle. The nearer the eccentricity is to zero, the closer the orbit is to being a circle, the nearer to one, the nearer the orbit is to a parabola.

Echo satellites
Two giant plastic balloons, inflated in space and used as *passive reflector* communications satellites. They were originally planned solely to study the effects of atmospheric drag, but they had a reflective aluminum coating added for communications purposes at the suggestion of the American communications engineer John Robinson Pierce (b. 1910). To the naked eye the Echo satellites appeared like bright stars drifting slowly across the sky. The Echo satellites were used for voice, teletype, and facsimile transmission; Echo 2 was used to exchange data with the Soviet Union, the first such joint East–West program.

Name	Launch date	Initial orbital elements			Notes
		perigee	*apogee*	*period*	
Echo 1	August 12, 1960	941 miles (1,514 km)	1,052 miles (1,693 km)	118 min.	Weight 166 lb. (75 kg), diameter 100 feet (30 m). Reentered May 24, 1968
Echo 2	January 25, 1964	642 miles (1,033 km)	816 miles (1,313 km)	109 min.	Weight 547 lb. (248 kg), diameter 135 feet (41 m). Reentered June 7, 1969

eclipse
Normally, the passage of one astronomical body into the shadow of another; but the term is also applied to the passage of the Moon in front of the Sun (a solar eclipse), though the event is more correctly termed an OCCULTATION of the Sun by the Moon.

Solar eclipses can occur only at new Moon, when the Moon comes between the Sun and Earth. The Sun is not eclipsed at each new Moon because the Moon's orbit is inclined by about $5°$ to the plane of the Earth's orbit. A solar eclipse can therefore happen only when the new Moon occurs near the two points where the lunar and terrestrial orbits cross (called the NODES). When the Moon is near its most distant from Earth (apogee), the tip of its conical umbra (dark central shadow) falls short of the Earth's surface by up to 20,300 miles (32,700 km). This produces a so-called annular eclipse, in which the dark disk of the Moon is surrounded by a bright ring (annulus) of sunlight. When the Moon is at its closest (perigee), its umbra covers a strip along the Earth's surface about 167 miles (269 km) wide. Outside this track, in the outer partial shade called the penumbra, observers over several thousand miles see a partial eclipse. Total solar eclipses last longest when the Sun is at its greatest distance (aphelion) and therefore appears smallest (early July), and the Moon is near perigee; this circumstance produced the longest eclipse of modern times, 7 minutes 14 seconds, on June 30, 1973.

Lunar eclipses occur when the Moon passes into the Earth's shadow, which it can do only at full Moon, when it is opposite the Sun in the sky. At the mean distance of the Moon, the Earth's shadow is about 5,700 miles (9,180 km) wide. A total lunar eclipse can last for up to 100 minutes, plus a further two hours during the Moon's passage in and out of the shadow.

Around this central umbra is the penumbra (diameter 10,200 miles or 16,400 km), in which the sunlight is only partly cut off by the Earth; taking this into account, a lunar eclipse as a whole can last up to nearly six hours. Because the atmosphere refracts sunlight into the Earth's shadow, the totally eclipsed Moon usually appears a dull copper color; variations are caused by changing atmospheric transparency. Lunar eclipses are rarer than solar eclipses, but, since they can be seen wherever the Moon is above the horizon, they are about equally frequent to any given observer. There are seldom more than two lunar eclipses in a year (1928 is an exception, with three), while there may be up to five solar eclipses.

Because the motions of the Earth, Moon, and Sun have been precisely calculated, eclipses can be accurately computed far into the past or future. A major work of this kind is the *Canon of Eclipses,* published in 1887 by the Austrian astronomer Theodor von Oppolzer (1841–1886), which contains tables of eclipses from 1208 B.C. to 2162 A.D.

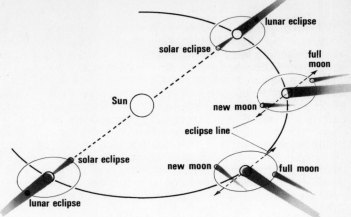

Eclipses can only happen when the plane of the Moon's orbit around the Earth and the Earth's orbit around the Sun coincide. At intermediate points, the new or full Moons are either above or below the Earth–Sun line and no eclipses appear.

Major Solar Eclipses until the year 2000

Date	Type	Duration of Totality	Central track
1977 October 12	Total	2 m 37 s	Pacific Ocean, Peru, Brazil
1979 February 26	Total	2 48	Pacific Ocean, U.S.A. Canada, Greenland
1980 February 16	Total	4 08	Atlantic Ocean, Congo, Kenya, Indian Ocean, India, China
1980 August 10	Annular		S. Pacific Ocean, Bolivia, Brazil
1981 February 4	Annular		Australia, New Zealand, Pacific Ocean
1981 July 31	Total	2 03	U.S.S.R., N. Pacific Ocean
1983 June 11	Total	5 11	Indian Ocean, East Indies, Pacific Ocean
1983 December 4	Annular		Atlantic Ocean, Equatorial Africa, Somalia
1984 May 30	Annular		Pacific Ocean, Mexico, U.S.A., Atlantic Ocean, Algeria
1984 November 22/23	Total	1 59	East Indies, S. Pacific Ocean
1985 November 12	Total	1 55	S. Pacific Ocean, Antarctic
1986 October 3	Annular/ Total	0 01	N. Atlantic Ocean
1987 March 29	Annular/ Total	0 56	Argentina, Atlantic Ocean, Congo, Indian Ocean
1987 September 23	Annular		U.S.S.R., China, Pacific Ocean
1988 March 18	Total	3 46	Indian Ocean, East Indies, Pacific Ocean
1988 September 11	Annular		Indian Ocean, south of Australia, Antarctic
1990 July 22	Total	2 33	Finland, U.S.S.R., Pacific Ocean
1991 January 15/16	Annular		Australia, New Zealand, Pacific Ocean
1991 July 11	Total	6 54	Pacific Ocean, Central America, Brazil
1992 June 30	Total	5 20	S. Atlantic Ocean
1994 May 10	Annular		Pacific Ocean, Mexico, U.S.A., Canada, Atlantic Ocean
1994 November 3	Total	4 23	Peru, Brazil, S. Atlantic Ocean
1995 April 29	Annular		S. Pacific Ocean, Peru, Brazil, S. Atlantic Ocean
1995 October 24	Total	2 05	Iran, India, East Indies, Pacific Ocean
1997 March 9	Total	2 50	U.S.S.R., Arctic Ocean
1998 February 26	Total	3 56	Pacific Ocean, Central America, Atlantic Ocean
1998 August 22	Annular		Indian Ocean, East Indies, Pacific Ocean
1999 February 16	Annular		Indian Ocean, Australia, Pacific Ocean
1999 August 11	Total	2 23	Atlantic Ocean, England, France, Central Europe, Turkey, India

Top right Known as the Seven Sisters, the Pleiades are a splendid example of an open galactic cluster. Their blue color reveals these stars as young type O or B objects. Wisps of gas and dust surrounding them reflect this color: the gas is not close enough to become "excited" and shine with its own characteristic colors.

Center right The most intriguing and important object in astrophysics—the Crab Nebula. Inside it is a pulsar, the spinning remnant of the star which exploded to create the whole nebula. The pulsar sustains the Crab's continued emission.

Below right One of the best known planetary nebulae in the sky, the Ring Nebula in Lyra. A medium sized telescope shows it as a pale ring, and long exposure photographs are needed to bring out the colors. The ring is in reality a shell of gas thrown off by the central star.

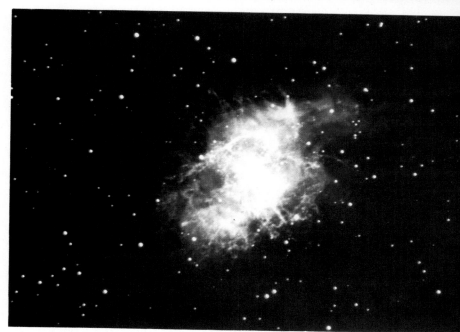

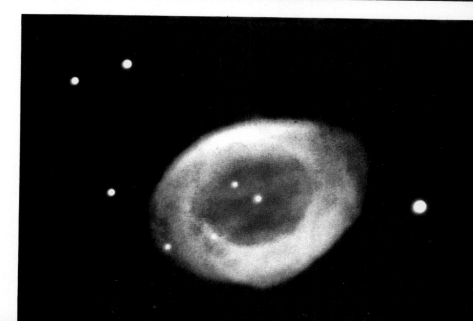

Above The brightest nebula in the sky, the Orion Nebula (M42) is easily visible to the naked eye as a greenish haze. Photographs reveal the colors due to hydrogen, oxygen, nitrogen and other gases glowing under the excitation of stars within the cloud.

Below The Lagoon Nebula (M8) in Sagittarius is one of the giant clouds of hydrogen in the Milky Way. The temperature of the gas in this nebula is about 7,000K, though it varies from point to point.

Lunar Eclipses until the year 2000

Date		Type	Duration of Totality	Moon overhead
1977	April 4	Partial		W. Brazil
1978	March 24	Total	1 h 30 m	Borneo
1978	September 16	Total	1 22	Indian Ocean
1979	March 13	Partial		Somalia
1979	September 6	Total	0 52	Polynesia
1981	July 17	Partial		N. Chile
1982	January 9	Total	1 24	Arabian Sea
1982	July 6	Total	1 42	Easter Island
1982	December 30	Total	1 06	Midway Island
1983	June 25	Partial		Pitcairn Island
1985	May 4	Total	1 10	Malagasy
1985	October 28	Total	0 42	Bay of Bengal
1986	April 24	Total	1 08	New Hebrides
1986	October 17	Total	1 14	Arabian Sea
1987	October 7	Partial (very small)		Venezuela
1988	August 27	Partial		Samoa
1989	February 20	Total	1 16	Philippines

Date		Type	Duration of Totality	Moon overhead
1989	August 17	Total	1 38	Brasilia
1990	February 9	Total	0 46	Bangalore
1990	August 6	Partial		W. Coral Sea
1991	December 21	Partial		Honolulu
1992	June 15	Partial		N. Chile
1992	December 10	Total	1 14	Sahara
1993	June 4	Total	1 38	New Hebrides
1993	November 29	Total	0 50	Mexico City
1994	May 25	Partial		S. Brazil
1995	April 15	Partial		Fiji
1996	April 4	Total	1 24	Gulf of Guinea
1996	September 27	Total	1 12	N.E. Brazil
1997	March 24	Partial		S. Colombia
1997	September 16	Total	1 06	Indian Ocean
1999	July 28	Partial		Tonga
2000	January 21	Total	1 24	West Indies
2000	July 16	Total	1 42	Coral Sea

The Moon blocks the Sun's brilliant disk from view during a total solar eclipse, revealing intricate structures in the Sun's chromosphere and lower corona. The symmetrical shape of the corona indicates that this photograph was taken near sunspot maximum, as do the large numbers of bright prominences.

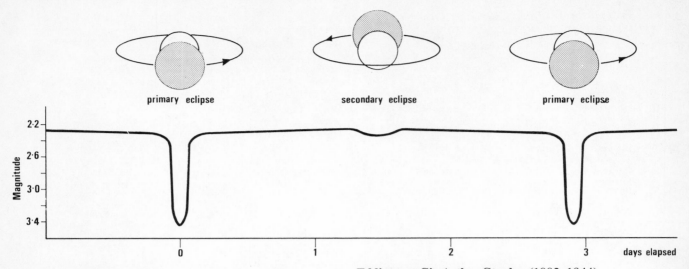

primary eclipse   secondary eclipse   primary eclipse

Algol is a typical eclipsing binary star. Its main drop in light (primary eclipse) occurs when the larger but cooler star passes in front of the smaller but hotter one. Secondary eclipse occurs when the larger star is behind the smaller; since the surface brightness of the larger star is so much lower than that of the smaller star, the drop in the total light at secondary eclipse is scarcely noticeable.

eclipsing binary

A system of two stars circling one another, with each periodically passing in front of the other and blocking off its light. The total light output of the star system therefore appears to vary periodically (see also VARIABLE STARS). ALGOL was the first such eclipsing binary to be noticed. There are two eclipses of Algol at each revolution; the deepest, so-called primary eclipse, occurs when the fainter yellow giant companion passes over the brighter blue dwarf star. Half an orbit later the blue star, which though brighter is smaller, blots out part of the yellow giant producing a shallow eclipse, the secondary minimum. The time taken for one star to move in front of another during eclipse reveals the size of the stars. Analysis of the light-curve of an eclipsing binary also reveals the angle at which the orbit of the stars is inclined to Earth. Varying X-ray sources discovered by recent scientific satellites (see X-RAY ASTRONOMY) are believed to be eclipsing binaries in which one of the components is a NEUTRON STAR or BLACK HOLE.

ecliptic

The plane of the Earth's orbit around the Sun. The path traced out by the Sun against the star background during the year corresponds to the plane of the ecliptic. Because of the tilt or inclination of the Earth's axis, the ecliptic is inclined to the celestial equator (the projection of the Earth's equator on the celestial sphere) at an angle known as the *obliquity of the ecliptic*. This angle slowly changes, due to the gradually changing inclination of the Earth's axis caused by the gravitational pulls of the Sun, Moon, and planets. At the beginning of 1976 it was 23° 26′ 32.66″, decreasing by about 0.45″ annually. The ecliptic takes its name from the fact that all lunar and solar eclipses occur when the Moon is on or near this plane.

Eddington, Sir Arthur Stanley (1882–1944)

British astrophysicist, noted for his pioneering investigations of the internal structure of stars, and his contributions to the understanding of relativity and cosmology. Eddington's early studies were of stellar distribution and motion, including an analysis of STAR STREAMING. In *Stellar Movements and the Structure of the Universe* (1914), he proposed that, contrary to general opinion, our Galaxy was only a small part of the total Universe. In 1917 Eddington established the basic theory of CEPHEID VARIABLE pulsations and in the early 1920s showed that the temperature of stars must rise to several million degrees at their centers to keep them from collapsing, and that the material inside them would be in the form of an ionized gas throughout. He developed a simple model for the interior of stars, showing how their gravity, gas pressure, and radiation pressure are interrelated. He found that radiation pressure increased so rapidly with mass that stars above about 50 solar masses could not exist because they would blow themselves apart. Eddington also championed the view that an atomic process of some kind was responsible for energy generation inside stars. From his theoretical work he calculated diameters for red giant stars which were confirmed by the measures of Albert MICHELSON. In 1924 Eddington announced the MASS-LUMINOSITY RELATION, which demonstrates the way the brightnesses of stars depend on their mass. Also in 1924 he predicted the ultra-high density of the WHITE DWARF companion of Sirius, which was confirmed when Walter S. ADAMS observed the gravitational red shift in its light. He summarized his work in *The Internal Constitution of the Stars* (1926).

Eddington is also noted for introducing the theory of relativity to the English-speaking public. He was early in grasping the significance of Einstein's work, and in 1919 led an expedition to Principe, an island off the west coast of Africa, to make the measurements at a total eclipse that first confirmed the prediction that starlight should bend as it passes through the Sun's gravitational field.

Effelsberg Radio Observatory

Location of the world's largest fully steerable radio telescope, 328 feet (100 m) in diameter. Effelsberg lies in the Eifel Mountains of West Germany, about

25 miles (40 km) west of Bonn. and is operated by the Max Planck Institute for Radio Astronomy in Bonn. The dish as a whole can operate at wavelengths down to 5 centimeters; the central part, 262 feet (80 m) wide, is surfaced to operate at wavelengths down to 1.2 centimeters. The telescope was put into operation in 1971.

Einstein, Albert (1879–1955)
German-born physicist, regarded as the greatest theoretical physicist of the century, and responsible for fundamental advances in a wide variety of fields. Einstein is most noted for his theory of RELATIVITY—a description of the structure of space and time. In his so-called Special Theory, published in 1905, Einstein replaced Newton's concepts of space, time, and the motion of bodies, which had gone unchallenged for more than two centuries, with a unified picture of *space-time.* The General Theory, which followed in 1916, was an even more radical departure, describing gravity in terms of such new concepts as the CURVATURE OF SPACE. Entirely new cosmological models have emerged from these ideas. Einstein's predictions were confirmed in 1919, when measurements at a total solar eclipse proved that starlight is bent on passing the Sun. In 1921 he was awarded the Nobel Prize in physics; this was in specific recognition of his explanation of the *photoelectric effect,* in which electrons are emitted from a surface struck by a beam of light. Einstein had demonstrated in 1905 that this could be explained by assuming that light acted as a stream of particles.

ELDO
The European Launcher Development Organization, set up in February 1964 to develop a space launcher based on the British Blue Streak strategic missile as a first stage. Other participants were Belgium, France, West Germany, Italy, and the Netherlands, with Australia providing the Woomera launch facilities. Although the consortium built two rocket designs, Europa I and Europa II, all launch attempts failed. The Europa program was subsequently canceled, and all European space activities were pooled in the EUROPEAN SPACE AGENCY. French desire for an independent launcher led to the development of the Ariane rocket.

electromagnetic radiation
The range of radiation, from gamma rays through the spectrum of visible light, to radio waves. Virtually all our information about the Universe, outside those parts of the solar system that can be sampled directly, reaches us in the form of electromagnetic radiation. Its EMISSION is the result of a change in speed or direction of charged particles, whether in the atom (producing SPECTRAL LINES) or on their own (as often for radio waves).

It can be thought of either as radiating waves, or as particles. Every charged particle has an electric and magnetic field associated with it, and as the particle is moved, the change in its position can be thought of as creating ripples in these fields. These ripples move at 186,000 miles (300,000 km) per second (see LIGHT, VELOCITY OF), the same for all forms of electromagnetic radiation. These ripples can also behave as particles, usually called *photons,* a term originally referring to light but now including all wavelengths.

The different forms of radiation are characterized by their WAVELENGTH or FREQUENCY. At the shorter wavelengths the particles are highly energetic, as demonstrated by the destructive nature of gamma rays and ultraviolet light. Longer wavelengths are increasingly less energetic.

Elektron satellites
Soviet satellites placed in orbit to study the Earth's Van Allen radiation belts. Two pairs were launched; one of each pair was sent to examine the outer section of the belt, the other the inner section. Elektrons 1 and 2 were launched in January 1964; Elektrons 3 and 4 followed in July of that year.

elements, abundance of
The chemical composition of the Sun and bright nebulae is revealed by analysis of their light (see SPECTROSCOPY). Taken as a whole, they have the so-called "present cosmic abundances" shown below. Although the Earth, Moon, and meteorites lack the light gases hydrogen and helium predominant elsewhere in space, the abundance of their other, heavier elements is remarkably similar to that in the Sun and other young stars. The oldest stars in our Galaxy are, however, deficient in elements heavier than helium. These stars often have a thousandth the heavy elements that the Sun has, and also slightly less helium. Since the surface layers of a star retain the cosmic abundance of the time at which it formed, it seems that the Milky Way originally condensed from a cloud of only hydrogen and helium. All the heavier elements (and some additional helium) have been produced by nuclear fusion in stars, and have been scattered throughout interstellar space by NOVA and SUPERNOVA explosions. Subsequent generations of

Present Cosmic Abundances of the Elements

Element	Symbol	Weight relative to hydrogen	Number of atoms to one million hydrogen atoms
Hydrogen	H	1	1,000,000
Deuterium (H_2)	D	2	100
Helium	He	4	73,000
Lithium	Li	7	0.002
Beryllium	Be	9	0.00003
Boron	B	11	0.001
Carbon	C	12	370
Nitrogen	N	14	120
Oxygen	O	16	680
Neon	Ne	20	110
Magnesium	Mg	24	33
Silicon	Si	28	31
Sulfur	S	32	16
Iron	Fe	56	26
All other elements			15

stars condensing from this gas are therefore progressively more enriched in the HEAVY ELEMENTS.

The helium present when the Milky Way formed (about 80 percent of the amount now observed) was produced from hydrogen in the very hot and dense early Universe, only 100 seconds after the BIG BANG in which the Universe originated. Deuterium (heavy hydrogen) can only have been formed in the big bang, since it reacts to form helium during the formation of stars. Lithium, beryllium, and boron are also destroyed in stars, and so are very rare in space. These three elements seem to be formed when carbon, nitrogen, and oxygen nuclei are broken up by the very fast COSMIC RAY particles (a process called SPALLATION).

ellipse
A shape somewhat like a squashed circle, produced by cutting a cone at an angle. An ellipse, being symmetrical, has two *foci*; the farther apart these are, the more flattened, or eccentric, the ellipse (see ECCENTRICITY).

ellipsoid
A solid surface of which any cross section is an ELLIPSE. So-called ellipsoidal mirrors for reflecting telescopes are in fact parts of prolate SPHEROIDS, since the cross section in the plane of the mirror is a circle and not an ellipse.

elongation
The angle between the Sun and a planet (generally Venus or Mercury), or between a planet and its satellite, as seen from Earth. *Greatest elongation* is the maximum angular separation.

emission
The production of ELECTROMAGNETIC RADIATION, including visible light and radio waves. Radiation is emitted when atomic particles (in practice, usually electrons) are accelerated or decelerated. Electrons can jump between one orbit and another in atoms when excited by the addition of energy. This produces the bright SPECTRAL LINES characteristic of nebulae. By a change in the direction of spin of the atomic particles, the 21-centimeter wavelength radio line of cold hydrogen gas is produced. Electrons moving freely in space may be decelerated or accelerated by close approaches to other particles. The radiation produced in this way is called BREMSSTRAHLUNG. Where electrons move at many different speeds, as in a hot object, the same process gives rise to the more familiar black body radiation, as emitted by stars or any hot bodies. Electrons may also be forced to move in curved paths by magnetic fields. This gives rise to SYNCHROTRON RADIATION.

emission nebula
A nebula that glows by the emission of light from its constituent gas (see NEBULA).

Enceladus
Third satellite of Saturn in order of distance from the planet, discovered in 1789 by William Herschel. It orbits Saturn every 32 hours 53 minutes at a distance of 147,950 miles (238,100 km). Enceladus has a diameter of about 375 miles (600 km).

Encke, Johann Franz (1791–1865)
German astronomer, who calculated in 1819 that a comet observed the previous year had the shortest period of any comet known; it is now called ENCKE'S COMET. He later deduced the masses of Mercury and Jupiter from the effects of their gravity on the comet's orbit. In 1825 Encke was appointed director of Berlin Observatory and began its reconstruction; the new observatory was opened in 1835. There he began the creation of a new star chart, which made possible the discovery by J. G. GALLE of the planet Neptune. Encke also analyzed in detail the observations of the transits of Venus in 1761 and 1769, deriving a distance for the Sun from Earth of just over 95 million miles. In 1837 Encke discovered a partial gap in Saturn's outer ring, now called *Encke's division*.

Encke's comet
The comet with the shortest known period—3.3 years. It is named for J. F. ENCKE, who in 1819 computed the orbit of a comet discovered the previous year by the French comet-hunter Jean Louis Pons (1761–1831), and identified it with comets observed in 1786, 1795, and 1805. In its orbit of the Sun, Encke's comet moves from inside the orbit of Mercury to about three-fourths the distance of Jupiter; it is a member of Jupiter's family of comets (see COMET). Encke's comet can be seen throughout its entire orbit, and it makes its next return to perihelion in 1977. The comet has a nucleus some 2 miles (3.2 km) across, with a mass of 100 billion tons. Since 1800 it has been slowly fading by 2 magnitudes a century; the gases it produces at each return have declined from 10 million tons to less than 1 million tons. Now decaying rapidly as an active object, it is expected to degas completely to form a dark, asteroid-like body within 60 to 100 years. Encke's comet is responsible for the TAURID meteor shower.

ephemeris
Collection of tables of the positions of a planet, satellite, or comet in the sky (plural: *ephemerides*). Planetary ephemerides can be calculated with confidence for many years ahead, as can the positions of most of their satellites, but the orbits of most comets are less certain because they can be observed over only a small part of their total path. Ephemerides for newly-discovered comets thus often require considerable revision as more observations are made available. The principal publications dealing with planetary and satellite ephemerides are the annual *American Ephemeris* and *Astronomical Ephemeris,* published in the United States and Britain respectively, with identical contents and produced in collaboration. Predictions for minor planets (asteroids) are issued by the Institute of Theoretical Astronomy at Leningrad.

epicycle
Device once used by astronomers who explained the movement of the planets by a combination of circles. An epicycle was a small circle on the rim of a bigger circle called the *deferent*. By adjusting the rate of movement of a planet around its epicycle, and the epicycle's rate of movement around the deferent, any required final motion could be reproduced. Such complex and clumsy systems were rendered

unnecessary when Johannes KEPLER showed in 1609 that the planets' orbits are actually ellipses.

epoch

A standard date to which star positions are referred. The effects of PRECESSION and NUTATION change the celestial coordinates of an object over time. The epoch gives a reference date on which to base the positions in a star catalog or atlas. Dates used currently are the years 1950 or 2000.

equation of time

The correction needed to account for the difference between mean solar time, as shown by a regularly running watch, and apparent solar time, as recorded by a sundial. The difference arises because the real Sun progresses across the sky at an irregular rate during the year, since the Earth's orbit is elliptical and its axis is tilted so that the Sun does not move along the celestial equator. Astronomers have invented a fictional MEAN SUN, which moves around the celestial equator at a constant rate, to serve as a basis for regular timekeeping. The greatest difference between time by the mean Sun and time by the real Sun occurs at the beginning of November, when the real Sun is due south about $16\frac{1}{2}$ minutes before noon, mean time. In February it is nearly 15 minutes late. The equation of time is zero about April 15, June 13, September 1, and December 25.

equator

The line around a rotating celestial body that is equidistant from the poles, defined as 0° latitude. The *celestial equator* is the projection of the Earth's equator on the CELESTIAL SPHERE, and from it the celestial equivalent of latitude, called DECLINATION, is measured.

equatorial mounting

A mounting for astronomical telescopes that has one axis (the *polar axis*) aligned with the axis of the Earth. The daily rotation of the Earth can be counteracted simply by driving the telescope about its polar axis at a rate of one revolution per day. Adjustment of the other axis (the *declination axis*) is made only during acquisition of the object and to compensate for guidance errors. Once the telescope is pointed at an object, the driving motor keeps the target in view as it moves across the sky. In contrast, the simple ALTAZIMUTH mounting, with axes in the vertical and horizontal planes, must be constantly adjusted in both

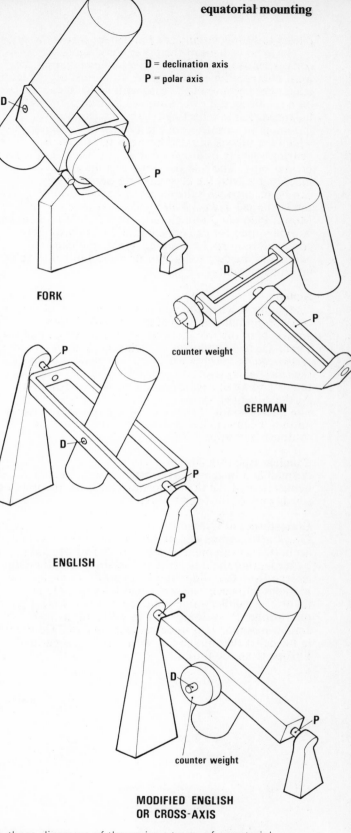

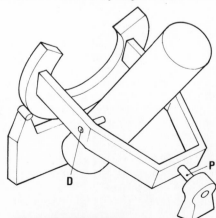

YOKE OR HORSESHOE

D = declination axis
P = polar axis

FORK

GERMAN

counter weight

ENGLISH

MODIFIED ENGLISH OR CROSS-AXIS

counter weight

In these diagrams of the various types of equatorial mounting, the polar axis in each case is shown by P, the declination axis by D. All are set to the same latitude— the angle which the polar axis makes with the horizontal is always the same.

planes to follow an object's movement across the sky.

The arrangement of the polar and declination axes can take various forms. The *German mounting* is the most compact type, and is used for almost all refracting telescopes because of the ease with which it can be set on a relatively tall pier; however, it is not very well balanced. Large reflecting telescopes are usually mounted on variations of the *English mounting,* in which the tube is situated between the two polar bearings, giving improved stability at the cost of much greater bulk. The celestial polar regions are inaccessible with the classical English mounting; they can be reached with the *modified English mounting,* or the U-shaped *Yoke mounting* like that used on the Mount Palomar 200-inch (508-cm). The *Fork mounting* is widely used for cameras, because it is reasonably compact, requires no counterweight, and can command the entire sky; but its use is limited to very large instruments.

equinox
The moment when the Sun crosses the celestial equator, and day and night are of equal length anywhere in the world (equinox means "equal night"). There are two equinoxes in each year. The *spring* or *vernal equinox* occurs when the Sun passes north of the celestial equator, on or near March 21; it marks the beginning of spring in the northern hemisphere. At the *autumnal equinox,* around September 23, the Sun passes south of the celestial equator and northern autumn begins. (The seasons are reversed for the southern hemisphere.)

Equuleus (the little horse)
A small and insignificant constellation in the northern hemisphere of the sky adjacent to Pegasus. Its brightest star is only of the fourth magnitude.

Eratosthenes of Cyrene (c.276–c.194 B.C.)
Greek astronomer who about 240 B.C. made the first accurate measurement of the Earth's size. He did so by measuring the difference in the altitude of the Sun as seen from two different places on Earth at the same moment. However, his work was later ignored in favor of a smaller value obtained by the Greek philosopher Poseidonius (c.135–c.50 B.C.) using a similar method. This underestimate inspired Columbus in his belief that only a few weeks' sailing separated Europe from the Indies.

Eridanus (the river)
A long constellation extending from the equator far into the southern sky. Its brightest star, Alpha Eridani, known as Achernar, is one of the brightest in the sky, of magnitude 0.47. Epsilon Eridani is a nearby star quite like the Sun, to which astronomers have listened for possible radio messages from other civilizations (see LIFE IN THE UNIVERSE). The star Omicron Eridani (also known as 40 Eridani), 15.9 light-years away, is a remarkable triple system; the main star resembles the Sun, the second star is a faint white dwarf, and the third an even fainter red dwarf. NGC 1300 is a well-known spiral galaxy in Eridanus.

Eros
An asteroid of the Amor group, also known as minor planet 433, discovered on photographs taken by Gustav Witt in Berlin and Auguste Charlois in Nice on August 13, 1898. It moves in a 1.81-year orbit between 1.13 and 1.8 a.u. from the Sun, inclined 11° to the ecliptic. Eros is a cigar-shaped body about 22 by 10 by 4 miles ($35 \times 16 \times 6$ km), rotating about its smallest axis every $5\frac{1}{4}$ hours. The irregular shape suggests it has been broken off a larger body. Eros has a reflectivity (albedo) of 14.2 percent and probably possesses a stony or stony-iron surface. In exceptional circumstances, Eros can pass quite close to the Earth, as in 1894, 1931, and 1975. In January 1975 the planet passed within 14 million miles (22.5 million km), appearing as a swiftly moving seventh-magnitude object. Until the advent of radar astronomy, observations of such close passages were used to define the exact value of the ASTRONOMICAL UNIT.

ERTS
Abbreviation for Earth Resources Technology Satellite, the original name for NASA's LANDSAT program. The series was renamed prior to the launch of Landsat 2 on January 22, 1975.

escape velocity
The speed an object must attain to escape from a gravitational field. Escape velocity depends on the mass of the body responsible for the gravitational field; the Earth, for instance, has an escape velocity at its surface of 6.95 miles (11.2 km) per second, while the less massive Moon has an escape velocity

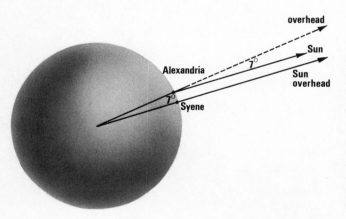

Eratosthenes knew that at noon on the summer solstice, the Sun appeared exactly overhead at a place in Egypt called Syene, near the modern Aswan. Yet on the same date at Alexandria, some way north, the Sun appeared about 7° from the vertical—roughly one fiftieth of a circle. Therefore the distance round the entire globe should be about fifty times the distance between Syene and Alexandria, which, in Greek units, was 5,000 *stadia*; hence the circumference of the Earth, according to Eratosthenes, was about 250,000 stadia. Modern research shows that the Greek *stadium* was equal to about one tenth of a mile, giving a figure for the Earth's circumference in modern units of 25,000 miles—almost exactly right.

of 1.5 miles (2.4 km) per second. A body with a small mass cannot retain an atmosphere, because gas particles of the atmosphere will move faster than the escape velocity. For an object already in circular orbit at a given distance, the escape velocity is $\sqrt{2}$ times the velocity required to maintain it in that orbit.

ESRO
The European Space Research Organization, an association of 10 countries—Belgium, Denmark, France, West Germany, Italy, Netherlands, Spain, Sweden, Switzerland, and the United Kingdom—set up in 1964 for cooperation in space research and satellite development. ESRO launched 184 sounding rockets for upper-atmosphere research, and built seven scientific satellites for launch by the United States. In 1975 ESRO merged with the European Launcher Development Organization ELDO to form the EUROPEAN SPACE AGENCY (ESA).

ESSA satellites
Series of meteorological satellites launched by the Environmental Science Services Administration. They superseded the earlier Tiros series, and were also known as the Tiros Operational Satellite (TOS) system. ESSAs 2, 6, and 8 used television cameras with automatic picture transmission (APT) to broadcast continuous cloud-cover photographs to any ground station. The other ESSA satellites stored their photographs for transmission on command to specific ground stations. The ESSA satellites were placed into near-polar orbits so that they covered the entire Earth each day; new satellites were orbited to replace previous ones as picture quality declined with age. In October 1970 the functions of the Environmental Science Services Administration were assumed by the newly created National Oceanic and Atmospheric Administration, which launched the subsequent NOAA series of WEATHER SATELLITES.

ether
The imaginary medium in which light was assumed to travel. The concept arose in the 19th century, when it was discovered that light travels in waves. By analogy with sound waves, which need air through which to travel, scientists supposed that light waves were transmitted through a medium called the ether. It was assigned curious properties: it had to fill all of space, be transparent and weightless, and allow bodies such as the Earth to move without any friction. By the turn of the century, the theory was abandoned because the MICHELSON-MORLEY EXPERIMENT had failed to confirm it, and because the behavior of light could be explained adequately without its use.

Eudoxus of Cnidus (c.400–c.350 B.C.)
Greek astronomer who produced the first detailed, if fanciful, system to account for the observed motions of the planets. Eudoxus was a pupil of the Greek philosopher Plato (c.427–c.347 B.C.) who, like PYTHAGORAS before him, believed that the heavens were perfect, and that celestial objects could thus move only in circles, which were considered the "perfect" shape. Eudoxus realized from his own observations that the planets did not move across the sky on simple circular paths. To explain the observed motions he introduced for each celestial object a nest of spheres, with the axis of each sphere set into the surface of the others. Eudoxus' scheme required a total of 27 spheres: one for the stars, three each for the Sun and Moon, and four each for the planets Mercury, Venus, Mars, Jupiter, and Saturn. The motions of the celestial objects could thus be accounted for by combinations of "perfect" motion, and the concept of heavenly spheres, particularly as later developed by ARISTOTLE, became firmly entrenched in astronomical thought. Eudoxus' scheme was in fact not totally successful in reproducing the motions of the celestial bodies, and his follower Callippus (c.370–c.300 B.C.) added another seven spheres, bringing the total to 34.

Europa
Smallest of the four main satellites of Jupiter, discovered in 1610 by Galileo. It orbits Jupiter every 3 days, 13 hours, 14 minutes, at a distance of 416,970 miles (671,050 km). Europa's diameter is about 1,900 miles (3,100 km), its density is three times that of water, and its mass two-thirds that of our Moon. Part of Europa's surface is covered with water frost, and detailed observations suggest that it possesses polar caps extending to about 30° from its equator. Europa is believed to be a rocky body, like the Moon.

European Southern Observatory
Astronomical observatory on Cerro La Silla, a mountain in northern Chile about 55 miles (90 km) northeast of the city of La Serena. The observatory, at an altitude of 8,000 feet (2,500 m), has as its major telescope a 142-inch (360-cm) reflector; other important instruments include a 39-inch (100-cm) Schmidt telescope and a 59-inch (150-cm) spectrographic telescope. Formally opened in 1969, the observatory is run by a consortium of Belgium, Denmark, France, West Germany, the Netherlands, and Sweden.

European Space Agency (ESA)
An organization of 11 European countries—Belgium, Denmark, Eire, France, West Germany, Italy, Netherlands, Spain, Sweden, Switzerland, and the United Kingdom—for space research, satellite applications, and rocket development. ESA began formal operation on May 31, 1975, as successor to the joint European Space Research Organization (ESRO) and the European Launcher Development Organization (ELDO).

The first ESA satellite was COS-B, launched for gamma-ray studies on August 9, 1975. The most adventurous of all ESA's projects is Spacelab, a two-part scientific laboratory to be flown in the cargo bay of the SPACE SHUTTLE, from which it will make astronomical and Earth observations. A pressurized module will allow scientists to control the operation of the instruments, which will be on an unpressurized pallet, giving a direct view of the sky. When not working in Spacelab, the crew of up to four will rest in the Orbiter. Each orbital flight of Spacelab will last from 7 to 30 days; after each flight it will return in the shuttle Orbiter, and will thus be used up to 50 times during a 10-year life. The payload can be modified for each flight, and will probably also include manufacturing tests and space medicine experiments.

Spacelab will be manned by American and European crews, each sharing its experimental facilities; it will be the only orbiting workshop available in the West during the early years of the shuttle's operation.

evening star
Popular term for the planet Venus when it appears shining brilliantly in the evening sky shortly after sunset. When Venus is rising in the morning sky before the Sun, it appears as the brilliant morning star.

exobiology
The study of life beyond the Earth (see LIFE IN THE UNIVERSE).

expanding Universe
The apparent recession of distant galaxies with speeds which increase proportionally to their distance. Because light from a receding object is shifted by the DOPPLER EFFECT toward the red end of the spectrum, the observational evidence for the expansion of the Universe is often referred to as the RED SHIFT. The relationship between a galaxy's red shift and its distance is called HUBBLE'S LAW. At very large distances, as the recession speed approaches the speed of light, the relationship breaks down. Different cosmological theories predict different amounts of breakdown (different values of the *deceleration parameter*), so that they could in principle be tested by observations, but in practice observations of the faint galaxies distant enough to show this breakdown are still too difficult to make with accuracy. Although we see all galaxies receding from us, we are not a center of repulsion. Actually, the expansion of the Universe resembles a spotted balloon being blown up: the distance between all spots is increasing, at a rate proportional to the distance between them (see also UNIVERSE).

Explorer satellites
Series of American scientific satellites, begun by the Advanced Research Projects Agency (ARPA) and transferred to NASA on its formation. Explorer 1 was the first U.S. satellite in orbit and discovered the Earth's Van Allen radiation belts. Explorer 6, nicknamed the paddlewheel satellite because of its outspread solar panels, transmitted the first crude television pictures of the Earth from orbit. A series of Interplanetary Monitoring Platform (IMP) spacecraft was later started within the Explorer program, as were the Small Astronomy Satellites (SAS). The Navy Solrad (solar radiation) and Injun satellites also came within the Explorer series. Other branches of the Explorer program are concerned with the structure of the atmosphere and with radio astronomy.

Explorer Satellites

Satellite	Launch date	Remarks
Explorer 1	January 31, 1958	First U.S. satellite in orbit. Discovered Van Allen belt. Reentered March 31, 1970
Explorer 2	March 5, 1968	Launch failure
Explorer 3	March 26, 1958	Radiation and micrometeoroid measurements

Satellite	Launch date	Remarks
Explorer 4	July 26, 1958	Radiation data
Explorer 5	August 24, 1958	Launch failure. (After this, launch failures were not given Explorer numbers)
Explorer 6	August 7, 1959	First television pictures of Earth
Explorer 7	October 13, 1959	Magnetic field and solar flare measurements
Explorer 8	November 3, 1960	Ionospheric research
Explorer 9	February 16, 1961	12-foot balloon to study atmospheric density
Explorer 10	March 25, 1961	Magnetic field measurements from near Earth to halfway to Moon
Explorer 11	April 27, 1961	Gamma ray measurements
Explorer 12	August 15, 1961	Radiation and solar wind measurements
Explorer 13	August 25, 1961	Micrometeoroid satellite
Explorer 14	October 2, 1962	Studied Earth's magnetosphere
Explorer 15	October 27, 1962	Radiation monitor
Explorer 16	December 16, 1962	Micrometeoroid measurements
Explorer 17 (AE-A)	April 2, 1963	Atmospheric Explorer
Explorer 18 (IMP A)	November 26, 1963	First Interplanetary Monitoring Platform; measured radiation between Earth and halfway to Moon
Explorer 19	December 19, 1963	12-foot balloon; identical to Explorer 9
Explorer 20	August 25, 1964	Ionospheric research
Explorer 21 (IMP B)	October 3, 1964	Magnetic field, cosmic ray, and solar wind studies
Explorer 22	October 9, 1964	Ionospheric and geodetic research
Explorer 23	November 6, 1964	Micrometeoroid measurements
Explorer 24	November 21, 1964	12-foot balloon to study atmospheric density
Explorer 25	November 21, 1964	Radiation data; same launch rocket as Explorer 24
Explorer 26	December 21, 1964	Radiation monitor
Explorer 27	April 29, 1965	Geodetic and ionospheric research
Explorer 28 (IMP C)	May 29, 1965	Studied magnetic field and radiation between Earth and Moon
Explorer 29 (GEOS-1)	November 6, 1965	First Geodetic Earth-Orbiting Satellite (GEOS). Carried flashing lights and laser reflectors for geodetic measurements
Explorer 30	November 18, 1965	Solar radiation satellite*

Satellite	Launch date	Remarks
Explorer 31	November 28, 1965	Ionospheric studies; complemented work of Canadian Alouette satellite launched with same rocket
Explorer 32 (AE-B)	May 25, 1966	Atmospheric Explorer
Explorer 33 (IMP D)	July 1, 1966	Magnetic field studies from Earth to beyond Moon. Lunar orbit intended, but failed
Explorer 34 (IMP F)	May 24, 1967	Measured radiation and magnetic field between Earth and halfway to Moon
Explorer 35 (IMP E)	July 19, 1967	In orbit around Moon; measured Earth's magnetic "tail"
Explorer 36 (GEOS-2)	January 11, 1968	Geodetic measurements
Explorer 37	March 5, 1968	Solar radiation satellite
Explorer 38 (RAE-A)	July 4, 1968	First Radio Astronomy Explorer, monitoring radio emissions from celestial and Earth sources
Explorer 39	August 8, 1968	Air Density Explorer; 12-foot balloon
Explorer 40	August 8, 1968	Injun Explorer, measured radiation†; launched by same rocket as Explorer 39
Explorer 41 (IMP G)	June 21, 1969	Magnetic fields and radiation measurements between Earth and halfway to Moon
Explorer 42 (SAS-A Uhuru)	December 12, 1970	First Small Astronomy Satellite, also known as Uhuru. Measured X rays from space
Explorer 43 (IMP I)	March 13, 1971	Studied Earth's magnetosphere out halfway to Moon
Explorer 44	July 8, 1971	Solar radiation satellite
Explorer 45	November 15, 1971	Magnetospheric research
Explorer 46	August 13, 1972	Meteoroid technology satellite
Explorer 47 (IMP H)	September 23, 1972	Measured radiation and magnetism at a distance about halfway to the Moon
Explorer 48 (SAS-B)	November 16, 1972	Measured gamma rays from space
Explorer 49 (RAE-B)	June 10, 1973	Radio Astronomy Explorer, in orbit around Moon, recording radio noise from Sun and Galaxy
Explorer 50 (IMP J)	October 25, 1973	Monitored solar flares and radiation from high orbit. Final IMP
Explorer 51 (AE-C)	December 15, 1973	Atmospheric Explorer

Satellite	Launch date	Remarks
Explorer 52	June 3, 1974	Hawkeye scientific satellite, studied interaction between solar wind and Earth's magnetic field
Explorer 53 (SAS-C)	May 7, 1975	Third Small Astronomy Satellite, studying X-ray sources
Explorer 54 (AE-D)	October 6, 1975	Atmospheric Explorer
Explorer 55 (AE-E)	November 19, 1975	Atmospheric Explorer, including upper-atmosphere ozone monitor

*Successor to the U.S. Navy Solrad series. Successful Solrad launches were: Solrad 1, June 22, 1960; Solrad 3, June 29, 1961; Solrad 6, June 15, 1963; Solrad 7a, January 11, 1964; Solrad 7b, March 9, 1965. The series was continued with Explorer 37 and Explorer 44. (Confusingly, the early Solrads are sometimes also called Grebs.)
†Previous announced successful Injun launches by U.S. Navy: Injun 1, June 29, 1961; Injun 3, December 12, 1962.

eyepiece

A lens or system of lenses for examining the image formed by a telescope's object glass or mirror. An eyepiece allows the eye to view the image from very close range; the magnification increases as the distance is reduced. The magnification can be expressed by dividing the FOCAL LENGTH of the telescope by the focal length of the eyepiece. Eyepiece focal lengths are normally quoted in millimeters, and values of between 4 and 40 mm are common. Most eyepieces consist of two spaced lenses or systems of lenses; that nearest the eye is called the *eye lens,* the other the *field lens,* because its purpose is to increase the field of view. Different eyepieces designs give different performance, and there are a number of well-known types, generally named for their designers. The main aspects to be considered are apparent field of view (the diameter, in degrees, of the circle of light seen by the eye); correction for SPHERICAL and CHROMATIC ABERRATION; eye relief, or the clearance between the face of the eye lens and the observer's eye; and freedom from scattered light and ghost images of bright objects caused by internal reflections.

Among simpler eyepieces are the *Huygenian* and *Ramsden* types, consisting of two separated plano-convex lenses. The *achromatic Ramsden,* with a double eye lens, is a common improvement. Among highly-corrected designs are the *orthoscopic* (a family of eyepieces of wide variation, but consisting generally of a single eye lens and a triple cemented field lens); the *monocentric* (a triple cemented lens), and the *Erfle* (a complicated system with three separate elements). The *Tolles* is an unusual but effective eyepiece design, consisting of a single short cylinder of glass with convex faces.

Dividing the apparent field of view of an eyepiece (generally no more than about 35–40° to avoid distortion) by the system's magnification gives the field of view actually seen through the telescope, the so-called *true field.*

F

facula

A bright patch seen near the edge of the Sun. Faculae usually occur near sunspots, but appear before them and also tend to outlive them. They are brighter and hotter than their surroundings high up in the solar atmosphere, so that the contrast is greatest at the

Sun's edge or *limb*. Faculae seem to correspond to regions of high magnetic field, which are presumably involved in the heating of the facular regions.

fireball

A meteor of magnitude −5 or brighter. Occasionally fireballs can be as bright as magnitude −20 to −25 (the full Moon is magnitude −12.7). They are produced by bodies as large as several feet across. Exploding or flaring fireballs, known as *bolides,* are common; they occur when the parent meteoroid disintegrates shortly after entering the atmosphere. Bodies producing fireballs of magnitude −10 or brighter are massive enough to lead to meteorite falls, but fewer than 1 percent reach the ground without disintegrating into dust.

Flamsteed, John (1646–1719)

First astronomer royal of England, appointed on March 4, 1675, by King Charles II. He was set the task of producing tables of the Moon's motion and star positions to enable seamen to determine their longitude, then the major problem in navigation. Finding longitude meant comparing the time locally, obtained by observing the Sun or stars, with a standard time reference. One suggestion for such a "standard clock" was the position of the Moon among the stars as it moved round its orbit. But the positions of objects in the sky were not then known precisely enough to provide the required accuracy, and so Flamsteed, a promising young astronomer, was given the task.

In 1676 the Royal Observatory was built for him in Greenwich Park, but he was given no instruments or assistants to accomplish the job. He borrowed and bought instruments and clocks, and began to teach private pupils to supplement his income. Flamsteed's ambition was to produce a major catalog of unprecedented accuracy. He was reluctant to part with his observations until they were complete, but Isaac Newton nevertheless arranged for some of the work to be published. The result was an inaccurate and incomplete catalog, many copies of which Flamsteed later destroyed.

The catalog on which Flamsteed's fame securely rests, the three-volume *Historia Coelestis Britannica* containing the positions of 2,935 stars, was published only posthumously, in 1725; it was the first major star catalog produced with the aid of a telescope, and is one of the foundation stones of modern astronomy. Flamsteed's catalog superseded that of the great Tycho BRAHE; it was far more extensive, and also six times as accurate. In 1729 a set of star maps based on the catalog was published, called *Atlas Coelestis.*

Flamsteed cataloged the stars in each constellation in order of increasing right ascension; other astronomers later numbered the stars to give the so-called Flamsteed numbers still used today. In 1694 he found a regular shift in position of the pole star which he thought was evidence of its PARALLAX; but its actual explanation was the ABERRATION OF STARLIGHT, which was later discovered by James BRADLEY. Flamsteed's planetary observations showed that the motions of Jupiter and Saturn are affected by each other's gravitational pull. And he deduced from observations of sunspots that the Sun's rotation period is about $25\frac{1}{4}$ days. The original building of the observatory at Greenwich is now called Flamsteed House.

flare

A brilliant burst of light occuring near a SUNSPOT. Flares are short-lived phenomena, lasting no more than a few hours, and are more common near SOLAR CYCLE maximum. They send out streams of particles which produce effects on Earth such as aurorae and radio blackouts. Flares appear to be produced by bursts of magnetic energy. They occur in the chromosphere and, being composed mainly of hydrogen, are best seen at the hydrogen wavelength known as the Hα spectral line.

An intensely bright solar flare of August 7, 1972, photographed in the red spectral line of hydrogen. This flare was one of many which occurred during that month, all associated with a particularly complex sunspot group. Several of these gave rise to extremely large disturbances in the Earth's magnetosphere.

flare star

Any star which suddenly increases its brightness and then fades to its former level. Some 500 are known. Flares typically last a few minutes and can be so intense as to multiply the brightness of the star a hundredfold. A flare star is usually a young star just beginning to burn hydrogen, and the flare activity represents its attempts to adjust its structure in order to achieve stability. The flares probably die away in intensity and frequency as the star becomes a stable dwarf, until they are as infrequent and as small as those occurring on the Sun (see also UV CETI STARS).

floccule

A striking bright or dark feature of the Sun's CHROMOSPHERE. Observations of the chromosphere at specific wavelengths emitted by hydrogen or calcium reveal a network of chains, roughly 20,000 miles (32,000 km) in diameter, which outline giant convection cells in the chromosphere called *supergranules*. In hydrogen light, the dark floccules are more prominent; in calcium light bright floccules outline the network. The floccules are themselves composed of smaller bright and dark mottles. The dark floccules are, in fact, not completely dark but show a "rosette" structure of bright and dark mottles. The flow of gas in the convection cells is thought to concentrate magnetic fields at the edge of the supergranules, thus forming floccules. Floccules are actually the same phenomena as SPICULES, only seen head-on directly in front of the Sun's disk.

flying saucers

See UFO.

focal length

The distance between the lens or mirror of an optical instrument and the image it forms of an object which is at infinite distance (as, in optics, all astronomical objects are assumed to be). The longer the focal length, the larger the scale of the image formed, but the smaller the field of view.

focal plane

The plane in which an optical instrument forms an image. Many systems, however, form their images in a curved field, rather than a plane, including the SCHMIDT TELESCOPE. To record such an image, a photographic plate must also be curved.

focal ratio

The focal length of a lens or mirror divided by its aperture; also termed *f* ratio. For an instrument of given aperture, a small focal ratio gives a shorter tube length, a wider field of view, and greater image brightness (it is said to be "faster," like a camera lens). Long-focus instruments give a larger but fainter image and are unwieldy. Typical focal ratios for visual instruments are *f*/5 to 8 for reflecting telescopes, and *f*/12 to 20 for refracting telescopes; compound and photographic instruments range from about *f*/2 to 30.

Fomalhaut

One of the brightest stars in the sky, of magnitude 1.16. It is a white star, 1.56 times the Sun's diameter, and 23 light-years away. Fomalhaut is the brightest star in the constellation of Piscis Austrinus.

forbidden lines

Bright lines which can appear in the spectra of certain objects such as gaseous nebulae, but which are not found in laboratory spectra. They are emission lines, caused when an atom's electron jumps from one orbit to another. In the denser gases familiar on Earth, the Electron is normally knocked out of such an orbit by a collision before it has a chance to jump. Therefore, the probability of observing such an energy jump under terrestrial conditions is very low—it is said to be forbidden. In thin gases, however, such as in a nebula, the chances of a collision are much reduced and the electron can remain in its orbit until ready to jump. Before the development of the concept of forbidden lines, two mysterious "elements" had been postulated to explain mysterious emission lines. "Nebulium" had unique green lines; it is in fact due to oxygen with two electrons missing. "Coronium," observed in the Sun's corona, produced emission lines now known to be forbidden lines principally of iron and nickel, heavily ionized.

Fornax (the furnace)

A small and faint constellation in the southern hemisphere of the sky adjacent to Eridanus, introduced by Nicolas Louis de Lacaille in the 1750s; he originally called it Fornax Chemica, the chemical furnace. Its main significance to astronomers is that it contains a dwarf elliptical galaxy, a member of our LOCAL GROUP, about 800,000 light-years away.

Fraunhofer, Joseph (1787–1826)

German optician and physicist, best known for charting the dark lines in the solar spectrum that are named after him (see FRAUNHOFER LINES). In 1814, while testing the optical properties of glass with the aim of improving the design and construction of ACHROMATIC (color-free) lenses, Fraunhofer noted a number of dark lines crossing the spectrum of the Sun. These had been discovered in 1802 by the English physicist William Hyde Wollaston (1766–1828), who had supposed they marked the boundaries between different colors. Fraunhofer systematically charted the lines, assigning the most prominent letters of the alphabet that are still used today. His instruments were sensitive enough to determine that the pattern of lines was the same for light from the planets (because they were reflecting sunlight), but was different for light from Sirius and other bright stars.

In the early 1820s Fraunhofer went on to study the phenomenon of diffraction, comparing the effects of prisms and diffraction gratings and calculating the wavelengths of certain spectral lines.

His unique theoretical knowledge and great practical skill made Fraunhofer the leading lens maker of his day, producing objectives remarkably free of spherical and chromatic aberration and coma. His most famous instruments were the $9\frac{1}{2}$-inch (24-cm) refractor at Dorpat and the $6\frac{1}{4}$-inch (16-cm) heliometer at Königsberg with which Friedrich BESSEL measured the parallax of 61 Cygni. A successor to Fraunhofer was George Merz (1793–1867), who in 1859 supplied the $12\frac{3}{4}$-inch (32.4-cm) Great Equatorial refractor to the Royal Greenwich Observatory.

Fraunhofer lines

The dark lines seen crossing the Sun's spectrum at certain wavelengths, caused by the absorption of light by the gases in its atmosphere. Over 500 were cataloged by Joseph FRAUNHOFER in 1815. All the Fraunhofer lines are ABSORPTION LINES; the Sun's atmosphere, up to a few hundred miles from its visible surface, is cooler than the photosphere itself, and atoms of various elements absorb light at various specific wavelengths. In this way, astronomers can analyze the chemical composition of other stars besides the Sun, because they all show dark Fraunhofer lines in their spectra. About 25,000 Fraunhofer lines have now been identified in the spectrum of the Sun.

frequency

The number of waves that pass a given point in a given period of time. In astronomy, the waves are of ELECTROMAGNETIC RADIATION (such as light or radio waves). Frequency is given in cycles per second, or *hertz* (Hz), after the German discoverer of radio waves, Heinrich Rudolf Hertz (1857–1894). The velocity of the waves divided by their WAVELENGTH will give the frequency.

G

Gagarin, Yuri Alekseyevich (1934–1968)

Soviet cosmonaut, the first man to fly in space. On April 12, 1961, he made one orbit of the Earth in the spaceship Vostok 1; his orbit had a maximum distance from Earth of 203 miles (344 km) and a minimum distance of 112 miles (190 km). The total flight lasted 108 minutes from lift-off to touchdown. Gagarin ate and drank during his flight and practiced writing under weightless conditions. He reported no problem in adjusting to weightlessness. For reentry the craft was aligned and the retro-rockets fired automatically. Despite some Soviet reports to the contrary, Gagarin ejected from the capsule at an altitude of 23,000 feet (7,000 m) and landed separately by parachute. Gagarin died on March 27, 1968, in a plane crash while training for another space mission.

galactic clusters

Shapeless star clusters found in the spiral arms of the Galaxy; they are also known as OPEN CLUSTERS, because their individual stars are scattered thinly within them, unlike the densely packed GLOBULAR CLUSTERS. Galactic clusters contain young, population I stars (see POPULATIONS, STELLAR). About 1,000 galactic clusters are known, though there may be a total of 18,000 in the Galaxy. The nearest 150, lying within 10,000 light-years, are strongly concentrated in the plane of our Galaxy, and outline three spiral arms, with the Sun located on the inside edge of one. The HYADES and PLEIADES are examples of galactic clusters. Typical galactic clusters contain some 100 stars and measure 10 light-years across. The stars in them will probably disperse over periods of billions of years. The giant star clouds known as STELLAR ASSOCIATIONS are sometimes also classified as galactic clusters.

galaxies

Systems of billions of stars bound together by their own gravity. Our own Milky Way is a typical galaxy. Other galaxies appear as fuzzy patches of light in small telescopes because of their great distances; as a result, they were originally called extragalactic nebulae and were cataloged in the same lists as star clusters and gaseous nebulae lying within the Milky Way. The three catalogs most commonly encountered are those of Charles MESSIER, published in 1784; the *New General Catalogue* (NGC) compiled by J. L. E. DREYER in 1888; and its supplement, the *Index Catalogue* (1895 and 1908). Galaxies listed in more than one can have different numbers: for example, the ANDROMEDA GALAXY is both M31 and NGC 224. Only three external galaxies are visible to the naked eye: M31 and the two MAGELLANIC CLOUDS.

Structure. Three different types of galaxies are found in the Universe. The Milky Way is an example of a SPIRAL GALAXY: these have a bright central region (nucleus) and a large spherical halo composed of old (population II) stars, with a disk of young stars, gas, and dust (population I) surrounding the nucleus. The disk is typically 100,000 light-years in diameter but only 2,000 light-years thick; its material is concentrated into two arms which spiral outwards from opposite sides of the nucleus. Some galaxies have a bar-like nucleus, and the spiral arms start at the ends of the bar. The individual stars and gas clouds in the disk rotate around the center of the galaxy in almost circular orbits, like the planets around the Sun. From the stars' velocities, the total mass of the galaxy can be calculated. A typical spiral is a hundred billion times as massive as the Sun. Since the Sun is a typical star, these galaxies must contain some hundred billion stars.

Elliptical galaxies lack spiral arms and appear as oval star systems consisting of old (population II) stars with no gas or dust clouds. They range in shape from circular (EO) systems, to flattened ovals three times as long as they are broad (E6). Even flatter galaxies are generally lens-shaped, and are classified SO because they seem to be intermediate in form between an elliptical and a spiral seen edge-on.

The smallest elliptical galaxies are no larger than the GLOBULAR CLUSTERS in the Milky Way, about a million times the Sun's mass. The most massive ellipticals (classified cD) are supergiants a hundred times heavier than our Galaxy; they are spheres about 300,000 light-years in diameter. Such elliptical galaxies are the largest and most massive single bodies known in the Universe.

The third class, *irregular galaxies,* are generally smaller than spirals. They contain both young and old stars, but lack any regular structure. The Magellanic Clouds, visible to the naked eye in the southern hemisphere, are intermediate in form between irregulars and small barred spirals, and orbit about our own Galaxy. Some irregulars appear to have been disrupted by explosions.

A few spiral and elliptical galaxies have small bright nuclei, and are known as SEYFERT or N-GALAXIES. These nuclei are often bright at infrared and radio wavelengths, and are apparently the sites of very violent explosions. Many very massive elliptical galaxies also have extended regions of radio emission on either side and are classified as RADIO GALAXIES.

Elliptical galaxy NGC 205, one of the companions of the
Andromeda galaxy. Unlike spiral galaxies, ellipticals
contain little dust or gas, and their smooth light distribution
arises from millions of old, red stars. The bright stars
visible on this galaxy photograph (and all others) are
foreground stars belonging to our own Milky Way.

The "**Whirlpool**" **galaxy** (M51) was the first in which
spiral structure was detected, by Lord Rosse in 1860.
It is comparable in size and appearance to our Milky Way
Galaxy, but is joined by a thin arm to a massive irregular
companion galaxy, bottom. Similar luminous bridges are
found to link other pairs of galaxies.

The apparently largest galaxy, left, of **Stephan's quintet** is
not associated with the other members, but is merely seen
superimposed on a more distant group of four galaxies.
The galaxy at the lower left is an elliptical, while the other
three seem to be distorted spiral galaxies, two of which are
interacting.

Formation and evolution. Galaxies were once thought to evolve from the spherical EO type, becoming progressively more elongated until they formed slowly unwinding spiral arms. The most favored theory today, however, is that the structure of a galaxy is determined at its initial formation; there is probably no evolutionary sequence from elliptical to spiral, or vice versa. Theories of galaxy formation generally assume that the stars formed during the gravitational collapse of a *protogalaxy,* a huge cloud of hydrogen gas. A slowly rotating protogalaxy collapses to a nearly spherical shape. The resulting high density of gas causes stars to form rapidly, creating an elliptical galaxy, all the stars of which are now old, and in which no gas remains. In the collapse of a faster-rotating protogalaxy, only some of the gas falls to the center to condense into stars immediately. The rest forms a rotating disk condensing more slowly, so that a significant amount of gas still remains, along with young stars only recently formed. This disk is unstable and forms arms which give the galaxy a spiral appearance. Irregular galaxies probably also originate in a fast-rotating gas cloud.

Clusters. Most galaxies are found in groups or clusters, gravitationally bound together. The Milky Way is in a small LOCAL GROUP which contains about 30 members, including the Andromeda galaxy. Larger clusters can have thousands of members. Many of these clusters appear spherical in form, and often surround a supergiant (cD) elliptical galaxy. The intrinsic brightness of cD galaxies is very similar from one cluster to another, and thus a cluster's apparent brightness is a good indicator of its relative distance.

Although the galaxies in a cluster orbit a common center of gravity, their actual motions across the sky are far too small to be seen directly. However, the DOPPLER EFFECT in their light can be measured, and reveals their orbital velocities. From these astronomers have calculated how much mass the average galaxy must have for the cluster to be gravitationally bound. The result is 10 to 100 times larger than is usually measured for galaxies. This indicates that there must be a large amount of invisible mass present.

Distances and velocities. Galaxies provide markers for cosmologists studying the expansion of the Universe. Individual stars can be seen in such nearby galaxies as M31. Measuring the brightness of these stars, particularly CEPHEID VARIABLES, allows astronomers to determine the distance of the galaxies. The Andromeda galaxy has been found to be about 2.2 million light-years away, yet on the scale of the Universe it is a near neighbor.

The nearest cluster of galaxies to the Local Group lies in the constellation Virgo. The brightest individual stars in this cluster can just be resolved, indicating a distance of 65 million light-years. The distance of more remote galaxies can be derived from the apparent magnitude of the very bright Type I SUPERNOVAE. These exploding stars always have the same intrinsic maximum brightness, and they can outshine their parent galaxy for a few days.

Astronomers find that the distance of a galaxy is directly proportional to the RED SHIFT in its spectrum; this shift is believed to be the result of the Doppler effect caused by the recession of the galaxies as the Universe expands. The red shift-distance relation is known as HUBBLE'S LAW.

Galileo lived in an age when to challenge the ideas which had been accepted since Greek times was regarded as heretical. He could use an acid tongue and sarcastic wit to great effect in arguments—which earned him many enemies. But his perseverance led to a revolution in thought.

Galileo Galilei (1564–1642)
Italian mathematician, astronomer, and physicist, responsible for a great series of contributions to modern scientific thought. While still a young man studying medicine at Pisa, Galileo discovered that a pendulum of a given length always takes the same time to complete one swing, whether that swing is large or small. Later in life he suggested applying the principle to clocks. Galileo was attracted to physics and was soon lecturing to the Florentine Academy on the hydrostatic balance and the center of gravity of solids; he returned to Pisa in 1589 as professor of mathematics. Here Galileo determined that all bodies fall at the same rate, although it is uncertain that he publicly demonstrated the fact from the Leaning Tower. In 1592 he moved to Padua, where he designed and made a calculating device known as a "military and geometrical compass" and wrote a short treatise on mechanics. Late in 1609 he heard of the invention of the telescope, built one himself, and was probably the first to survey the heavens telescopically. His discoveries of the phases of the planet Venus, of mountains on the Moon, and of the four largest satellites of Jupiter, helped convince him that the views of COPERNICUS were right. He published his results early in 1610 in *Sidereus Nuncius* (The Starry Messenger).

Later in 1610 Galileo returned to Florence as the Grand Duke's mathematician and philosopher, and published a book about his observations of sunspots. All the evidence obtained with the telescope and from his investigations into physics convinced Galileo that the age-old doctrines about the nature of the Universe were in error, and he set about writing his famous *Dialogue Concerning the Two Chief World Systems—Ptolemaic and Copernican,* published in

Florence in 1632. A witty, penetrating discussion of ancient and modern views, it raised a storm of controversy. Unfortunately Galileo, who had a fiery temper and a sarcastic wit, had already made many enemies, and his opponents had the book banned. Galileo was brought before the Inquisition and forced to recant his belief in the Copernican theory.

Galileo was treated leniently, but had to live the rest of his life under virtual house arrest, moving to Arcetri on the outskirts of Florence in 1633, the year after his trial. Although he kept clear of controversy, he continued his research, and in 1638 a new book of his was published at Leiden in the Netherlands from a manuscript smuggled out of Italy.

This last work, *Discourses Concerning Two New Sciences,* discussed the principles of mechanics. Together with the physical principles published in the *Dialogue,* it demolished old and outdated ideas and laid the foundations of the new mathematical physics. At the same time, Galileo's telescopic observations of the heavens opened up a new dimension in astronomy. By observation, experiment, and the use of mathematics, Galileo was a founder of modern scientific method in the physical sciences.

Galle, Johann Gottfried (1812–1910)
German astronomer, discoverer of the planet Neptune. In 1835 Galle became assistant to Johann ENCKE at the Berlin Observatory. On September 23, 1846, the French mathematician Urbain LEVERRIER asked Galle to look for a new planet whose position he had just calculated. That same night Galle and Heinrich Louis d'Arrest (1822–1875), a discoverer of comets and asteroids, set to work using the new star chart being compiled by Encke, and spotted the planet almost immediately; it was subsequently named Neptune. Galle was interested in comets (he discovered three) and asteroids, and spent much of his career computing their orbits. His studies helped prove the connection between the orbits of comets and certain meteor showers. In 1872 he suggested measuring the distances of certain asteroids during their close approaches to establish an accurate scale of the solar system; this method has been used successfully to yield a precise value for the ASTRONOMICAL UNIT.

gamma-ray astronomy
The study of radiation emitted by celestial objects at wavelengths shorter than X rays (i.e. less than 0.1 angstrom). These rays are absorbed by the Earth's atmosphere, and can be studied only by rockets and satellites. The first gamma-ray satellites were the Vela satellites, launched by the United States and intended to monitor nuclear explosions. They began operating in 1967, and immediately detected bursts of gamma rays from space. There are four or five bursts a year, which seem to come from random directions.

This discovery was completely unexpected, and the origin of the bursts is still disputed. They usually last only a few seconds, and must come from an object only a few light-seconds across, about the size of the Sun. Their random distribution means that the gamma-ray sources are either within a few hundred light-years of the Sun, or else are far outside our Galaxy. Objects of intermediate distance would be concentrated toward the plane of the Milky Way,

like the fainter stars in the sky.

Gamma-ray sources lying relatively near the Sun could be produced by collisions of comets with compact NEUTRON STARS, or by giant stellar flares a billion times more powerful than solar flares. Stellar explosions (SUPERNOVAE) in distant galaxies could also produce these short bursts.

Sources producing a continuous stream of gamma rays have also been located. The first such source was the CRAB NEBULA, the remains of a supernova within the Milky Way. The gamma rays originate in the very small dense neutron star (or PULSAR) which formed in the explosion. They are found to "flash" 30 times a second as the rapidly rotating star sweeps a beam of radiation past the Earth.

The center of our Galaxy is another gamma-ray source, and satellites have also found a general "background" of gamma rays which arrive continuously from all directions. This background may be due to very many weak sources scattered randomly, like the background observed by X-RAY ASTRONOMY.

Large satellites for gamma-ray research, such as the American SAS B and the European COS B, have now been launched and should lead to rapid advances in gamma-ray astronomy.

Gamow, George (1904–1968)
Russian-born American astrophysicist, best known for his association with the BIG-BANG theory of cosmology. Gamow proposed that the matter of the Universe originally existed in a primordial state called the "ylem," and that helium and perhaps other elements were formed from the ylem shortly after the big bang that started the Universe's expansion. The theoretical work was performed with the American physicist Ralph Asher Alpher (b. 1921); together with Hans BETHE, they published the famous Alpher-Bethe-Gamow paper (α, β, γ) in 1948. They predicted the existence of a BACKGROUND RADIATION in the Universe, left over from the big bang. Gamow also made pioneering studies of energy-generation inside stars, and the evolution of normal stars to the red giant stage.

Ganymede
The main moon of Jupiter, 3,275 miles (5,270 km) in diameter, and one of the largest moons in the solar system (Titan and Triton may be larger); it was discovered by Galileo in 1610. Ganymede orbits Jupiter once every 7 days 3 hours 42.5 minutes at a distance of 665,120 miles (1,070,400 km). Its mass is about twice that of our Moon, and its density roughly twice that of water; one model suggests that Ganymede may be a muddy ball coated with a layer of frozen water about 300 miles (480 km) deep. Observations show that about half its visible surface is covered by water frost. Visual mapping and space probe photographs have revealed dark markings, possibly due to dust or craters on its surface.

Gauss, Carl Friedrich (1777–1855)
German mathematician and astronomer who made fundamental contributions to celestial mechanics (the study of orbital motion and gravitational interaction). His most striking achievement came in 1801, when he "rediscovered" on paper the asteroid

Ceres. It had recently been located by Giuseppe PIAZZI, but it then disappeared behind the Sun, and the combined efforts of astronomers were unable to relocate it. Gauss had devised a method for working out an object's orbit from only three good positional measurements; he had also invented the mathematical technique known as "least squares," which can be used to fit a smooth curve to observations of uneven quality. With these techniques, Gauss calculated the orbit of Ceres and told observers where to find the asteroid. In 1807 Gauss became director of the observatory at Göttingen, where he spent the rest of his life. His *Theoria motus* of 1809 contained his work on orbital motion and gravitational perturbation, which were used by later mathematicians in tracking down Neptune from its effects on the motion of Uranus. Gauss also turned his attention to geodesy and terrestrial magnetism.

gegenschein
A faint, diffuse patch of hazy light close to the ecliptic, directly opposite the Sun in the sky; the word is German for "counterglow." The gegenschein was first noted by the Danish astronomer Theodor Brorsen in 1854. The glow is about 20° across at maximum, produced by sunlight reflected from dust particles, like the ZODIACAL LIGHT. It is much fainter than the Milky Way.

Gemini (the twins)
Constellation of the zodiac, in the northern hemisphere of the sky, best seen in the northern winter. The Sun is in Gemini at the SUMMER SOLSTICE, passing through the constellation from late June to late July. It is best known for its two brightest stars, CASTOR and POLLUX. It contains a bright star cluster, M35 (also known as NGC 2168), containing about 120 stars, and 2,600 light-years away.

Gemini project
American space program to practice rendezvous and docking techniques and to gain experience of long-duration space missions. The Gemini missions showed that men could live safely in space for periods long enough to reach the Moon and return, and that the docking operations required for Moon missions could be successfully accomplished. The achievements of the Gemini program established American leadership in the space race.

Gemini hardware. The Gemini project was born late in 1961, while the first Mercury flights were underway. The capsule had to be bigger than Mercury and it had to be maneuverable, with considerable on-board control. To assist with rendezvous and docking, and to make pinpoint landings possible, a microcomputer was installed, the forerunner of the Apollo on-board computers without which the Moon landings could never have been made.

The crew capsule in Gemini, called the reentry module, was conical, with an extended nose called the rendezvous and recovery section, which contained rendezvous radar and landing parachutes. Also in the nose were 16 thrusters for attitude control during reentry; these added considerably to the accuracy of Gemini splashdowns. The base diameter of the reentry module was 7 feet 6 inches (2.3 m), and the diameter of the extended nose

The first close rendezvous in space was between Geminis 6 and 7 in December 1965. This view, taken through the window of Gemini 6, shows the whole of Gemini 7 in its orbital mode, including the white-painted equipment section. This part carried the attitude control rockets, and was jettisoned before reentry.

39 inches (1 m). The conical crew section was 5 feet 10½ inches (1.8 m) long; including the rendezvous and recovery section, it was 11 feet (3.35 m) long. Behind the reentry module was the adapter module, in two parts: the retrograde section, containing retro-rockets; and the equipment section, containing fuel cells for electricity generation, and attitude control rockets for in-flight maneuvering. Before reentry, the equipment section was jettisoned to reveal the retro-rockets, which later were also jettisoned to leave the heat shield at the base of the reentry module. The base diameter of the adapter module was 10 feet (3 m), and its total length 7 feet 6 inches (2.3 m). The complete Gemini weighed about 8,000 lb. (3,600 kg).

Astronaut access to the capsule was through two hatches, one above each crew seat. Each hatch had a crescent-shaped forward-facing window. Gemini spacecraft were launched by a two-stage Titan II rocket. In the event of a launch failure, the astronauts would be catapulted to safety by ejector seats. For spacewalks, both astronauts donned suits, the craft was depressurized, and one of the hatches was opened. An astronaut either left the capsule entirely, or performed a "stand-up" EVA by looking out the hatch.

The missions. Gemini was tested in orbit before manned flights began. The complex nature of the missions demanded a new control center; this was the Manned Spacecraft Center (now Johnson Space Center) in Houston. Gemini 4 was the first mission not controlled from the Cape.

The Gemini series accomplished the first orbit change by a manned spacecraft (Gemini 3), the first space docking (Gemini 8), and various duration records for spaceflight and extravehicular activity. Considerable photography of the Earth's surface and weather patterns was also accomplished during the Gemini missions, paving the way for later Earth resource surveys.

Mission	Launch date	Results
Gemini 1	April 8, 1964	Unmanned orbital test flight
Gemini 2	January 19, 1965	Unmanned suborbital flight to test reentry heat shield
Gemini 3	March 23, 1965	Virgil I. Grissom and John W. Young made 3 orbits of Earth. First manned spacecraft to change orbit
Gemini 4	June 3, 1965	James A. McDivitt and Edward H. White; White became first American to walk in space, maneuvering with a hand-held jet gun for 21 minutes. 62 orbits
Gemini 5	August 21, 1965	Leroy G. Cooper and Charles Conrad made 8-day, 120-orbit flight
Gemini 7	December 4, 1965	Frank Borman and James A. Lovell made record-breaking 14-day, 206-orbit flight
Gemini 6	December 15, 1965	Walter M. Schirra and Thomas P. Stafford made first space rendezvous, maneuvering with Gemini 7. 15 orbits
Gemini 8	March 16, 1966	Neil A. Armstrong and David R. Scott made first space docking, with an Agena target vehicle. A stuck thruster caused the spacecraft to roll dangerously, and Gemini 8 undocked for an emergency splashdown. 7 orbits
Gemini 9	June 3, 1966	Thomas P. Stafford and Eugene A. Cernan; intended docking with Agena target vehicle frustrated by a shroud that failed to jettison. Cernan performed a total of 2 hours 7 minutes of EVA. 45 orbits
Gemini 10	July 18, 1966	John W. Young and Michael Collins rendezvoused and docked with Agena target vehicle, and used its engine to boost themselves into a new orbit of apogee 476 miles (766 km). Disengaged from first Agena and then docked with Agena vehicle used in Gemini 8, which had been parked in a new orbit. Collins retrieved a micrometeoroid detector from the side of the Agena during a 30-minute EVA. 43 orbits
Gemini 11	September 12, 1966	Charles Conrad and Richard F. Gordon docked with Agena target vehicle and used its propulsion system to boost themselves into a new orbit with a record-breaking apogee of 850 miles (1,368 km). Gordon attached a tether to the Agena during a spacewalk; Gemini undocked and kept station with the tethered Agena. 44 orbits
Gemini 12	November 11, 1966	James A. Lovell and Edwin E. Aldrin docked with Agena target vehicle. Aldrin performed a total of 200 minutes of stand-up EVA, photographing a solar eclipse. On the third day Aldrin worked for 129 minutes on Agena. 59 orbits

Geminid meteors

One of the principal annual meteor streams, first detectable about December 7 each year. The Geminids reach a maximum rate of about 55 meteors an hour on December 13–14, then rapidly decline within a day. The Geminid radiant is 1° west of Castor in Gemini at maximum, but moves eastward at 1° a day; the radiant reaches its greatest altitude at 2 A.M. The stream follows an elliptical 1.63-year orbit, ranging from 0.129 a.u. to 2.63 a.u. from the Sun. The Geminids are the strongest of the meteor showers, but they have no ascertainable parent comet.

geocentric

Term meaning Earth-centered. A *geocentric orbit* is an orbit about the Earth. *Geocentric coordinates* are positions of celestial objects as measured from the center of the Earth. The *geocentric system* was the ancient belief that the Earth was the center of the Universe.

geostationary

Term describing an orbit in which a satellite appears to hang stationary over a point on the Earth's equator, frequently used for COMMUNICATIONS SATELLITES. It is also termed a synchronous orbit, because the satellite's orbital period is synchronized with the Earth's rotation. A geostationary orbit is circular, 22,300 miles (35,900 km) above the equator. In practice, slight errors in positioning the satellite, together with the perturbing effects of the Sun's and the Moon's gravity and the irregular gravitational field of the Earth, all tend to pull the satellite off station. It is therefore not absolutely stationary, but describes a slight figure-of-eight motion in the sky. Ground stations must therefore make slight tracking movements to follow the path of the satellite, and its position must be corrected occasionally by small gas thrusters on board.

giant star

A large bright star; the very biggest and brightest stars are called SUPERGIANTS, but there is no firm dividing line. The brightness of a star depends on its surface temperature and size. A giant star is brighter than a dwarf star of the same temperature; a brightness difference of five magnitudes corresponds to a diameter difference of ten times. Stars enter a giant stage toward the end of their evolution; they pass through it relatively quickly, so that giants are a hundred times rarer than main sequence stars. Yet their brightness allows them to be easily spotted: three-fourths of the hundred brightest stars are giants, including a sprinkling of supergiants (see also BLUE GIANT; RED GIANT).

gibbous

Term describing the phase of the Moon or a planet between half and full illumination.

Glenn, John Herschel (b. 1921)

First American to orbit the Earth. He circled the Earth three times in his craft *Friendship 7* on February 20, 1962, landing after a total flight time of 4 hours 55 minutes 23 seconds. His altitude varied between 100 miles (161 km) and 163 miles (262 km). During the second orbit Glenn saw what he termed

globular cluster

"fireflies," bright specks around the spacecraft which were later realized to be flakes of paint illuminated by the Sun. Glenn controlled the orientation of his capsule by "flying" it with a joystick. During reentry the retro-rockets were not jettisoned for fear that the heat shield had come loose; it was a false alarm caused by a faulty warning light. Glenn, who was selected as an astronaut in 1959, retired from the space program in 1964. He later entered politics and was elected to the U.S. Senate from Ohio in 1974.

globular cluster

A spherical-shaped cluster of old stars. Globular clusters are distributed in a halo around the center of our Galaxy. Some 125 globular clusters are known in our Galaxy, with between 100,000 and 10 million stars in each. The average diameter of a globular cluster is about 100 light-years. The two brightest clusters are easily visible to the naked eye and were once mistakenly cataloged as stars. They are Omega Centauri and 47 Tucanae, both too far south to be seen from mid-northern latitudes. The brightest globular cluster which can be distinguished further north is M13, also known as the Great Cluster in Hercules. It is just visible to the naked eye.

Globular clusters swarm about the center of our Galaxy, with half the known number lying in the constellations Scorpio and Sagittarius. From this distribution, Harlow SHAPLEY in 1917 deduced the true size of the Milky Way system and the Sun's position toward its edge. Similar globular clusters are found distributed about the center of other galaxies including the Andromeda Galaxy and M/04 in Virgo, known as the Sombrero Hat. Most globular clusters lie well away from the plane of the Galaxy, unlike the much younger GALACTIC CLUSTERS which are concentrated in the spiral arms.

Globular clusters are among the most ancient objects in the Galaxy; the oldest are estimated to be 13 billion years of age, and they all contain population II stars (see POPULATIONS, STELLAR), such as RR LYRAE VARIABLES. Globular clusters are believed to be the first parts of a galaxy to form as a giant cloud of gas collapses into a disk. They are believed to orbit the center of the Galaxy on elliptical paths like comets around the Sun. Some "tramp" globulars not bound to any galaxy exist in our LOCAL GROUP.

globule

A circular or oval dark nebula made of dust, seen silhouetted against a background of stars or a bright nebula. Large globules are several light-years in diameter, and contain as much mass as several dozen stars. The smallest globules, however, are only about the size of the solar system, and contain about as much mass as the Sun. These are sometimes termed *Bok globules,* after the Dutch-born American astronomer Bart Jan Bok (b. 1906), who first drew attention to them in 1947. Approximately 100 globules are known, but there must be many more that remain unseen because they are projected against dark areas of sky or are hidden behind larger nebulae. Globules may be clouds collapsing to create star clusters or individual stars.

Goddard, Robert Hutchings (1882–1945)

American rocket pioneer, who built and launched the

The globular cluster M13 in Hercules, visible as a fuzzy patch to the naked eye, is revealed in a telescope as a ball of millions of stars some hundred light-years across. These clusters, which surround our Milky Way Galaxy in a spheroidal cloud, are by far its oldest members.

world's first liquid-fueled rocket. Like the other two visionaries of spaceflight, Konstantin TSIOLKOVSKY and Hermann OBERTH, Goddard worked out the theory of rocket propulsion very much on his own and could see that rockets provided the only way of traveling in space. Goddard was a professor of physics at Clark University at Worcester, Massachusetts. He took out his first rocket patents in 1914; they included designs for propulsion systems and for a form of multistage rocket. He eventually registered over 200 patents, and in 1960 the federal government finally paid $1 million for use of these patents to his widow and the Guggenheim Foundation, which had supported him.

During World War I Goddard developed a tube-launched rocket that was to become the bazooka. But his real interest since boyhood was spaceflight. Goddard had written a paper in 1916 for the Smithsonian Institution, which awarded him a small grant; the paper was published in 1919 under the title *A Method of Reaching Extreme Altitudes*. In it, Goddard outlined the value of rockets for upper atmosphere research, and even dwelt on the possibility of firing rockets to the Moon.

This notion brought him considerable publicity, some of it hostile; Goddard thereafter continued his work in seclusion. On March 16, 1926, he made history by flying the world's first liquid-propellant rocket. Using liquid oxygen and gasoline, it flew for 2.5 seconds, rising 41 feet (12.5 m). In 1929 the aviator Charles A. Lindbergh visited Goddard, and later arranged for grants from the Daniel Guggenheim Fund. With these, Goddard moved to a ranch near Roswell, New Mexico, in 1931, where he continued his experiments on a larger scale.

A 1936 publication, *Liquid-Propellant Rocket Development,* summarized his work, which included the development of new stabilization and combustion techniques that helped propel his rockets more than a mile in height at near-supersonic speeds. He then turned to the development of propellant pumps to force the liquids into the combustion chamber, launching rockets 22 feet (6.6 m) in length, 18 inches (45 cm) in diameter, and weighing 450 lb. (200 kg) or more.

With the help of only four assistants, Goddard had developed most of the basic components needed for successful long-range rocketry. There was little official interest, except during World War II when he supervised development of jet-assisted takeoff for aircraft. He died before he could return to his peacetime activities—by then, the potential of the rocket as a long-range weapon had been demonstrated by the German V-2. Perhaps the greatest tribute was paid to him by Wernher von Braun when questioned shortly after the war. "Don't you know about your own rocket pioneer?" he asked. "Dr. Goddard was ahead of us all."

Goddard Space Flight Center

Facility of the National Aeronautics and Space Administration (NASA) for space science research and satellite tracking, established in 1959 at Greenbelt, Maryland. The Center has been responsible for over half NASA's Earth satellites, including members of the Explorer series, the Orbiting Astronomical Observatory, Orbiting Geophysical Observatory, and Orbiting Solar Observatory programs, Landsat, and various weather and communications satellites. Goddard scientists have also sent experiments aboard satellites of other agencies. Goddard Space Flight Center is the headquarters of STADAN, the worldwide Space Tracking and Data Acquisition Network for tracking unmanned satellites, and it is also the switching center for the manned and deep-space tracking networks operated by the Johnson

Robert Goddard with his first liquid-fueled rocket near Auburn, Mass, in 1926. The Goddard Space Flight Center at Greenbelt, Md, is named after him.

Space Center and Jet Propulsion Laboratory. The Center is named for the American rocket pioneer Robert H. Goddard.

Gold, Thomas (b. 1920)

Austrian-born American astronomer, who with Hermann BONDI proposed the famous STEADY-STATE THEORY of the Universe in 1948. In a later cosmological theory he proposed that if the Universe ceased expanding and began to contract, time might also run in reverse. More recently, he has provided the standard theoretical model of a PULSAR, describing it as a rotating neutron star with a strong magnetic field, emitting lighthouse-beams of energy. He is also known for his prediction that the surface of the Moon is covered in dust.

Gould, Benjamin Apthorp (1824–1896)

American astronomer, the first director (1870) of Cordoba Observatory in Argentina, where he initiated production of the *Cordoba Durchmusterung,* a southern extension of the BONNER DURCHMUSTERUNG star catalog. The *Cordoba Durchmusterung* was eventually published in 1930 under the editorship of Gould's successor, the American astronomer Charles Dillon Perrine (1867–1951), who discovered two moons of Jupiter and 13 comets. The catalog covers the sky from −23° to the south pole, listing 613,953 stars to the 10th magnitude. Gould established the existence of a band of hot, young stars around the sky inclined at an angle of 20° to the Milky Way; this band, apparently first noted by John Herschel, is now called *Gould's belt*. It is a flattened system of stars within about 1,000 light-years of the Sun which has been expanding for about 40 million years; the stars of the Pleiades cluster are part of Gould's belt. In 1849 Gould founded the *Astronomical Journal*.

granulation

A mottling effect on the Sun's visible surface (the PHOTOSPHERE) caused by columns of gas rising from the Sun's hot interior like water boiling in a pan. This mottling is also termed the "rice-grain" effect. Granules are between about 200 and 1,000 miles (300–1,500 km) in diameter, with bright centers where the hot gas rises and darker rims where cooler gas descends. Granules are in constant motion and exist for only a few minutes. This continual seething is believed to transmit energy into the upper layers of the Sun's atmosphere.

Supergranules are similar but independent large-scale convection phenomena in the chromosphere.

gravity

A fundamental property of matter, which produces a mutual attraction between all bodies. Although gravity is by far the weakest of the known forces of nature, on the astronomical scale it usually overwhelms all other forces, determining the motions of planets, stars, galaxies, and even the Universe. Two theories of gravity have been widely accepted. The first was proposed by Isaac NEWTON, and described gravity as a force acting across the space between two bodies, reducing in intensity with the square of the distance (the inverse-square law of force). With this theory

A closeup view of the Sun's photosphere, showing the fine network of granules, each only 600 miles (1,000 km) across. The bright centers of these roughly hexagonal cells are sites where hot material carried from the interior reaches the surface; the dark boundaries are regions of cooling, descending gas.

Newton was able to account for the orbits of the planets in our solar system. The other theory, proposed by Albert Einstein, is called general RELATIVITY, and treats gravity not as a force, but as a distortion in the geometry of space and time. This remarkable identification of gravity with geometry is founded upon the *equivalence principle* first discovered by Galileo in the 17th century. Anyone who accelerates rapidly in an automobile experiences a pushing force that feels exactly like a gravitational force. This equivalence of the forces of acceleration and gravity is also responsible for the weightlessness which astronauts experience in orbit. Although they are still within the Earth's gravity, its force is neutralized by the acceleration in free fall as they orbit the Earth. The apparent weightlessness arises because all the surrounding objects are falling equally fast through space together, a phenomenon which suggests that gravity is better regarded as a property of space and time, rather than of matter itself.

The extreme feebleness of gravity means that it is difficult to perform convincing laboratory experiments to understand its true nature. One example of this concerns *gravity waves*, which are predicted from theory by analogy with electromagnetic waves. Gravity waves might be produced as "ripples" from supernova explosions, or by the swallowing of matter into black holes, and would travel outward at the speed of light. In 1969 the American physicist Joseph Weber (b. 1919) claimed to have detected gravity waves coming from the center of our Galaxy. To detect these waves, Weber had set up two large aluminum cylinders that would be slightly distorted whenever a wave passed through them. The expected effect would be so weak it would be necessary to measure changes in the length of the bar of less than one atomic nucleus. Other researchers have been unable to confirm Weber's claims, although experiments to detect gravity waves continue.

Great Bear
See URSA MAJOR.

great circle
A line on the surface of a sphere which divides it into two equal hemispheres; on the Earth, lines of

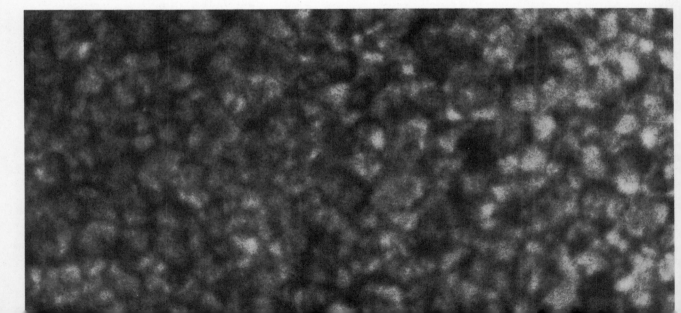

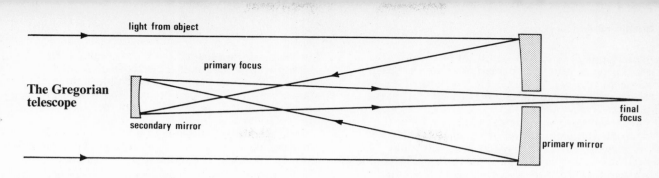

The Gregorian telescope

light from object

primary focus

secondary mirror

primary mirror

final focus

longitude are great circles, but lines of latitude (except for the equator) are not. The horizon produces an approximate great circle when projected onto the CELESTIAL SPHERE (the imaginary sphere carrying the celestial objects and rotating around the Earth once a day).

Green Bank Observatory
See NATIONAL RADIO ASTRONOMY OBSERVATORY.

greenhouse effect
The warming of a planet by its atmosphere. Short-wavelength radiation such as visible light reaches the surface of the planet, which warms up and radiates heat. Some of this heat is absorbed by the atmosphere, which acts as an insulating blanket around the planet. The effect is the same as in a greenhouse. Without an atmosphere, the temperature of the Earth would be about $-30°C$, rather than the average $16°C$ observed. The atmosphere of Venus is mostly carbon dioxide, a strong heat absorber. This, along with some traces of water vapour, produces unusually high temperatures on Venus, in what is sometimes termed the runaway greenhouse effect.

Greenwich Mean Time (GMT)
Local time as calculated at the longitude of the Royal Greenwich Observatory (the *Greenwich meridian*). Greenwich Mean Time became the international time reference in 1884, but in 1972 it was superseded by ATOMIC TIME. GMT is often called Universal Time (UT) by astronomers.

Greenwich Observatory
See ROYAL GREENWICH OBSERVATORY.

Gregorian telescope
The earliest proposed form of reflecting telescope, suggested by the Scottish astronomer-mathematician James GREGORY. The image formed by a concave primary mirror is refocused by a much smaller concave secondary mirror, which reflects the light back through a central hole in the primary. The original design, published in 1663, called for a PARABOLOID figure on the primary mirror and a so-called prolate ELLIPSOID secondary, although different combinations are possible. The Gregorian telescope did not become a practical proposition until the 18th century, when instrument makers were able to make the mirror shapes required. The Gregorian design gives an upright image, but the tube length of a given system is about 50 percent longer than that of the equivalent CASSEGRAIN TELESCOPE, and the design has fallen into disuse. Its former popularity was due to the fact that the concave secondary mirror, which can be made to form a real image for testing purposes, was easier to make than the convex

secondary of a Cassegrain, which cannot be tested directly. Improved techniques have removed this objection to the Cassegrain system.

Gregory, James (1638–1675)
Scottish mathematician and astronomer, who in 1663 published the first design for a reflecting telescope, now known as the GREGORIAN TELESCOPE, which, however, he was unable to construct. In 1668 Gregory proposed that the distances of the stars could be estimated by comparing their brightness to that of the Sun. Using this method he calculated Sirius to be 83,190 times the distance of the Sun. Newton amended his method in 1685 and arrived at a more accurate value of 950,000 solar distances; the true figure is 550,000. James' nephew, David Gregory (1659–1708), was an early champion of Newton's theory of gravity. In 1695, David Gregory suggested that chromatic aberration could be eliminated by combining lenses of different composition; this was the first proposal for an ACHROMATIC lens.

Grissom, Virgil Ivan "Gus" (1926–1967)
American astronaut, first man to make two space flights. His first flight, a suborbital mission on July 21, 1961, was the second space launching in the Mercury series. On March 23, 1965, he and John Young flew the first manned spacecraft in the Gemini series. Grissom was also scheduled to command the first of the three-man Apollo flights. But during a simulated countdown at the launchpad on January 27, 1967, a fire in the spacecraft killed him and fellow astronauts Roger Bruce Chaffee (1935–1967) and Edward WHITE.

Grus (the crane)
A constellation of the southern hemisphere of the sky, introduced by Johann Bayer in 1603. Its brightest star, Alpha, known as Alnair, is a white star of magnitude 2.16.

H

H I, H II regions
Areas of interstellar space containing hydrogen gas. In H I regions the hydrogen is in its so-called neutral form—cool and un-ionized. In H II regions it is hot and ionized.

H I clouds emit no visible light, although they can produce dark absorption lines in light passing through them, but the cool (about 100°K) hydrogen atoms emit radio waves at a wavelength of 21 centimeters, as first predicted by H. C. van de Hulst in 1944. This 21-cm line is produced when the electron in the hydrogen atom changes its direction of spin. Hydrogen is the principal constituent of interstellar matter, and 21-cm observations of H I regions provide considerable information about the structure of our MILKY WAY Galaxy. The motions of the gas can be studied from the slight Doppler shift in the 21-cm line.

H II regions (gaseous NEBULAE) contain hotter material, at a temperature of about 10,000°K, in which hydrogen atoms exist as IONS, with the electrons no longer bound to the protons. Interactions between free electrons and protons produce radio noise over a broad range of wavelengths. It was this radiation that was observed coming from the Milky Way by the first radio astronomers. Certain optical spectral lines are also emitted, produced by atoms in which the electron has not gained enough energy for IONIZATION. The ionization is caused by ultraviolet radiation from very hot stars; the ORION NEBULA, for example, is ionized by four central stars (the Trapezium). These stars have condensed from the H II region which they now ionize. H II nebulae are typically a thousand times denser than the interstellar H I clouds, and are 10 to 100 light-years across.

Hale, George Ellery (1868–1938)
American astronomer, a major pioneer of the experimental branch of astronomy called astrophysics. He founded the Yerkes, Mount Wilson, and Mount Palomar observatories, and was responsible for the construction of the world's largest optical telescopes. As a young man in 1889, he invented the spectroheliograph, a device for photographing the Sun's prominences and other solar phenomena; this success turned his interest permanently toward solar physics. Hale's father, a wealthy manufacturer, supported his scientific interests and built him a private observatory. A 12-inch (30-cm) refractor was installed there in 1891, and Hale began regular observations with his spectroheliograph. In 1892 he moved to the University of Chicago, where he used his organizational talent to set up the Yerkes Observatory, opened in 1897, and containing the world's largest telescope, a 40-inch (102-cm) refractor with which he continued his work on the Sun and stellar spectroscopy. In 1895 Hale began the *Astrophysical Journal,* still one of the major outlets for research papers.

In 1904, with a grant from the Carnegie Foundation, Hale founded the Mount Wilson Solar Observatory, on a peak near Pasadena, California. There in 1905 he proved that sunspots were cooler areas on the Sun, and in 1908, with a tower telescope 60 feet (18.3 m) high feeding underground instruments, found that intense magnetic fields are associated with sunspots. Hale's advances produced the first understanding of these strange solar blemishes, which had puzzled astronomers for centuries. Hale then made the even more important discovery that the magnetic polarity of sunspots is reversed in each successive 11-year solar cycle. In 1912 a 150-foot (45.7-m) tower telescope was completed on Mount Wilson. Observations with this instrument led Hale to predict an overall solar magnetic field, eventually confirmed in the 1950s by H. D. and H. W. BABCOCK.

In 1908 a 60-inch (152-cm) reflecting telescope was set up on Mount Wilson, using a glass disk donated by Hale's father. This allowed stellar spectra to be studied in considerably greater detail than ever before.

But Hale soon persuaded the businessman John D. Hooker to provide funds for a 100-inch (254-cm) mirror, which was installed on the mountain in 1917 and was responsible for a major step forward in knowledge of the Universe through the work of Edwin HUBBLE.

Hale then began thinking of a still larger project. Once the existence of other galaxies had been proved with the 100-inch, Hale wanted an instrument of greater power to study them in detail: a 200-inch (508-cm) reflector. The Rockefeller Foundation provided $6 million in funds, and Mount Palomar was picked as the site for the new observatory. Hale had died by the time the 200-inch opened in 1948; in 1970 the Mount Wilson and Palomar Observatories were renamed the HALE OBSERVATORIES in his honor.

Hale Observatories
Name since 1970 of the Mount Wilson and Palomar Observatories, founded by George Ellery HALE and operated by the Carnegie Foundation and the California Institute of Technology. Mount Wilson, where Hale began observations in 1904, is 20 miles (32 km) northwest of Los Angeles. The observatory, at an elevation of 5,705 feet (1,742 m), has tower telescopes 60 feet (18.3 m) and 150 feet (45.7 m) high for solar studies, a 60-inch (152-cm) reflector opened in 1908, and the famous 100-inch (254-cm) Hooker telescope opened in 1917, named after the Los Angeles businessman John D. Hooker who in 1906 provided $45,000 to purchase the mirror. The observatory on Mount Palomar, 50 miles (80 km) northeast of San Diego, is at an elevation of 5,597 feet (1,706 m). Its main telescope is the 200-inch (508-cm) Hale reflector, opened in 1948. Another major instrument is a Schmidt telescope with a 72-inch (183-cm) mirror and 48-inch (122-cm) corrector plate, which took the famous Sky Survey photographic atlas. In 1970 a 60-inch (152-cm) photometric telescope was installed on Mount Palomar.

In 1969 the Big Bear Lake Solar Observatory, operated by Hale Observatories, opened 100 miles (160 km) east of Los Angeles, at an elevation of 6,700 feet (2,042 m). The Hale observatories also operate the Las Campanas observatory, a southern outstation in Chile.

Hall, Asaph (1829–1907)
American astronomer, discoverer of the two moons of Mars during the planet's close approach to the Earth in 1877. On August 11 of that year Hall first glimpsed Deimos, the smaller and more distant moon; he confirmed his sighting on August 17, and the same night found Phobos. Hall made his discoveries with the 26-inch (66-cm) refractor of the U.S. Naval Observatory in Washington, where he was chief observer. He went on to become an authority on the satellites of the solar system. Hall was also an assiduous observer of double stars, and in 1892 showed that the two stars of 61 Cygni were in orbit around each other.

Halley, Edmond or Edmund (1656–1742)
Second English astronomer royal, computer of the orbit of the famous comet that now bears his name. Halley was inspired by the star-cataloging activities of John FLAMSTEED at Greenwich to perform similar

work for the southern stars. In 1676 he traveled to St. Helena, an island in the South Atlantic, and on his return two years later published a catalog of 341 stars, the first southern catalog made with the aid of a telescope, which established his scientific reputation. Halley became a friend and admirer of Isaac Newton, and in 1684 encouraged him to publish his work on gravitation and motion. The outcome was the famous *Principia,* which Halley paid for and saw through the press.

Halley contributed to several branches of science, but his most famous work came in 1705, with the publication of *Synopsis of Cometary Astronomy,* in which he computed the orbits of 24 comets. Three of them—the comets of 1531, 1607, and 1682—were so similar that he concluded they must all be the same body, moving around the Sun with a 76-year period. He predicted a return in 1758, though he did not live to see it confirmed. It is now named HALLEY'S COMET in his honor.

In 1679 Halley had first suggested that observations of the transits of Venus across the face of the Sun could be used to measure the scale of the solar system. In 1716 he made specific suggestions for the predicted transits of 1761 and 1769, which were observed (long after his death) throughout the world. In 1718 Halley found that the stars Sirius, Aldebaran, and Arcturus had moved perceptibly from their positions as recorded in Ptolemy's star catalog, thus revealing their PROPER MOTIONS.

Halley's major work had therefore already been completed when in 1720, at the age of 64, he was appointed astronomer royal in succession to Flamsteed. However, he had long been interested in the motion of the Moon, and as early as 1684 had observed regular deviations from its predicted motion. Despite his age, Halley spent the remainder of his life assiduously observing the Moon through one entire SAROS cycle of 18 years. These observations were eventually published in 1749 as additions to tables of the Moon and planets he had prepared in 1719, but had kept from publication. He died before he could analyze these observations, but they proved of enormous value to later astronomers in calculating the complex nature of the Moon's motion.

Halley's comet
The first periodic comet to be identified, and the most prominent of its type. When Edmond HALLEY calculated the orbit of the comet of 1682, he noted that it was very similar to those of the comets of 1531 and 1607. He concluded that all three were the same object, moving in an elliptical 76-year orbit, and he predicted a return in about 1758. More accurate figures were later computed by the French mathematician Alexis Claude Clairaut (1713–1765). The comet was first seen on its return by the German amateur astronomer Johann Georg Palitzsch (1723–1788) on Christmas Day, 1758. The comet reached perihelion on March 13, 1759, returned again in 1835, and was particularly well placed for observations in 1910. It will return to perihelion in February 1986, but is expected to be a disappointing object since it will be badly placed for observation. Halley's comet moves between 0.59 and 35.3 a.u. of the Sun (from between the orbits of Mercury and Venus to beyond the orbit of Neptune). The comet has

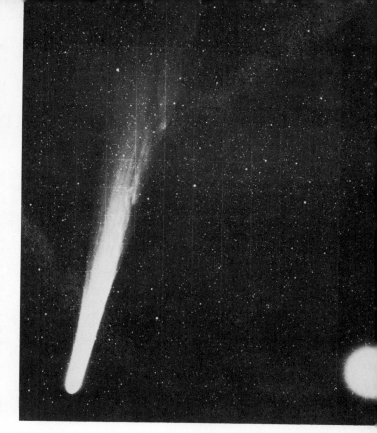

Halley's comet, close in the sky to Venus, **right**, on its last approach to the Sun in 1910. The comet takes 76 years to complete its highly elliptical orbit, traveling far beyond Neptune, and is due to return in 1986, when astronomers have predicted that it may not be as spectacular as in previous appearances.

now been identified at 21 returns prior to 1531, the earliest being in 87 B.C. It has a mass of about 50 trillion tons and develops a 20-million-mile (32-million-km) tail at perihelion. The Earth passed through the tail in 1910. The comet is thought to be the parent source of meteors in the Eta Aquarid and Orionid meteor streams.

Harvard College Observatory
Astronomical facility of Harvard University, founded in 1839, which in 1973 joined with the SMITHSONIAN ASTROPHYSICAL OBSERVATORY to form the Harvard-Smithsonian Center for Astrophysics. At its Agassiz Station it operates a 61-inch (155-cm) reflecting telescope and a 60-ft (18.3-m) radio telescope. It also has a radio astronomy station at Fort Davis, Texas, and shares in the operation of the 4,550-foot (1,387-m) Boyden Observatory, 15 miles (27 km) east of Bloemfontein, South Africa, founded in 1927 as the southern station of Harvard. This contains a 60-inch (152-cm) reflector and 32/36-inch (81/91-cm) Schmidt-Baker camera.

harvest Moon
The full Moon nearest to the autumnal equinox in late September, when the Moon seems to rise at about the same time each night for several nights in succession.

heavy elements

In astronomy, all the chemical elements except HYDROGEN and HELIUM. Astronomers often refer to heavy elements as "metals," even though they include such nonmetallic elements as carbon and oxygen. All have been built up by nuclear fusion reactions inside stars. They are scattered through space by nova and supernova explosions, and stars subsequently formed from the debris are richer in heavy elements than were preceding generations. The variation in the abundance of elements among stars thus provides a clue to the evolution of stars and galaxies. The oldest stars in our Galaxy, which occur in the globular clusters and in the extended "halo" of the Galaxy, have only a thousandth the proportion of heavy elements found in younger stars like the Sun (see ELEMENTS, ABUNDANCE OF).

heliacal rising and setting

The closest observable rising or setting of a star to its conjunction with the Sun. The heliacal risings of bright stars such as Sirius were used by ancient astronomers as the basis of their calendar; the stars acted as time-markers of the Sun's yearly journey around the sky. The term is now sometimes also used to denote the simultaneous rising or setting of a star with the Sun.

heliocentric

Term meaning Sun-centered. A *heliocentric orbit* is an orbit about the Sun. The *heliocentric system* was the belief that the Sun is the center of the Universe. Copernicus' heliocentric theory superseded the geocentric (Earth-centered) system of the Greeks, although modern astronomy has shown that the Sun is only the center of the solar system, not of the Universe.

heliometer

A device for measuring the angular distance between objects in the sky. The first successful heliometer was made in 1754 by John DOLLOND. The heliometer's OBJECT GLASS is divided into two equal D-shaped halves that can slide along the join, producing a split image. When the two halves form a circular lens, a normal image is produced; but as they slide past each other two sets of images are produced, one by each semicircular half of the lens. The angular separation of two objects is found by measuring the amount by which the lenses have to be moved in order to superimpose their two images. The heliometer derives its name from the original purpose—to measure the diameter of the Sun. An advantage of the heliometer over ordinary micrometers is that it can easily measure angular distances larger than the field of view of the eyepiece, and for many purposes it can produce more accurate results. Although the measurement of large angular distances on photographs has made the heliometer itself obsolete, double-image micrometers using the heliometer principle are often used with ordinary telescopes. The German astronomer Friedrich BESSEL used a $6\frac{1}{4}$-inch (15.8-cm) heliometer to obtain the first reliable PARALLAX measurement of a star in 1838.

Helios probes

Joint space program of Germany and the United States to study the Sun and interplanetary space. The first Helios probe was launched from Cape Canaveral on December 10, 1974, into an elliptical orbit that takes it to within 28 million miles (45 million km) of the Sun every 192 days, closer than any previous probe; the first such approach was on March 15, 1975. An identical craft was launched on January 15, 1976, for a close approach to the Sun of within 27 million miles (43 million km) on April 17.

heliostat

An arrangement of two plane mirrors which feeds the image of a celestial object into a stationary telescope. One mirror is equatorially mounted (on an axis parallel to that of the Earth), and reflects light onto the other mirror which directs the beam into the telescope. The advantage of this system over the COELOSTAT is the ease with which different objects can be brought into view in quick succession. On the other hand, the field of view slowly rotates as the Earth turns, while the field of the coelostat is fixed in orientation. Another form of heliostat, more correctly called a *siderostat,* uses a single equatorially-mounted plane mirror to reflect the light toward one of the celestial poles and directly into the telescope, which is set up parallel to the Earth's axis (see COUDÉ FOCUS).

helium in space

Helium is the second-lightest and second most abundant element in space; about one helium atom exists for every ten hydrogen atoms in the Universe. Helium is several hundred times more abundant than any of the other elements (the so-called HEAVY ELEMENTS). It is formed from hydrogen by nuclear reactions in the center of normal stars like the Sun, and may be ejected into space by NOVA or SUPERNOVA explosions. But only about 15 percent of interstellar helium has been created in this way, the largest part was probably produced in the BIG-BANG explosion from which the Universe is believed to have originated. The oldest (population II) stars in the Milky Way Galaxy, for example, contain almost the same proportion of helium as do the younger stars; but they lack the heavier elements created since the big bang in stars and stellar explosions (see also ELEMENTS, ABUNDANCE OF; HYDROGEN IN SPACE).

Heracleides Ponticus (c.388–c.315 B.C.)

Greek astronomer, who made the first known suggestion that the Earth rotates on its axis. Heracleides also proposed that since Mercury and Venus never appear to stray far from the Sun, they could actually be in orbit around it. This was the earliest suggestion that not all objects revolved around the Earth. The ideas of Heracleides, though ignored at the time, had an influence on Renaissance astronomers such as Copernicus and Tycho Brahe.

Hercules

A major constellation of the northern hemisphere of the sky, best seen in summer. Alpha Herculis, also known as Ras Algethi, is a double star, the primary of which is a red supergiant 500 to 600 times the Sun's diameter, and one of the largest stars known; it varies irregularly between third and fourth magnitude. The fainter star, a yellow giant of magnitude 5.4, is also a SPECTROSCOPIC BINARY of period 51.6 years. Hercules contains two globular clusters. One is M13 (NGC 6205), one of the brightest globular clusters visible, about 22,500 light-years away and containing some 300,000 stars. The other globular, M92 (NGC 6341), is roughly 36,000 light-years away.

Hermes

An asteroid of the Apollo group, discovered by Karl Reinmuth at Heidelberg in November 1937. Hermes is about $\frac{1}{2}$ mile (0.8 km) in diameter. The asteroid passed within 500,000 miles (800,000 km) of the Earth's sunlit side (only twice the distance of the Moon), and observations were only possible for five days. The 2.1-year orbit which carries Hermes between 0.62 and 2.7 a.u. of the Sun is therefore poorly determined and the asteroid has been lost. It retains its preliminary designation of Minor Planet 1937 UB and has never been formally numbered. Hermes passes closer to the Earth than any other known asteroid and should be relocated during a future encounter.

Herschel family

A family of British astronomers of German origin, who made pioneering contributions to astronomy.

Sir William Herschel (1738–1822) was trained as a musician, coming to England in 1757, and becoming an organist at the resort of Bath. He was an avid amateur astronomer, and being dissatisfied with commercially available telescopes he began to construct his own reflectors. Herschel had a flair for grinding and polishing the metal mirrors then in use; when he wanted larger ones than those readily available, he cast his own metal blanks at home.

William Herschel was an indefatigable observer who regularly surveyed the night sky. In 1781, during his third survey of the heavens, he found an object that turned out to be Uranus, the first planet to be discovered since prehistoric times. Herschel became famous; he was granted a royal pension as astronomer to King George III, and forsook music for astronomy. He moved from Bath to be near Windsor and the king, finally settling at Slough where he spent the rest of his life. Unlike his contemporaries, who concentrated their attention on the Sun, Moon, and planets, Herschel occupied himself with the stars, which required large mirrors to gather increasing amounts of light as he probed ever-further into space. In 1783 he built a reflector with an aperture of 18 inches (46 cm); this proved to be of fine optical quality and became Herschel's favorite instrument. In 1785 the king financed the construction of a gargantuan telescope with a mirror 48 inches (122 cm) in diameter. Completed in 1789, it was the largest telescope in the world and a sight important visitors to England were taken to see. However it proved cumbersome, and although Herschel made a number of discoveries with it, he still preferred the 18-inch telescope.

Herschel tried to determine stellar parallax by observing the changing separation between two stars close together in the sky. Although he did not measure any parallaxes in this way, he did discover that many such pairs were actually two stars in orbit about each other. In 1782, 1785, and 1821, he published positions

Sir William Herschel, **top**, the ''father of stellar astronomy''— at 56. His fame was widespread by this time, though he was not knighted until 1816, when he was 78. The discovery of Uranus, which made him famous, was made from his back garden in Bath in 1781.

Caroline Herschel, **below left**, at the age of 97, a year before her death in 1848. Her work in discovering comets earned her recognition in her own right. From her notes and diary, we have a detailed record of her brother William's life.

The famous photograph of Sir John Herschel, at the age of 75, by family friend Mrs Julia Cameron. Herschel had a keen interest in photography: he had invented his own method, essentially the modern process, in 1839 within days of hearing of the early Daguerrotype system. He made the first glass photographic negative, of his father's great ''Forty Foot'' reflector, just before it was dismantled in that year.

and other details of a total of 848 double stars. Herschel also devised methods to determine stellar brightness and found that stars differ widely in their intrinsic luminosity.

Herschel spent most of his time cataloging all celestial objects visible in his telescopes. He made a particular study of hazy objects or nebulae, recording a total of 2,500 of them. When his 48-inch telescope resolved some into separate stars, he speculated on whether an even larger telescope would show them all to be merely clusters of stars. Further observations convinced him, correctly, that this was not so; some nebulae were patches of what he called a "shining fluid."

In 1783 Herschel determined that the Sun was moving through space toward a point in the constellation Hercules (see SOLAR APEX). Perhaps his most significant contribution was an analysis of more than 3,000 selected areas of the sky, which suggested that the stars are arranged into a circular, lens-shaped slab—our Galaxy.

Caroline Herschel. In 1772 William brought his sister Caroline Lucretia Herschel (1750–1848) over from Germany, and in due course she became her brother's devoted and untiring assistant. Herschel observed every night if the sky was suitable, employing watchers on overcast nights to notify him as soon as the clouds cleared; Caroline was always at his side, helping him and writing up the results by day. She herself discovered no less than eight comets and revised the star catalog of John FLAMSTEED. She was awarded many honors for her own astronomical research, which would certainly have been more extensive had she not spent so much of her time helping William, with whom she remained until his death.

Sir John Herschel. William Herschel's son John Frederick William Herschel (1792–1871) refurbished the 18-inch and took it to South Africa where, between 1834 and 1838, he made extensive surveys of the southern skies. He also observed at Slough, cataloging stars, measuring their brightnesses with a new device of his own, and discovering 525 nebulae. He devised new ways of determining binary star orbits, and cooperated in making accurate observations of the difference in longitude between London and Paris.

John's interests were extremely broad. He experimented in photography, coining the words "positive" and "negative," and helped the photographer William Henry Fox Talbot (1800–1877) in his research. He also made useful contributions in the fields of biology, botany, and meteorology. An able administrator as well as research scientist, John Herschel was closely connected with the Royal Society and was instrumental in forming and running what is now the Royal Astronomical Society. He undertook a host of public duties, and was a great popular writer on science, particularly astronomy. Although his second son, Alexander Stewart Herschel (1836–1907), made important meteor observations, establishing the radiant of several streams, after John's death the great scientific contributions of the Herschel family effectively came to an end.

Hertzsprung, Ejnar (1873–1967)
Danish astronomer, who showed that the color and luminosity of stars are related, and discovered the division of stars into giants and dwarfs. These discoveries were first outlined in papers published in 1905 and 1907, and were confirmed in 1913 by the work of the American astronomer Henry Norris RUSSELL. The Hertzsprung-Russell diagram, a plot of stars' luminosity against their spectral types, is now named for the two men. Hertzsprung discovered on his graphs that most stars fell into a broad band, now called the MAIN SEQUENCE, which indicated the relationship between color and brightness for the majority of stars. The group of giant stars, however, lay above this main sequence in an area separated by what is called the *Hertzsprung gap*.

Hertzsprung made a study of the stars of the Pleiades, separating members of the cluster from foreground stars by their different motions. In 1911 he discovered that the star Polaris was a CEPHEID VARIABLE, and in 1913 determined the distance of the Small Magellanic Cloud by comparing the brightness of the Cepheid variable stars in it with the brightness of nearby Cepheid stars in our own Galaxy.

Hertzsprung-Russell diagram
A graph on which the color or temperature of stars is plotted against their brightness. It is named for the astronomers who independently developed it in the early part of this century: the Dane Ejnar HERTZSPRUNG and the American Henry Norris RUSSELL. The position of a star on the H-R diagram reveals its physical nature and the stage it has reached in its evolution. The diagram is useful for reading off the intrinsic brightness (ABSOLUTE MAGNITUDE) of a star when only its spectrum is known; by comparing the star's absolute magnitude with its observed apparent magnitude, its distance can be readily computed.

Features of the H-R diagram. Most stars lie in a band stretching diagonally across the diagram and termed the *main sequence*. Stars on the main sequence are ordinary stars in the prime of their lives, like the Sun. A star's position on the main sequence depends on its mass: the more massive stars lie toward the upper end, while the smallest stars lie near the bottom. The Sun, being average, is placed about halfway along. Above and to the right of the main sequence are stars which are much brighter than the main-sequence stars of the same temperature; these are giants, and represent a later stage in a star's evolution, when it has exhausted its hydrogen fuel and begun to swell in size. Eventually the Sun, like all main-sequence stars (except, perhaps, the smallest and coolest) will evolve off the main sequence to become a giant star. Above the giants is a scattering of supergiants, formed when stars much larger than the Sun evolved off the main sequence. The more massive stars evolve faster, and thus move off the main sequence first; astronomers can estimate the age of a star cluster (whose stars all formed together) by noting the point at which stars have evolved off the main sequence to become giants. Certain types of inherently variable stars, such as the CEPHEID VARIABLES, are found in specific regions among the giants and supergiants on the H-R diagram. Eventually, stars like the Sun are believed to puff off their outer layers into space, leaving the small but intensely hot core which forms a WHITE DWARF; these are found at the bottom left of the H-R diagram.

History of the H-R diagram. Russell, in his 1913

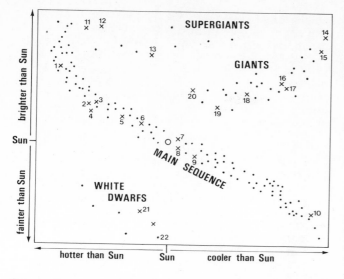

Schematic representation of the Hertzsprung-Russell diagram: the dots represent schematically placed stars, crosses represent actual stars. Brightnesses and temperatures of stars are shown relative to the Sun (circle). Main sequence stars are: 1.Spica; 2.Vega; 3.Castor; 4.Sirius; 5.Altair; 6.Procyon; 7.Alpha Centauri; 8.Tau Ceti; 9.Epsilon Eridani; 10.Barnard's star. Giants and supergiants: 11.Rigel; 12.Deneb; 13.Polaris; 14.Betelgeuse; 15.Antares; 16.Mira; 17.Aldebaran; 18.Arcturus; 19.Pollux; 20.Capella. White dwarfs: 21.Sirius B; 22.Procyon B.

graph, plotted the absolute magnitude against spectral class for nearby stars whose distances he measured from their PARALLAX; he was plotting a star's brightness against its temperature. Hertzsprung, working on the Hyades and Pleiades star clusters in 1911, plotted the apparent magnitude of the stars in each cluster against their COLOR INDEX. (This plot is sometimes distinguished from Russell's version by the name *color-magnitude diagram.*) Since the stars in each cluster are at virtually the same distance, their apparent magnitudes are in direct proportion to their absolute magnitudes; and, like spectral class, color index measures temperature. Hertzsprung and Russell's diagrams were thus essentially the same. The measurement of stellar brightnesses and colors has now become far more refined, and the H-R diagram is established as a major analytical tool of astrophysics.

Hevelius, Johannes (1611–1687)
Polish astronomer, best known for his famous atlases of the Moon and stars. Hevelius' volume of 1647, *Selenographia,* contained his Moon map, the first major attempt of its kind, notable for showing the Moon's libration (rocking motion). Hevelius' map also introduced the first lunar nomenclature, though few of his names for lunar features survive today, those of Riccioli having been adopted in preference. Hevelius observed the Moon and planets with telescopes of his own construction, up to 150 feet (46 m) long and supported from masts by a complex system of pulleys and stays. To make star observations he built copies of the instruments used by Tycho

Brahe, which did not involve optics. His best-known work is his catalog of 1,564 stars, the most extensive of its day, published posthumously in 1690. He engraved accompanying maps which introduced seven now-familiar constellations—Canes Venatici, Lacerta, Leo Minor, Lynx, Scutum, Sextans, and Vulpecula.

Hewish, Antony (b. 1924)
British radio astronomer, discoverer of the rapidly flashing radio sources known as PULSARS. Hewish and his colleagues at the Mullard Radio Astronomy Observatory in Cambridge, England, first detected rapidly pulsating radio signals in mid-1967, with an instrument intended to measure how radio signals from quasars flickered as they passed through gusts in the stream of atomic particles flowing from the Sun. In early 1968 Hewish's team announced that pulsars were probably tiny rotating NEUTRON STARS, predicted by physicists but never previously detected, in which protons and electrons have been crushed together to form neutrons. Hewish was corecipient of the 1974 Nobel Prize in physics for his discovery.

Hipparchus of Nicaea (fl. 146–127 B.C.)
Greek astronomer, considered the greatest figure of ancient astronomy. His most notable discovery was that the Earth wobbles like a spinning top in space over a period of about 26,000 years, an effect which he termed PRECESSION. He deduced this effect from a regular shift between the positions of stars that he cataloged and those recorded by earlier observers; the shift meant that the Sun preceded the stars in its yearly motion around the sky, hence the name.

Hipparchus measured the length of the year from the Sun's motion to an accuracy of within six minutes. He abandoned the complex system of multiple spheres proposed by ARISTOTLE to explain the movement of heavenly bodies, and turned instead to the geometrical devices involving circles that the Greek APOLLONIUS had suggested. Hipparchus found that the Sun's motion could be explained by assuming it followed an off-center circle, known as an *eccentric,* around the Earth. Hipparchus also investigated the motion of the Moon, a complex problem that has occupied astronomers ever since, and was unable to account for it with total success. His work did, however, lead to major advances in the prediction of eclipses of the Sun and Moon.

Hipparchus also left an important legacy in his catalog of 850 stars, completed in 129 B.C.; it was during its compilation that he discovered precession. Hipparchus divided the stars in his catalog into six classes of brightness, from first magnitude (the brightest) to sixth magnitude (faintest visible to the naked eye), a system still used today. He collated observations of the planets to aid later astronomers in calculating their motions; the observations existing in his time were too sketchy for him to work out their orbits accurately. The work of Hipparchus had a great effect on PTOLEMY, much of whose great book the *Almagest* is actually based on Hipparchus.

Horologium (the pendulum clock)
A faint and unremarkable constellation adjacent to Eridanus in the southern hemisphere of the sky, introduced in the 1750s by Nicolas Louis de Lacaille.

Hoyle, Sir Fred (b. 1915)

British astronomer, best known for his support of the STEADY-STATE THEORY of cosmology, which holds that the Universe has appeared essentially the same at all times, and that matter is being continuously created as the Universe expands. This idea was proposed in 1948 by Thomas GOLD and Hermann BONDI; Hoyle then developed the theory in mathematical terms, modifying Einstein's theory of general relativity. Hoyle has also made fundamental contributions to our understanding of the structure and evolution of stars, notably the evolution of Sun-type stars into red giants; much of this work was presented in his famous 1955 book *Frontiers of Astronomy*. Collaborating with Geoffrey and Margaret BURBIDGE and the American astrophysicist William Alfred Fowler (b. 1911), he described how chemical elements can be built up by nuclear reactions inside stars, eventually being scattered into space by supernova explosions (see ELEMENTS, ABUNDANCE OF).

H-R diagram

Abbreviation for HERTZSPRUNG-RUSSELL DIAGRAM, on which stars are plotted according to spectral type and magnitude.

Hubble, Edwin Powell (1889–1953)

American astronomer who proved that galaxies of stars exist outside our own Milky Way, and discovered that the Universe is expanding. These advances have been compared in fundamental importance to the establishment of the heliocentric theory of the solar system three centuries earlier. Hubble's first observations, at Yerkes Observatory, were of the fuzzy patches called nebulae. In 1917 he came to the conclusion that some nebulae were patches of gas within our Galaxy, while the spiral-shaped nebulae were probably far more distant. In 1919 Hubble joined George Ellery HALE at Mount Wilson Observatory, and with the 60-inch (152-cm) telescope there showed that the gaseous nebulae in our own Galaxy were made luminous by stars embedded in them.

Then he turned to other nebulae with the new 100-inch (254-cm) telescope. In 1923 he found a CEPHEID VARIABLE star in the outer regions of the Andromeda nebula, M31; this and other stars showed conclusively for the first time that M31 was in fact a separate star system, far beyond our own.

Hubble went on to study numerous other galaxies, introducing in 1925 the now-standard classification system of spiral galaxies, barred spirals. ellipticals, and irregulars (see GALAXIES). In 1929, when comparing the distances of galaxies with their radial velocities (speed of movement deduced from spectroscopic observations), he found that the galaxies seemed to be receding with speeds that increased with distance. This relationship, termed HUBBLE'S LAW, revealed that the Universe is apparently expanding. Hubble's law was confirmed over even greater distances by the American astronomer Milton LaSalle Humason (1891–1972).

Hubble found that galaxies were distributed evenly over the sky except for a "zone of avoidance" along the plane of the Milky Way, caused by obscuring clouds of dust. Hubble helped plan the 200-inch (508-cm) telescope at Mount Palomar.

Hubble's constant

A measure of the rate at which the Universe is expanding. HUBBLE'S LAW shows that a galaxy's speed of recession depends on its distance; Hubble's constant is the figure that relates velocity to distance, and is determined by observation. Current figures put Hubble's constant at about 10 miles per second per million light-years (55 km/sec/megaparsec). Galaxies, in other words, recede at a rate of 10 miles per second for every million light-years of their distance. Some theories of cosmology predict that the rate of expansion of the Universe may change over time. A rough measure of any such change can be found by comparing the local expansion rate with the rate at very distant regions of the Universe. Because of the time the light from these distant regions has taken to reach us, we see them as they appeared many billions of years ago. Current results suggest scarcely any measurable change, which seems to indicate that the expansion of the Universe will continue indefinitely. If the expansion of the Universe has been constant over time, then the inverse of Hubble's constant yields the age of the Universe—that is, the time since expansion began from the BIG BANG. This period is 18 billion years.

Hubble's law

The relationship between a galaxy's distance and its speed of recession, announced in 1929 by Edwin HUBBLE. Hubble found that distant galaxies were receding at speeds which increased in proportion to their distance: the farther the galaxy, the faster its recession. This result, termed Hubble's law, shows that the Universe is apparently expanding. Like spots on an inflating balloon, every galaxy moves away from every other, and no galaxy is at the center of expansion; therefore, the expansion of the Universe appears everywhere the same. It follows that the rate at which two galaxies move apart is proportional to their separation—galaxies twice as far apart recede at twice the speed. Hubble confirmed this simple model by demonstrating that the RED SHIFT in a galaxy's light (which measures the recessional velocity) was proportional to the apparent brightness of the galaxy (which is dependent on its distance). Hubble's law received an immediate explanation in the theory of relativity, which regards the expansion of the Universe as a uniform expansion of space itself. The actual rate of expansion is given by HUBBLE'S CONSTANT.

Huggins, Sir William (1824–1910)

English astronomer, a pioneer of stellar spectroscopy. His first observations, published in 1863, showed that the stars were made of incandescent gas like the Sun, and that although stars differed in composition they still contained elements that were familiar on Earth. In 1864 he found that, although some of the fuzzy patches in the sky called nebulae were groups of distant stars, others were actually clouds of glowing gas. He was unable to identify the bright spectral lines emitted by the gaseous nebulae and attributed them to an unknown substance he called "nebulium;" it was not until 1927 that the American astronomer Ira S. BOWEN finally proved that these lines were caused by known elements under the rarefied conditions of space (see FORBIDDEN LINES). The

Italian comet observer Giovanni Battista Donati (1826–1873) had taken the first spectrum of a comet in 1864, showing it to be made of gas; Huggins went on to find evidence of carbon compounds in comets. He also observed the clouds of hot gas at the Sun's edge called prominences, and discovered a cloud of hot gas that had been expelled from a nova that flared up in the constellation Corona Borealis in 1866. In 1868 he measured the first RADIAL VELOCITY for a star, Sirius, by observing the shift in position of its spectral lines caused by its speed of recession.

Huygens, Christiaan (1629–1695)
Dutch scientist, best known in astronomy for his description of the rings of Saturn as a swarm of particles orbiting the planet like tiny moons. He gave this explanation in 1655, the same year that he discovered Saturn's largest moon, Titan. Although Saturn's rings had been seen before, no one had fully understood their true nature. Huygens built his own telescopes, the best of their time. In 1659 he discovered the first markings on Mars, notably the dark area called Syrtis Major. His optical studies led him in the 1650s to invent the two-lens Huygenian EYEPIECE, still popular today. He is also credited with first suggesting that light travels as a wave, which allowed him to give explanations of refraction and reflection far better than any provided by the theory that visualized light purely as a stream of particles. Among his many other scientific advances, Huygens invented the first successful pendulum clock in 1656, thus aiding astronomers in their problem of accurate timekeeping.

Hyades
A V-shaped cluster of stars in the constellation Taurus. The Hyades are about 148 light-years away; they are the nearest cluster to the Sun. They are moving through space at a speed of about 27 miles (43 km) per second. The distance to the Hyades is found by observing the motions of the stars (see MOVING-CLUSTER METHOD). The result is extremely important, because it is used as the first step for determining distances to most other stars and galaxies. The Hyades itself contains about 200 known stars spread over a diameter of about 12 light-years. Accompanying the cluster is a stream of about 200 other scattered stars together with the Praesepe cluster ("the Beehive"). The age of these stars is estimated at 600 million years.

Hydra (the water snake)
The largest constellation in the sky. It extends from just north of the celestial equator, in its head region, into the southern hemisphere. Its brightest star, Alpha Hydri, also called Alphard, is an orange giant star of magnitude 2.16.

hydrogen in space
Hydrogen, the lightest element, is the most abundant substance in the Universe. It occurs in three forms in interstellar space: neutral atoms (H I), ions (H II), and molecules (H_2). Hydrogen atoms are distributed between the stars in our Milky Way Galaxy. Although the average density is about one atom to 5 cubic centimeters, individual areas of cold hydrogen (H I REGIONS) vary considerably. Cold hydrogen gas emits the 21-centimeter radiation observed in RADIO

ASTRONOMY. Areas of luminous hydrogen gas (H II regions) have densities some thousand times greater; their electrons have been stripped from the atoms by the ultraviolet radiation from the hot stars condensing within them. Hydrogen molecules (H_2) are rare in interstellar space, because they are broken into individual atoms by ultraviolet radiation. However, satellite observations at ultraviolet wavelengths have shown that some do occur, and the study of interstellar molecules suggests that H_2 is common in very dense interstellar clouds where there is enough dust to shield the molecules from ultraviolet radiation. Most of the hydrogen in dense clouds may be in this form, with a density of more than a million molecules per cubic centimeter.

Ordinary hydrogen atoms consist of an electron orbiting a proton. The heavier isotope, deuterium (symbol D), is also found in space. It possesses a neutron attached to the central proton, and is twice as heavy as ordinary hydrogen. Deuterium is easily destroyed by nuclear reactions in stars, and the deuterium in space cannot have been made in stars by nuclear fusion reactions, like the heavy elements. It was probably formed very soon after the BIG-BANG explosion in which our Universe originated, when the whole Universe was at a temperature of 1 billion degrees K. The observed abundance of deuterium seems to indicate that the Universe contains a relatively low density of matter, and that there is insufficient gravitational attraction to prevent the expansion of the Universe from continuing indefinitely.

Hydrus (the lesser water snake)
A constellation near the south pole of the sky, introduced in 1603 by Johann Bayer. Beta Hydri is a star similar to the Sun.

hyperbola
A symmetrical curve which never closes in on itself. A hyperbola is formed by cutting a cone at an angle steeper than either of the cone's sides. A hyperbola looks like a PARABOLA whose arms diverge, instead of becoming parallel at infinity. The orbit of an object passing through the solar system, but never captured by the Sun, would be a hyperbola.

hyperboloid
A solid shape formed by rotating a HYPERBOLA around its central axis (the line about which it is symmetrical). The mirrors of certain types of reflecting telescopes are occasionally hyperboloid.

hypergolic propellants
Rocket fuels that are self-igniting; they burn spontaneously when mixed, thus eliminating the need for an ignition system. Hypergolic fuels were used on the main engine in the Apollo service module, and in the lunar module.

Hyperion
A satellite of Saturn, the eighth in order of distance from the planet; it was discovered in 1848 by W. C. and G. P. BOND. Hyperion is about 300 miles (480 km) in diameter and orbits Saturn every 21 days 6 hours 38 minutes at an average distance of 921,350 miles (1,483,000 km).

I

Iapetus

Second most distant moon of Saturn, discovered in 1671 by Giovanni Domenico CASSINI. Iapetus orbits Saturn every 79 days 7 hours 56 minutes at a distance of 2,212,200 miles (3,560,200 km). It has a density three times that of water, indicating it is probably a rocky body like the Moon; the diameter of Iapetus is about 1,000 miles (1,600 km). The satellite probably presents the same face to Saturn at all times. Its trailing side appears six times brighter than its other side; it is believed that Iapetus is mostly snow-covered, with the dark side representing a large patch of bare rock.

IAU

Abbreviation for the INTERNATIONAL ASTRONOMICAL UNION.

Icarus

An asteroid of the Apollo group, discovered by Walter BAADE at Mount Palomar in June 1949. Icarus, also known as minor planet 1566, is 0.87 miles (1.4 km) in diameter. It moves in an exceptionally eccentric ellipse passing from within 0.19 a.u. of the Sun (where it must be heated to a dull red), out to as far as 2.0 a.u. Inclined at 23° to the ecliptic, the 1.1-year orbit brings Icarus close to the Earth every 19 years. In June 1968 it passed within 4 million miles (6.4 million km) of the Earth, appearing as a fast moving 12th-magnitude object. It will pass close to the Earth again in 1987. Icarus is a near-spherical body with a $2\frac{1}{4}$-hour rotation period, the shortest of any asteroid. It has a reflectivity (albedo) of 17.8 percent and a stony or stony-iron surface composition. Icarus is probably the rocky central core of an extinct and degassed periodic comet nucleus rather than a true asteroid (see ASTEROIDS).

image intensifier

An electronic device which increases the brightness of a faint image. Image intensifiers are now becoming widely used in professional astronomy; they reduce photographic exposure times and reveal faint detail never before visible. Low levels of light are amplified by using a *photoemitter,* a layer of material which emits electrons when struck by light. The emitted electrons, which represent the image, are accelerated by electric fields through an evacuated tube, where they strike a phosphor screen, like a television screen. The electrons cause the screen to glow, reproducing the original image. Electromagnets or charged plates are placed around the tube to focus the electrons on the phosphor screen. A variant is the *electronographic tube,* or electronic camera. In this case, no phosphor is used, and the electrons strike the photographic emulsion directly. In other forms of image tubes, the image on the photoemitter is allowed to build up and is then scanned as in a television camera. This allows a remote display on a television screen and permits "time exposures" to increase light collection.

inclination

The angle between the plane of a particular orbit and a reference plane, which for objects in the solar system is usually the plane of the Earth's orbit. For objects orbiting the Earth, the reference plane is more usually the plane of the Earth's equator.

Indus (the Indian)

An insignificant constellation near the south pole of the sky, named in 1603 by Johann Bayer. The star Epsilon Indi, 11.2 light-years away, is one of the closest stars to us, and slightly cooler than the Sun.

inferior conjunction

The instant at which Mercury or Venus is in a direct line between the Earth and Sun (see CONJUNCTION).

inferior planet

A planet with an orbit closer to the Sun than Earth, i.e. Venus and Mercury.

infrared astronomy

The study of radiation from space between the wavelengths of red light (7800 Å) and about 1 millimeter (where microwave radio astronomy begins). This includes the range of wavelengths we feel as heat. An object which releases most of its energy in the infrared is cooler than our Sun, which has a surface temperature of 6,000°K and radiates most strongly in the visible part of the spectrum.

The infrared sky. Because the brightness of a star at a particular wavelength depends on its temperature, eyes sensitive to infrared radiation would see a very different pattern of constellations from those we are familiar with. The pattern of bright stars "seen" by an infrared detector depends on exactly which wavelengths it is most sensitive to.

The first survey of the sky in the infrared was made at Mount Wilson in the mid-1960s, and used detectors most sensitive at 2.2 microns, about four times the wavelength of yellow light. The survey covered 75 percent of the sky, and revealed 20,000 sources, most of them stars with surface temperatures of 1,000 to 2,000°K. The brightest 5,500 of these sources made up the first catalog of infrared stars.

Very few of these bright infrared stars are visible at optical wavelengths, and very few of the brightest visible stars emit strongly in the infrared. The 300 brightest infrared stars are spread at random across the sky, as are the brightest stars at visual wavelengths. But the faintest 300 infrared stars in the Mount Wilson catalog were concentrated strongly in the plane of the Milky Way, particularly in the direction of the galactic center.

This distribution indicates that these sources are more than 1,000 light-years distant, and must therefore be relatively powerful. They are probably bright, hot stars, whose light has passed through interstellar dust clouds on the way to our telescopes. The dust scatters the shorter wavelengths, but the longer wavelengths can penetrate through the dust, so that the infrared light from the star is detected. This is similar to the way dust in the atmosphere scatters blue light from the Sun at sunrise and sunset, but allows red light to penetrate.

Dust in our Galaxy is concentrated in the plane of the Milky Way and obscures the galactic center at optical wavelengths. Although ordinary light is 99.9999999 percent absorbed, the absorption at 2 microns wavelength is only 90 percent. Infrared studies thus provide a means of penetrating the obscuring dust, and are of great importance in understanding the structure of our Galaxy.

The galactic center. Many galaxies have a bright central nucleus, visible with optical telescopes, but such a nucleus in our Galaxy would be obscured by dust. Radio and infrared observations, however, show an extended source of energy at the galactic center. The ANDROMEDA GALAXY also shows a peak of infrared emission at its nucleus, and a comparison indicates that the center of our Galaxy contains millions of ordinary stars packed into a space only a few light-years across. At the core, stars are 200 times closer together than they are near the Sun, so the night sky on a planet orbiting one of those stars would be 40,000 times brighter than ours.

There appears to be a point source of very intense infrared radiation at the galactic center itself, only $\frac{1}{3}$ light-year in diameter but radiating as much energy as 300,000 Suns. Its nature is not known; possibly it is a giant BLACK HOLE.

In some galaxies with very bright optical nuclei (SEYFERT GALAXIES) the infrared emission from the nucleus comprises most of the total radiation from the galaxy. Some QUASARS are also strong infrared sources. In both cases the radiation is strongest at long wavelengths (about 70 microns). Further sky surveys at these wavelengths will probably reveal many similar sources.

Very cool objects. Much of the early excitement of infrared astronomy centered on the prospect of detecting PROTOSTARS, clouds of gas and dust, at only a few hundred degrees K, early in the process of collapsing to form stars. These hopes have not been realized, but the search for very cool objects has revealed some interesting phenomena.

Most of the cool stars detected so far (with temperatures below 1,700°K) are long-period variables, perhaps related to the cool variables visible optically (such as MIRA); but a few of the sources are very cool but unvarying, quite unlike anything observed before. An early discovery was a bright infrared star in Cygnus, as bright at 2 microns as Vega (the fourth-brightest star in the visible sky), and brighter at 20 microns than any object except the Sun. The spectrum of this source corresponds to a temperature of 1,000°K. Although it may simply be a cool young star, there is a possibility that it is a very bright supergiant star, of a kind never before observed, surrounded by obscuring dust.

Infrared detectors. Infrared radiation is absorbed by water vapor in the atmosphere, and so it is important to observe from high altitudes. Almost any optical reflecting telescope can be adapted for infrared studies, using detectors sensitive at infrared wavelengths (such as lead sulfide PHOTOELECTRIC CELLS) mounted at the focus, and usually cooled by liquid nitrogen to improve sensitivity. Using the best optical telescopes is wasteful since observations at the longer infrared wavelengths do not require the precision of optical mirrors. Infrared observations can be made at twilight, however, when the telescope cannot be used optically.

The first Mount Wilson infrared survey was made with a specially-built instrument with a 62-inch (157-cm) mirror; but infrared detectors are so lacking in sensitivity that an optical instrument would need only a 2-inch (0.8 cm) mirror to make an equally sensitive survey of the sky. Most large telescopes have been used for infrared work, but new infrared "flux collectors" of large aperture are now being planned. The biggest such instrument under construction is the 152-inch (390-cm) British infrared telescope on Mauna Kea in Hawaii.

Intelsat

The International Telecommunications Satellite Corporation, an organization set up by a number of nations in 1964 to produce and operate satellites for

Satellite	Launch date	Remarks
Intelsat I (Early Bird)	April 6, 1965	Stationed over Atlantic
Intelsat II-A (Lani Bird)	October 26, 1966	Failed to achieve synchronous orbit
Intelsat II-B (Pacific 1)	January 11, 1967	Stationed over Pacific
Intelsat II-C (Atlantic 2)	March 22, 1967	Stationed over Atlantic
Intelsat II-D (Pacific 2)	September 27, 1967	Stationed over Pacific
Intelsat III-A	September 18, 1968	Launch failure
Intelsat III-B	December 18, 1968	Stationed over Atlantic
Intelsat III-C	February 5, 1969	Initially located over Pacific, but moved to Indian Ocean when malfunction reduced capacity
Intelsat III-D	May 21, 1969	Stationed over Pacific; completed first commercial global comsat system
Intelsat III-E	July 26, 1969	Intended to be stationed over Atlantic, but entered improper orbit due to launch-vehicle malfunction
Intelsat III-F	January 15, 1970	Stationed over Atlantic
Intelsat III-G	April 23, 1970	Stationed over Atlantic
Intelsat III-H	July 23, 1970	Intended to be stationed over Pacific, but failed to enter synchronous orbit
Intelsat IV-A	January 26, 1971	Stationed over Atlantic
Intelsat IV-B	December 20, 1971	Stationed over Atlantic
Intelsat IV-C	January 23, 1972	Stationed over Pacific
Intelsat IV-D	June 13, 1972	Stationed over Indian Ocean
Intelsat IV-E	August 23, 1973	Stationed over Atlantic
Intelsat IV-F	November 21, 1974	Stationed over Pacific
Intelsat IV-G	May 22, 1975	Stationed over Indian Ocean
Intelsat IVA-A	September 25, 1975	Stationed over Atlantic
Intelsat IVA-B	January 29, 1976	Stationed over Atlantic

international communications; the countries contribute to the cost and share the profits of operating a global satellite system. The Corporation's satellites are named Intelsat, and there have been several different designs since the launching of the first, Early Bird, in 1965. Early Bird, based on the experimental SYNCOM satellites, was the only Intelsat I type of satellite. Stationed in synchronous orbit above the Atlantic, it provided the first commercial satellite link between North America and Europe, carrying 240 two-way telephone conversations, or one television channel. The 85-lb. (38.5-kg) satellite covered only the North Atlantic with its antenna. The succeeding 190-lb. (86-kg) Intelsat II series, of which there were four satellites, each covered a wider area, although they carried the same number of circuits.

The first global system arrived with satellites of the Intelsat III series, which were put into orbit above the Atlantic, Pacific, and Indian Oceans. Each Intelsat III could handle up to 1,200 simultaneous telephone conversations, or four television channels, and their antennae covered all the Earth visible to them. They weighed 332 lb. (150.5 kg) each. Their successors, of the advanced Intelsat IV design, weigh 1,585 lb. (719 kg), and include spot-beam antennae to cover small areas of high traffic, as well as antennae to cover the entire globe. Their capacity is 5,000 telephone circuits or 12 color-television channels. Intelsat IV-A, a modified design, incorporates more spot-beam antennae to give a capacity of up to 9,000 telephone circuits. As each new generation of satellites has been orbited, their predecessors have been retired or put on standby. With the growth in international communications traffic, particularly across the North Atlantic, satellites of an entirely new generation— Intelsat V—will be needed by the early 1980s.

interference

An effect which occurs when two waves of the same wavelength are combined. If the waves coincide, peak for peak and trough for trough (*in phase*), they produce a wave of twice the height, or *amplitude*; but if the peaks of one coincide with the troughs of the other (*out of phase*) they cancel out each other completely. These conditions are known respectively as *constructive* and *destructive* interference. All ELECTROMAGNETIC RADIATION, including visible light, consists of waves, and can undergo interference.

An optical interference filter consists of several thin layers of transparent material coated on glass. Light reflected within the layers interferes with itself. The thickness of the layers is designed so that only a narrow range of wavelengths passes through without destructive interference. The rest of the light is reflected back. These filters are often used in narrow band PHOTOMETRY, to measure the magnitudes of stars at different wavelengths.

Two waves not exactly in or out of phase combine to give a wave whose amplitude is intermediate between constructive and destructive interference. This amplitude can thus be used to measure the phase difference between the waves. The INTERFEROMETER is based on this principle.

interferometer

A pair of receiving devices, such as radio or optical telescopes, linked to make the equivalent of a much larger receiver. Interferometers are most important in radio astronomy, where the long wavelengths mean that a dish several miles across would be needed to give a resolution equal to that of an optical telescope (about 1 arc second). In practice, two radio telescopes observing the same source are linked and their outputs combined electronically. Unless the source is directly overhead, the radio waves must travel further to one telescope than to the other. The waves received are thus to some extent out of step with each other. As the Earth rotates this *phase difference* changes, and the combined signal varies in amplitude (see INTERFERENCE). The rate of change depends on the position of the source, which can thus be measured very accurately. For a radio source that is extended over a part of the sky, different parts of the source will be in or out of phase at the same moment. The combined signal is an average of all these, and so its total variation is less than that from a point source, giving clues to the radio source's structure.

A shorter wavelength, or a longer baseline, reveals finer details. Radio telescopes separated by tens of miles can be linked by telemetry. In *very long baseline interferometry* (VLBI), which uses telescopes in different continents, the outputs of each are recorded on magnetic tape and combined later. A resolution of a few millionths of an arc-second can be achieved by VLBI. By using many fairly short interferometers and combining the results in a computer it is possible to make detailed maps of radio sources. This technique of APERTURE SYNTHESIS effectively creates a telescope as big as the separation of the longest interferometer used. A common technique is to use the Earth's rotation to simulate the effect of a large dish.

Optical interferometers can help resolve details too small for ordinary optical telescopes to distinguish. In 1920 the American physicist Albert MICHELSON used a stellar interferometer attached to the Mount Wilson 100-inch (254-cm) telescope to measure the diameters of some large nearby stars by interferometry. At Narrabri Observatory, about 250 miles (400 km) northwest of Sydney, Australia, is a stellar interferometer consisting of two reflectors 21 feet (6.5 m) in diameter, each made of 251 smaller mirrors, moving on a track 600 feet (200 m) in diameter. This optical interferometer has measured the diameters of many nearby bright stars.

International Astronomical Union (IAU)

An organization containing several thousand leading astronomers, founded in 1919 for international cooperation in astronomical research; it grew out of an international solar union founded by George Ellery Hale. The IAU holds a general assembly every three years, at which recent findings are discussed, and which decides on such matters as the standardization of constellation boundaries and the nomenclature of astronomical objects. The IAU has a telegrams bureau at the Smithsonian Astrophysical Observatory, Cambridge, Massachusetts, which notifies observatories of objects such as novae and newly discovered comets that require immediate additional observations.

interstellar absorption

The dimming of light by dust particles in space. These

dust grains, about .00001 centimeter in size, scatter some of the light from stars and galaxies that would otherwise reach the Earth. Red light is less affected than blue, so that distant objects appear redder as well as fainter. From the degree reddening, the total amount of light absorbed can be determined, and the apparent magnitude which the object would have in the absence of absorption can then be calculated. This revised magnitude must be used if the object's distance is to be found from the DISTANCE MODULUS.

Interstellar dust is concentrated in the plane of the Milky Way. The long "rift" down the middle of the Milky Way, seen best in the constellation Cygnus, is due to the greater depth of dust in that direction, which absorbs more of the light from distant stars. Dark nebulae, such as the COALSACK in the Southern Cross, which were once thought to be holes through the Milky Way, are in fact denser clouds of dust which block off distant stars in patches.

Interstellar absorption lines, seen in the spectra of distant stars, are caused by atoms of elements such as calcium or sodium in the interstellar gas.

interstellar molecules

Many simple molecules are found in clouds of gas and dust in the Milky Way, and are revealed by their characteristic radio emission lines. A few molecules can also be detected optically; methylidyne and cyanogen were found in the late 1930s by the absorption lines they produced in the spectra of starlight. Since the late 1960s, however, the development of microwave radio techniques has revealed over three dozen molecules, some of them fairly complex and containing up to nine atoms, in the interstellar material of our Galaxy. A few compounds, including formaldehyde, have also been detected in other galaxies.

Detection. The radio emission from these molecules is produced by changes in their spin and oscillation. These motions produce a characteristic set of radio wavelengths by which each compound can be identified. It is not always straightforward to assign an observed radio line to a particular molecule (there are several molecules producing emission at 3.4 millimeters, for example), and so it is important that molecules are identified by two or more lines whenever possible. Some hydroxyl, water, and methyl alcohol sources are small and very powerful. They appear to be caused by natural MASERS.

Distribution. About half the interstellar molecules so far detected have been found in only one or two sources, such as the ORION NEBULA. Only hydroxyl radical, formaldehyde, and carbon monoxide have been found distributed widely around our Galaxy. Suitable conditions for their formation in space seem to occur in dark dust clouds and GLOBULES, H II REGIONS, emission nebulae, and certain infrared stars similar to MIRA-type variable stars. The reasons why different molecules are found in different sources are far from fully understood.

Interstellar chemistry. The existence of fairly complex molecules in detectable quantities indicates that some interstellar clouds have densities greater than previously suspected. Atoms can only form molecules if they come into contact; to produce the observed molecular abundances, clouds must have densities of at least 10 million atoms per cubic

centimeter. The density of these clouds also prevents the rapid destruction of molecules, since the dust particles shield out the interstellar ultraviolet radiation which would otherwise break the molecules up.

Leaving aside the chemically unreactive substance helium, most of the molecules are made up of the four most common elements: hydrogen, nitrogen, carbon, and oxygen. The next most common atoms, sulfur and silicon, are also found in some molecules. But the molecules observed are not those which would be produced by random sticking together of the common atoms; the formation of organic (carbon-containing) compounds seems to be particularly favored by the conditions in interstellar space. The relative abundances of molecules also vary from source to source. Astronomers believe that molecules build up on the surface of dust grains in the cloud, rather than by simple collisions between the gas atoms. Interstellar molecules provide good evidence that the material of the solar system is typical of our whole Galaxy. The presence of organic molecules in dense interstellar dust clouds may also be significant for the origin of LIFE IN THE UNIVERSE.

invariable plane

The plane of average spin, both orbital and axial, of the solar system. It is obtained by adding up the total angular momentum of each planet. The invariable plane is inclined by $1° 43'$ to the plane of the Earth's orbit.

Io

Second-closest satellite of Jupiter, discovered in 1610 by Galileo. Io orbits Jupiter every 42 hours $27\frac{1}{2}$ minutes at a distance of 262,070 miles (421,760 km). Io's diameter is 2,273 miles (3,658 km) and its density 3.48 times that of water. Io is orange in color, with orange-brown polar caps similar in color to Jupiter's red spot. It is surrounded by a tenuous atmosphere of sodium, apparently released from salty deposits on its surface; an ionosphere has also been detected. Radio astronomers discovered in 1964 that sudden radio bursts from Jupiter are linked with the position of Io in its orbit, and it is probable that the satellite interacts with Jupiter's magnetic field to cause occasional discharges of particles from Jupiter's radiation belts.

ions

Atoms which have either lost or gained electrons, and thus have acquired a positive or a negative charge, respectively. Stars and much of the interstellar gas are composed of positive ions and the electrons they have lost, a mixture known as a PLASMA.

The process by which ions are formed is termed *ionization*. It usually refers to the loss of electrons by atoms in high-speed collisions in a hot gas. The voltage required to ionize an atom is its *ionization potential*; for hydrogen this is 13.6 volts. Removal of successive electrons from multi-electron atoms becomes progressively more difficult; the second ionization potential is thus greater than the first, and so on.

ionosphere

The region of the Earth's atmosphere containing atoms and molecules that have been *ionized* (had electrons removed) by solar radiation, making them electrically charged. The ionosphere actually consists of several layers at altitudes extending from about 30 miles (50 km) to about 300 miles (480 km) where it merges with the MAGNETOSPHERE. It extends through the regions of the atmosphere known as the mesosphere and thermosphere (see EARTH). Because it contains electrically charged particles, the ionosphere can reflect radio waves of certain wavelengths.

The ionosphere is divided into three main regions: the D, E, and F layers. The D layer, below about 55 miles (90 km), is weakly ionized and plays little part in reflecting radio waves; it largely vanishes at night. The E layer, between 55 and 100 miles (90 and 160 km), contains more strongly ionized molecules, and the F layer above has the greatest concentration of ionized atoms. Variations in the ionization of the E and F layers are closely linked to changes in solar activity, including flares, the 27-day rotation of the Sun, and the roughly 11-year SOLAR CYCLE. These two layers are also known as the Heaviside-Kennelly and Appleton layers, respectively. The F (Appleton) layer is sometimes further subdivided into F1 and F2 layers centered at about 110 miles (180 km) and 160 miles (250 km) in altitude (the heights at which many spacecraft orbit). Probing the ionosphere by sounding rockets and satellites has been of major importance in understanding the ionosphere.

ISIS satellites
International Satellites for Ionospheric Studies, two joint Canadian-U.S. satellites for ionospheric research, continuing the Canadian ALOUETTE satellite program.

J

Jansky, Karl Guthe (1905–1950)
American communications engineer, the founder of radio astronomy. Working at the Bell Telephone Laboratories, Jansky began to investigate the causes of interference in long-distance telephony in 1931. He tracked some of them down to natural sources such as thunderstorms, but he also found another source of radio noise that seemed to move with the stars. In 1932 he identified the source as lying in the direction of the constellation Sagittarius, which is toward the center of our Galaxy. Listening at a wavelength of 15 meters, Jansky was the first to detect the radio emission of the gas lying between the stars of the Milky Way. Jansky never followed up these first radio astronomy observations, and it was left to the radio amateur Grote REBER to pursue them. The unit of strength of cosmic radio waves has been named the jansky in his honor; it is equal to 10^{-26} watt per square meter per hertz.

Janssen, Pierre Jules César (1824–1907)
French pioneer of solar physics, inventor of the spectrohelioscope, an instrument for examining the Sun in the bright lines of the spectrum emitted by certain atoms. At a total solar eclipse in 1868 Janssen took spectra of the bright flames called prominences seen at the Sun's edge, and found that they emitted bright lines due to the gas hydrogen. Janssen realized that these bright lines could be observed when the Sun was not eclipsed, and developed an instrument (the spectrohelioscope) that allowed him to observe the

Sun and its prominences at a particular hydrogen wavelength. Janssen's observations showed that the bright envelope seen around the Sun at eclipses and named the CHROMOSPHERE is in fact a cloud of hot hydrogen. Also at the 1868 eclipse, Janssen observed puzzling spectral lines which the English astronomer Norman LOCKYER interpreted as being due to a previously unknown element he named helium; only in 1895 was this element detected on Earth. In 1862 Janssen had discovered that absorption of light by the Earth's atmosphere produced dark lines in the solar spectrum, and he suggested (correctly) that the same effect in light reflected from planets would allow study of the composition of their atmospheres. He attempted to make such analyses, but with little success. In 1876 the French government granted him an observatory site at Meudon near Paris, where he set up a 33-inch (83-cm) refractor and used it to take a pioneering atlas of solar photographs.

Janus
Tenth satellite of Saturn, discovered in December 1966 by the French astronomer Audouin DOLLFUS. The satellite was found when the rings of Saturn were seen edge-on, so that it was not hidden by glare. Janus orbits Saturn every 17 hours $58\frac{1}{2}$ minutes at a distance of 98,500 miles (159,000 km), which is just at the edge of Saturn's rings; the satellite cannot normally be seen because of the rings' brilliance. Janus has a diameter of about 200 miles (320 km).

JD
Abbreviation for JULIAN DATE, a system of day numbering commonly used in astronomy.

Jeans, Sir James Hopwood (1877–1946)
British astrophysicist, best known for his support of the theory that the solar system originated from the Sun's encounter with a passing star. From his study of rotating bodies, Jeans concluded that a fast-spinning star would either split in two, forming a binary system, or spin off a stream of gas into space (he thought that something similar might be occurring, on a large scale, to account for the spiral shapes of galaxies). He could find no intermediate case in which a cloud of gas around the star would form into planets, and he concluded that the so-called nebular hypothesis of the solar system's origin, as advanced by KANT and LAPLACE, was untenable. Instead, in 1919, he developed the idea put forward by the Americans Thomas Chrowder Chamberlin (1843–1928) and Forest Ray Moulton (1872–1952) that a passing star drew out a cigar-shaped filament of material from the Sun that condensed into planets. This famous theory, however, is no longer accepted (see PLANETS). One remarkably foresighted suggestion of Jeans, made in 1928, was that matter is continually created in the Universe, an idea which is now of great prominence in cosmology (see CONTINUOUS CREATION; STEADY-STATE THEORY). From 1928 onward Jeans devoted himself to the popularization of astronomy, producing such well-known books as *The Universe Around Us, The Mysterious Universe,* and *The Stars in their Courses.*

Jet Propulson Laboratory (JPL)
A division of the National Aeronautics and Space

Administration (NASA), situated in Pasadena, California. JPL was set up in November 1944 for missile development, and became part of NASA at the end of 1958. It is operated for NASA by the California Institute of Technology. JPL supervises the Deep-Space Network of NASA tracking stations, which include its own 210-foot (64-m) antenna at Goldstone. JPL scientists are also engaged in unmanned space projects to explore the solar system. America's first satellite, Explorer 1, was developed at JPL, and the laboratory has since been responsible for the Ranger, Surveyor, and Mariner series of probes. The Orbiter section of the Viking spacecraft was also the work of JPL. NASA's Space Flight Operations Facility, the mission control center for unmanned Moon and space probes, is situated at JPL. The laboratory operates an astronomical observatory at Table Mountain, California.

Jodrell Bank

British radio-astronomy observatory near Macclesfield, Cheshire, operated by the University of Manchester; it is also known as the Nuffield Radio Astronomy Laboratories. It was founded after World

The 250-foot diameter Mark I radio telescope at Jodrell Bank was for many years the largest fully steerable dish in the world. Bearings in the housings at either side allow the dish to tip, while the whole structure rotates on a circular rail track not visible here.

War II by Bernard LOVELL, and began by using radar equipment to study cosmic rays and meteors. In 1952 construction began on the famous 250-foot (76-m) dish, which was the world's largest fully steerable radio telescope until the opening of the 328-foot (100-m) EFFELSBERG dish in Germany in 1971. The great Jodrell Bank dish was completed in 1957 in time to track Sputnik 1, which put it in the world's headlines. Jodrell Bank has also received signals from other space probes, but its main work is in radio astronomy, in such areas as the detection of radio emission from flare stars, the measurement of radio-source diameters, and the location of large numbers of pulsars. In 1964 an elliptical-shaped dish, 125 by 83 feet (38 × 25 m), was installed at Jodrell Bank for observations at shorter wavelengths.

Johnson Space Center

A NASA installation near Clear Lake, southeast of Houston, Texas. Opened in 1963, the facility was originally known as the Manned Spacecraft Center: it was renamed in 1973 in honor of former President Lyndon Johnson. The center houses Mission Control, which directs manned space missions after lift-off. The first mission to be controlled from Houston was Gemini 4 in 1965; previous manned flights had been controlled from Cape Canaveral. In addition to the planning and direction of space missions, Johnson Space Center is also responsible for the design and management of spacecraft, and for astronaut selection and training: astronauts practice maneuvers such as spacecraft docking in full-scale simulators at JSC. Also housed at JSC is the Lunar Receiving Laboratory, where Moon rock samples are stored and studied. Rice University operates the Lunar Scientific Institute at JSC under a NASA grant.

Jones, Sir Harold Spencer (1890–1960)

Tenth British astronomer royal, who made precise calculations of the distance between the Earth and the Sun (the ASTRONOMICAL UNIT), thus establishing the accurate scale of the solar system. The calculations were based on the close approach of the asteroid EROS in 1930–1931, of which Spencer Jones organized worldwide observations. Finding the distance between two objects orbiting the Sun provides a baseline for determining the scale of the entire solar system; before the advent of radar, close approaches of asteroids were the most accurate way of achieving this. After 10 years' work, Spencer Jones in 1941 announced that the Sun's distance is almost exactly 93 million miles, a figure that has been scarcely altered since. Spencer Jones was appointed astronomer royal in 1933. He showed that the Earth's speed of rotation is not truly constant, and in 1938 introduced quartz-crystal clocks to supplant purely astronomical observations as an accurate basis of timekeeping; in 1936 he introduced a system of time distribution by telephone. After World War II, Spencer Jones supervised the removal of the Royal Observatory from its centuries-old site at Greenwich, London, to a new and better site at Herstmonceux, Sussex.

Julian date

Chronological system which acts as a standard reference for logging historical events recorded on the different calendars of various civilizations. The Julian date is particularly useful to astronomers for coordinating ancient records of eclipses and other astronomical events. The system was devised in 1582 by the French scholar Joseph Justus Scaliger (1540–1609), and named for his father, Julius Caesar Scaliger; it has no connection with the Julian calendar. Scaliger chose January 1, 4713 B.C. as a starting date, from which Julian days are numbered consecutively. Thus, January 1, 1975, is JD 2,442,414 and January 1, 1980 is JD 2,444,240. Julian days begin at noon. The system is still used in astronomical computing for such purposes as predicting eclipses or plotting the light-curve of a variable-brightness star.

Juno

The third asteroid to be discovered, also known as minor plant 3. Juno was found by the German astronomer Karl Ludwig Harding (1765–1834) on March 29, 1807. It orbits the Sun every 4.36 years, moving from the inner portion of the asteroid belt at 1.98 a.u. out to its fringes at 3.36 a.u.; the orbit is inclined at 13° to the ecliptic. Juno is an irregular object some 140 miles (226 km) across, with a mass of about 3×10^{16} tons. It rotates in $7\frac{1}{4}$ hours with a pole near Gamma Camelopardalis. Juno is a reddish asteroid with a reflectivity (albedo) of 19 percent. Its spectrum suggests a stony-iron surface composition akin to the SIDEROLITE meteorites.

Juno rockets

American space launchers used for early satellites and probes. The Juno I was a Jupiter-C rocket modified by adding a solid-fuel top stage with satellite attached. This combination launched the Explorers 1, 3, and 4 satellites. The Juno II was based on the JUPITER ballistic missile (not to be confused with the smaller Jupiter-C, which was a development of the Redstone missile). Juno II had the same solid-fuel upper stages as Juno I, contained within a jettisonable nose cone. Juno II launched the Pioneer 3 and 4 probes, as well as Explorers 7, 8, and 11.

Jupiter

The fifth planet in order of distance from the Sun, and the largest planet in the solar system, with $2\frac{1}{2}$ times the mass of all the other planets combined. The planet's dimensions are enormous. Its equatorial diameter is 88,730 miles (142,800 km) or 11 Earth diameters. Its volume is equal to 1,319 Earths, but with a mass of 1.9×10^{24} tons, its density is only 1.33 the density of water, a quarter that of the Earth. Jupiter moves at 8 miles (13 km) per second in an elliptical path averaging 483,631,000 miles (778,329,000 km) from the Sun, taking 11.9 years to complete one orbit.

Jupiter spins on its axis faster than any other planet. Its rotational period of 9 hours 50 minutes 30 seconds is particularly remarkable for a body of such size; a point on Jupiter's equator spins around at 22,000 miles (35,000 km) per hour. Its rate of rotation is so rapid that the planet is slightly flattened at the poles and bulging at the equator; its equatorial diameter is nearly 5,500 miles (8,800 km) greater than its polar diameter. The rapid rotation, flattening at the poles, and unusually low density, all suggest that Jupiter is not a solid planet like the Earth, but a largely fluid body.

Surface. Small telescopes reveal a regular pattern of light zones and dark belts running parallel to Jupiter's equator, usually yellow to reddish-brown in color. Larger instruments show a wealth of fine detail which can change in a few hours. The relative prominence of the various belts varies over a period of years, while the region around the south equatorial belt can undergo sudden and rapid upheavals, darkening and developing many dark and light patches and spots. Such outbreaks occur at intervals of about 6 to 10 years, and particularly active disturbances, such as that of 1971, spread right around the planet before slowly subsiding. Detailed photographs taken by Pioneer spacecraft in 1974 and 1975 show that the equatorial regions are remarkably devoid of fine structure, but consist instead of large-scale swirls and plumes of cloud, formed because the planet's rapid

Blue light photograph of Jupiter taken by Pioneer from 1,121,000 miles shows the complexity of its cloud structure. Although the Great Red Spot is not in this picture, many small white spots are visible, as well as the light zones and dark belts parallel to the equator.

rotation has "smeared" the individual convection cells of hot gas rising from below the clouds. Away from the equator, the alternating cloud bands and darker belts become progressively more ragged and full of fine detail. Close to the poles there are no organized belts at all, and the convection cells are visible as a mass of dark and light spots, swirls, eddies, and disorganized mottlings.

The most remarkable feature on Jupiter, and its only permanent surface marking, is the RED SPOT, discovered in 1666. Smaller, temporary, red spots have been seen at high latitudes. The only other features to rival the longevity of the great red spot are three small white ovals close to the south temperate belt, first glimpsed in the 1920s and recorded intermittently ever since.

Atmosphere. Jupiter is remarkable in that it emits almost twice as much heat as it receives from the Sun. This may be heat left over from its formation. There is a strong GREENHOUSE EFFECT which keeps the temperature of the upper atmosphere at about $120°$ to $138°K$ at both equator and pole, even on the dark side of the planet. The deeper parts of the atmosphere are probably stagnant, with turbulence and "weather" occurring only in the top 40 to 50 miles (64 to 80 km). Water condenses to form clouds at low levels,

ammonia condenses at intermediate levels, and methane forms the highest clouds. The light zones on Jupiter are dense cloud layers at about $130°K$, formed from upwelling currents of gas. Dark objects, especially the belts, are actually the denser, warmer ($230°$ to $270°K$) and lower regions of the atmosphere. Occasionally, very dark small features have shown temperatures as high as $310°K$, representing transient windows into even lower regions.

The main colors on Jupiter are probably due to various combinations of ammonia and hydrogen sulfide. These can produce a wide variety of green, yellow, orange, red, and brown colorations carried into the visible levels by updrafts. The brightest red areas, such as the red spot, may be due to the presence of red phosphorus. When the atmosphere is disturbed, large amounts of colored material rise to the cloud tops where they are blown around and produce a general darkening. The great equatorial disturbances at 6 to 10-year intervals are probably caused by the release of heat which has slowly built up below a temperature inversion in the low weather levels.

The rotation periods of Jupiter's features vary for different latitudes. At high latitudes, the major white cloud zones rotate with a period of 9 hours 55 minutes, similar to that of 20 MHz radio noise detected from Jupiter. The radio noise is thought to be caused by activity below the weather zone, so that its rotation also represents the rotation period of the dense main body of Jupiter. Materials in the region of the dark belts, however, show rotation periods which can vary by as much as six minutes. The 20,000-mile-(32,000-km)-wide equatorial region of Jupiter's clouds

101

rotates five minutes faster than the dense core, demonstrating the fluid nature of the visible surface.

Jupiter is rich in the building blocks of living organisms—all the molecules thought to have been involved in the beginnings of life on Earth. At the bottom level of the weather zone, the temperature reaches 300° to 350°K, at which there would be a plentiful supply of liquid water. It is conceivable that forms of life have evolved here, possibly simple forms of stress-adapted bacteria.

Below the clouds. Jupiter consists mainly of hydrogen, with a much smaller amount of helium. Heavier elements are present in combination with hydrogen. The volatile products form the weather zone, while at greater depths the planet is believed to consist of a hydrogen/helium mix with a rocky-metallic core some 6,000 miles (10,000 km) across at the center. The central temperature must be about 30,000°K, with a pressure of 100 million Earth atmospheres. The hydrogen mix changes from a dense fluid into a metallic state about 15,000 miles (25,000 km) below the surface of the planet. The slow change of fluid to metal, or perhaps the slow gravitational contraction of the entire central core, could be the source of Jupiter's excess heat flow. The depth of Jupiter's atmosphere is about 600 miles (1,000 km). Unlike the planets nearer the Sun, Jupiter has retained a large proportion of volatile materials and is closer to the Sun in total composition.

Magnetosphere. Jupiter possesses a strong magnetic field and intense radiation belts, stretching out to 90 planetary radii. A dipole magnetic field about 10 times as strong as the Earth's field, with north and south poles reversed, relative to the Earth, is found close to the planet. Beyond this is a non-dipole field about three times stronger. Jupiter's radiation belts, containing trapped high-energy protons and electrons, are some 10,000 times stronger than the Earth's Van Allen belts. They emit strong bursts of radio waves partly influenced by Io, the innermost of the large moons of Jupiter. Io orbits within the radiation belts, and material eroded from its surface forms a huge cloud of hydrogen and sodium which encircles Jupiter.

Moons. Jupiter has at least 14 moons. The four brightest—Io, Europa, Ganymede, and Callisto— are termed the Galilean satellites, after their discoverer. The only other man to discover four Jovian moons was the American astronomer Seth Barnes Nicholson (1891–1963). Other discoverers of Jovian moons were E. E. Barnard, the American astronomer Charles Dillon Perrine (1867–1951), and the British astronomer Philibert Jacques Melotte (1880–1961). Most recently, Charles T. Kowal of Hale Observatories has conducted a fruitful search for faint extra satellites. Jupiter's moons fall into three main groups.

The innermost five have periods from 12 hours to 17 days and move in near-circular orbits in the equatorial plane of the planet. The next group of four have periods of from 239 to 260 days, moving in elliptical orbits inclined at about 28° to the planet's equator. The final group of four are tiny bodies with periods of 631 to 758 days, moving in retrograde ellipses at great distances from Jupiter. The two outer groups are probably captured asteroids, and must contain other as yet undiscovered members.

Near Jupiter's north pole the orderly atmospheric belts and zones break up into scalloped and oval features, well shown in this Pioneer 11 photograph from 373,000 miles. The completely disordered clouds at top right probably represent the tops of individual hurricane-like storms, each hundreds of miles across.

Moons of Jupiter

Moon		Mean distance from Jupiter (km)	Orbital period (days)	Diameter (km)	Discoverer and date
V	Amalthea	181,000	0.498	240	Barnard, 1892
I	Io	421,760	1.769	3,659	Galileo, 1610
II	Europa	671,050	3.551	3,100	Galileo, 1610
III	Ganymede	1,070,400	7.155	5,270	Galileo, 1610
IV	Callisto	1,882,600	16.689	5,000	Galileo, 1610
XIII	Leda	11,100,000	239	15	Kowal, 1974
VI	Himalia	11,477,600	250.566	100	Perrine, 1904
X	Lysithea	11,720,250	259.219	20	Nicholson, 1938
VII	Elara	11,736,700	259.653	30	Perrine, 1905
XII	Ananke	21,200,000	631*	20	Nicholson, 1951
XI	Carme	22,600,000	692*	20	Nicholson, 1938
VIII	Pasiphae	23,500,000	744*	20	Melotte, 1908
IX	Sinope	23,600,000	758*	20	Nicholson, 1914

(The characteristics of Jupiter's 14th moon, discovered by Kowal in 1975, are not yet known.)
*Retrograde

Jupiter rocket

American medium-range ballistic missile, which in 1958 was installed in Italy and Turkey. The missile was 58 feet (17.7 m) high, 105 inches (2.7 m) wide, and weighed 110,000 lb. (50,000 kg). It was successfully test fired on March 1, 1957, making it the first operational U.S. missile of its range. A Jupiter missile blasted the monkeys Able and Baker on a suborbital ride from Cape Canaveral on May 28, 1959, in preparation for later manned launches. To aid in testing components for the Jupiter, the smaller REDSTONE rocket was modified to produce the so-called Jupiter-C design. The Redstone was lengthened and had a revolving drum attached at the top. The drum contained two concentric rings of solid-fuel rockets, which comprised the second and third stages of the vehicle; the drum was spun by small jets to even out the thrust of the solid-propellant rings. Nose cones for reentry into the atmosphere were tested on Jupiter-C firings. A small solid-fuel fourth stage with a satellite attached turned Jupiter-C into the launcher named Juno I. It was this rocket that launched the first American satellite, Explorer 1 (see JUNO ROCKETS). Confusingly, the Jupiter military missile also had extra stages added to it to act as a satellite launcher. In this configuration it was called Juno II.

K

°K

The symbol for degrees Kelvin, the temperature scale measured from ABSOLUTE ZERO, the coldest temperature possible. A Kelvin degree is equivalent to a centigrade degree.

Kant, Immanuel (1724–1804)

German philosopher who, though not known as a scientist, did contribute to modern theories of the origin of the solar system. In his 1755 book *General History of Nature and Theory of the Heavens* (subtitled *Concerning the Structure and the Mechanical Origin of the Whole Universe, According to Newtonian Principles*) he described our solar system as part of a lens-shaped system of stars, the Galaxy, and suggested that the fuzzy patches called nebulae seen by astronomers were in fact other galaxies. Kant proposed that the Sun and planets arose from the condensation of diffuse material in the Universe, as a result of gravitational attraction between particles. This is now called the nebular hypothesis, and was further developed by the French mathematician the Marquis de LAPLACE.

Kapteyn, Jacobus Cornelius (1851–1922)

Dutch astronomer who measured positions and brightnesses of 454,875 stars on photographs taken at the CAPE OBSERVATORY, to produce the *Cape Photographic Durchmusterung*, a major catalog of southern-hemisphere stars and the first large-scale photographic survey in astronomy. Kapteyn began work in 1886, and his results appeared in three volumes between 1896 and 1900. He studied the distribution of stars in space, finding that giant stars are a thousand times less common than fainter, Sun-type stars. Kapteyn concluded that the Sun was near-centrally placed in a rotating, lens-shaped Galaxy of stars;

however, the obscuring effect of gas prevented him from seeing the true extent of our Galaxy, and we now know that the Sun is far from the center. Kapteyn found that the motions of stars with respect to the Sun seem to cluster in two opposite directions, a phenomenon known as STAR STREAMING. It is now known that this effect is due to the Galaxy's rotation.

Kapustin Yar

Soviet rocket-launch site about 60 miles (100 km) southeast of Volgograd and northwest of the Caspian Sea. It was originally an intercontinental missile test site, and was later modified for space use. Early Soviet sounding rockets were launched from Kapustin Yar, including modified V-2s that carried animals. The first orbital launch was Cosmos 1 in 1962. Subsequent space launches from Kapustin Yar have also been of small Cosmos scientific satellites. Although the second space launch site to be used, it is far less important than either TYURATAM or PLESETSK.

Kennedy Space Center (KSC)

NASA installation at Cape Canaveral, Florida, responsible for the preparation and launching of manned and unmanned space rockets. Most U.S. space missions use the KSC facilities at Cape Canaveral, although some are launched from facilities operated by KSC at the Western Test Range in Lompoc, California. Headquarters of KSC are on Merritt Island, which is the site of the Vehicle Assembly Building (VAB) 525 feet (160 m) tall, inside which the giant Saturn rockets were assembled for Apollo and Skylab missions, and which is now modified for Space Shuttle assembly. Rockets assembled in the VAB are fired from one of two pads at nearby launch complex 39. Kennedy Space Centre covers 84,000 acres. (*See picture p. 104.*)

Kepler, Johannes (1571–1630)

German mathematician and astronomer, who discovered that the orbits of the planets are ellipses, not circles, and derived the three fundamental laws of planetary motion known as KEPLER'S LAWS. Kepler was destined for a career as a Lutheran minister, but after being appointed a teacher of mathematics at Graz, he pursued the subject of mathematical astronomy for the rest of his life. Kepler accepted the novel theory of Copernicus, which placed the Sun at the center of the Universe. Kepler tried to see if a series of regular geometric solids would fit between the spheres which were thought to carry the planets around the sky. This ingenious attempt, published as the *Mysterium Cosmographicum* (The Cosmographic Mystery), in 1596, brought him into contact with the famous Danish astronomer Tycho BRAHE. In 1600 Kepler moved to Prague to work with Tycho, and on the latter's death in 1601 succeeded him as Imperial Mathematician. From an analysis of Tycho's observations, Kepler found that Mars moves in an elliptical orbit at varying speeds depending on its distance from the Sun. These momentous findings, which firmly established the truth of the Copernican scheme and banished forever the ancient notion that planetary motions must be based on circles, were published in his book of 1609, *Astronomia Nova* (New Astronomy).

Kepler had an underlying belief that God had

The Vehicle Assembly Building dominates the Kennedy Space Center, Fla. Here the Saturn 5 rocket which launched Apollo 11, the historic first manned Moon flight, is moving along the 3½ mile Crawlerway to Launch Complex 39A.

constructed the Universe according to a divine plan. The idea in his *Mysterium Cosmographicum* was one reflection of this, but he now returned to the question in the light of new studies of Tycho's observations which showed that the motion of all planets resembled that of Mars. Kepler found a relationship between the size of a planet's orbit and the time it takes to complete a circuit of the Sun. He also discovered relationships between the orbital speeds of the planets and the notes in a musical scale, developing a scheme of divine musical harmony which he published as *Harmonices Mundi* (Harmonies of the World) in 1619. At the same time he devised a physical theory to account for elliptical motion and published all his planetary results in *Epitome Astronomiae Copernicanae* (Epitome of Copernican Astronomy) between 1618 and 1621. In 1627 he also published the *Rudolphine Tables,* which gave planetary positions based on Tycho's observations.

Kepler also studied and wrote on the subject of optics. After he had used one of Galileo's telescopes, he published a design of his own that later became more widely adopted. He observed and wrote about the nova of 1606, another proof of how mistaken were the old views about the changelessness of the heavens.

Kepler's laws
Three laws governing the motions of the planets in their orbits around the Sun, derived by Johannes KEPLER from the observations of the Danish astronomer Tycho Brahe. Kepler found the premise that the planets orbit in circles at a uniform speed untrue whatever scheme of planetary motion he tried. He therefore boldly broke with tradition and tried

various oval orbits, finally discovering that only elliptical orbits along which the planets traveled at varying speeds would fit the facts. His three laws state:

(i) that every planet orbits the Sun in an ellipse, with the Sun itself at one focus of that ellipse;

(ii) that the radius vector (line from the Sun to the planet) sweeps out equal areas within the ellipse in equal times;

(iii) that a fixed ratio exists between the time taken to complete an orbit (the orbital period) and the size of the orbit, and that this ratio is the same for every planet. The ratio may be expressed as P^2/a^3, where P is the orbital period and a the semi-major axis of the ellipse (the distance from the center of the ellipse to a point on the ellipse nearest to one focus). If P and a are known for one planet, then a may be found for any other planet whose period is already known. From this the scale of the solar system can thus be calculated.

Kirkwood gaps
Areas in the asteroid belt in which few or no asteroids orbit. They are named for the American astronomer Daniel Kirkwood (1814–1895), who first noted them in 1857. He pointed out that they lie at exact fractions of the orbital period of Jupiter (the orbital periods are thus said to be COMMENSURABLE); asteroids in these regions would therefore be swept into another orbit by Jupiter's regular gravitational pull. Kirkwood also explained the Cassini division in Saturn's rings by commensurability of the ring particles with the orbital periods of Saturn's satellites.

Kitt Peak National Observatory
Observatory located in the Quinlan Mountains 56 miles (90 km) southwest of Tucson, Arizona, at an elevation of 6,770 feet (2,064 m), opened in 1960 and operated by the Association of Universities for

Research in Astronomy (AURA). Kitt Peak hosts 14 telescopes, the largest single concentration of astronomical instruments in the world. Its main telescope is a 158-inch (400-cm) reflector, opened in 1973. The McMath Solar Telescope is the world's largest solar telescope, with a 60-inch (150-cm) main mirror and a 300-foot (91.4-m) focal length. Among the observatory's other telescopes is an 84-inch (210-cm) reflector used for optical and infrared observations; a 50-inch (130-cm) reflector for infrared studies; two 36-inch (91-cm) reflectors; two 16-inch (40-cm) reflectors; and a 24-inch (61-cm) solar telescope. In addition, the University of Arizona's STEWARD OBSERVATORY operates a 90-inch (230-cm) and 36-inch (91-cm) reflector on the same site. Also on the mountain is a 52-inch (140-cm) reflector operated jointly by the University of Michigan, the Massachusetts Institute of Technology, and Dartmouth College; and a 36-foot (10.75-m) radio telescope of the National Radio Astronomy Observatory.

Komarov, Vladimir Mikhailovich (1927–1967)
First Soviet cosmonaut to make two space flights, and the first man to be killed during a space mission. Komarov commanded the three-man Voskhod 1 flight launched on October 12, 1964. The capsule, a Vostok modified to hold three men, remained in space for a day. On April 23, 1967, he was launched alone in Soyuz 1, the first of a new generation of Soviet manned spacecraft. But during this test flight the capsule became difficult to control, and Komarov had to make an emergency reentry after 18 orbits. According to official reports, the parachute became twisted, apparently due to the spin of the spacecraft, and the Soyuz crashed to Earth, killing Komarov.

Korolev, Sergei Pavlovich (1907–1966)
The anonymous Russian "Chief Designer of Carrier Rockets and Spacecraft." As such, he was responsible for the development of the first Soviet intercontinental missile, which in modified form became the basic Soviet space launcher; he also supervised the design of the first Soviet artificial satellites and space probes, as well as the Vostok, Voskhod, and Soyuz manned capsules. In 1931 Korolev was a founder member of the Moscow Group for the Study of Rocket Propulsion called GIRD, where he headed work by pioneers such as Friedrich TSANDER and Mikhail TIKHONRAVOV on the first Soviet liquid-fueled rockets. In 1933 this organization was merged with the work of rocketeers in Leningrad to form the Rocket-Science Research Institute (RNII), where work progressed on simple missiles. Imprisoned by Stalin in 1938, Korolev was set to work during World War II to develop jet-assisted takeoff for aircraft. Freed after the war, he began to develop improved versions of the V-2, using them for atmospheric sounding and animal flights. In 1954 his group embarked on the construction of the first Soviet intercontinental missile, which in 1957 was turned to space use with Sputnik 1. Korolev masterminded all Soviet space activities from that date until his death.

Kuiper, Gerard Peter (1905–1973)
Dutch-born American solar-system expert. In 1944 he discovered the atmosphere of Saturn's satellite Titan, and in 1948 found that the Martian atmosphere is made of carbon dioxide. In 1948 he discovered the fifth moon of Uranus, Miranda, following it in 1949 with the discovery of Nereid, the second moon of Neptune. Kuiper measured the diameter of Pluto in 1950 and found it to be similar to that of Mars, which was much smaller than had previously been imagined. He proposed that the solar system formed from a turbulent cloud of dust and gas around the Sun, in which small regions, called protoplanets, grew dense enough to contract into planets. This idea was inspired by his work on double stars, and he viewed the solar system as a "failed" twin-star system. This led him to suggest that planetary systems exist around at least one star in a hundred, making them far more common than supposed; this view is now widely supported. Kuiper was also interested in the surface features of the Moon and compiled several atlases of lunar photographs. In 1960 he founded the Lunar and Planetary Laboratory at the University of Arizona.

L

Lacaille, Nicolas Louis de (1713–1762)
French astronomer, called the "father of southern astronomy" for his pioneer mapping of the southern skies. From 1751 to 1753 Lacaille made observations from the Cape of Good Hope, mapping nearly 10,000 stars and introducing 14 new constellations: Antlia, Caelum, Circinus, Fornax, Horologium, Mensa, Microscopium, Norma, Octans, Pictor, Pyxis, Reticulum, Sculptor, and Telescopium. He measured the positions of the Moon, Mars, and Venus, simultaneously with similar observations in Europe, in order to determine more accurately the scale of the solar system. From his observations he arrived at a value for the Sun's distance that was only 10 percent too small.

Lacerta (the lizard)
An inconspicuous constellation sandwiched between Andromeda and Cygnus, introduced by the Polish astronomer Johannes HEVELIUS.

Lagrange, Joseph Louis (1736–1813)
French mathematician, who in 1772 discovered what are now known as the LAGRANGIAN POINTS. Lagrange made studies of the overall stability of the solar system, showing that no long-term changes could be expected in the planets' orbits. Together with LAPLACE, he dominated the field of mathematical astronomy for a generation. Lagrange later headed the commission that established the metric system of weights and measures.

Lagrangian points
Five points in space at which a very small body can remain in a stable orbit with two very massive bodies. The points were first recognized by Joseph Louis LAGRANGE and are rare cases in which the relative motions of three bodies can be computed exactly. In the case of a massive planet orbiting the Sun in a circular path, the first stable point (L_1) lies on that orbit diametrically opposite the large planet itself. The points L_2 and L_3 are both on the Sun-planet line, one closer to the Sun than the planet and the other farther away. The remaining two points, L_4 and L_5,

are positions on the planet's orbit that each form an equilateral triangle with the planet and the Sun. Clouds of gas and dust are believed to collect at the Lagrangian points in the Moon's orbit around the Earth, and the TROJAN asteroids are found at the Lagrangian points in the orbit of Jupiter.

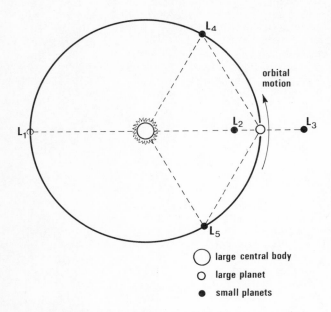

The five positions to which a small body can move in fixed relative position with a massive planet orbiting a large central body.

Landsat

Two identical NASA satellites for surveying the Earth's resources, originally called Earth Resources Technology Satellites (ERTS). Both Landsats were launched into orbits 567 miles (912 km) above the Earth, circling from pole to pole 14 times a day. As the satellite orbits, its sensors scan a strip 115 miles (185 km) wide. After 252 orbits (18 days) the satellite has completed one survey of the entire Earth and is back over its starting point. The satellite passes over the Earth's equator at about 9.30 A.M. local time on each orbit.

ERTS was developed from the Nimbus weather satellite, which it resembles. It possesses two main sensor systems: a so-called multispectral scanner (MSS), which scans the Earth at four wavelengths, two in the visible region and two in the infrared; and three television cameras, which take pictures simultaneously in three wavelengths, from which color images can be assembled. Different features of the Earth are prominent in different wavelengths, enabling experts to examine large areas of forest and crops, monitor water and air pollution, and study marine resources and biology. Several countries, led by Brazil and Canada, have built ground stations to receive pictures direct from Landsat as it passes overhead. Landsats 1 and 2 were launched in 1972 and 1975. A third Landsat, containing improved sensors, is scheduled for launch late in 1977.

Langley Research Center

A NASA installation at Hampton, Virginia, for research and development in aeronautical and astronautical technology. The installation is named for the American aviation pioneer Samuel Pierpont Langley (1834–1906); it was established in 1917 as Langley Field, a center for aircraft development. Langley came under NASA when that body was set up in 1958, and provided experienced personnel for many of NASA's other branches, notably the Manned Spacecraft Center (now JOHNSON SPACE CENTER) in Houston, Texas. Langley scientists helped develop the high-flying rocket planes that were the precursors to manned spaceflight. The Mercury, Gemini, and Apollo spacecraft were all developed at Langley, where wind-tunnel tests solved the problem of a capsule's reentry into the atmosphere. Facilities also allow structural testing of rocket models to study their behavior in flight. Simulators at Langley permitted astronauts to practice docking and lunar landings in full-scale mock-ups. The Langley Research Center was responsible for the Lunar Orbiter series of unmanned spacecraft and the Viking program.

Laplace, Pierre-Simon, marquis de (1749–1827)

French mathematical astronomer, formulator of the so-called nebular hypothesis to explain the solar system's origin. In his *Exposition du système du monde* (1796), he described the solar system as having formed from a giant cloud or nebula of material. As this cloud shrank, it threw off rings of material that condensed to form the planets; at the center was formed the Sun. Laplace suggested that the many spiral-shaped nebulae observed in space were examples of such a process in action, although we now know the spiral nebulae to be enormous galaxies of stars. Laplace's hypothesis was similar to an idea put forward some time earlier by the German philosopher Immanuel KANT. In outline it is similar to modern ideas of the solar system's origin.

Laplace published a massive five-volume *Traité de mécanique céleste* between 1799 and 1825, summarizing work on gravitational astronomy since Newton. With his contemporary LAGRANGE, he made many important investigations in the field of celestial mechanics, including a major analysis of the Moon's motion which led to the construction of new and more accurate tables. Laplace showed that there were limits to the possible variations of inclination, eccentricity, and size of the planets' orbits, demonstrating that no major changes could be expected from interactions among the planets.

Large Space Telescope (LST)

An astronomical telescope planned to be orbited by the SPACE SHUTTLE in the early 1980s. The LST is a remote-controlled observatory in space, receiving commands from Earth and returning data by radio link. The telescope is planned to have a main mirror of 95 inches (240 cm) aperture. Orbiting above the blurring effects of the atmosphere, the LST will achieve a level of performance far higher than that of the best Earth-based telescopes; it will be able to discriminate detail 10 times as fine, and objects 100 times as faint. The telescope will also be able to function at wavelengths the Earth's atmosphere prevents from reaching instruments on the ground. Astronauts will visit the LST from time to time to replace equipment and make repairs; this will give it an operating life of at least 10 years, and possibly as long as that of a ground-based observatory.

laser

A device for producing an intense beam of light; the name is an acronym for *Light Amplification by Stimulated Emission of Radiation*. Invented in 1960, lasers are optical forms of MASERS, and work on similar principles. A laser tube contains a helium-neon mixture (gas laser) or cylindrical ruby crystal (solid-state laser), whose atoms are excited. Light of a particular wavelength stimulates these excited atoms to emit light at that same wavelength, producing a precisely aimed beam of very intense light by reflections from mirrors at each end of the tube.

One notable astronomical application of the laser is its use in plate-scanning machines to provide small, brilliant spots of light for measuring photographic density. More spectacularly, the later Apollo missions carried a laser altimeter in the command module, which worked by timing the return of a laser beam reflected from the lunar surface. Intense laser beams have even been directed from major observatories on the Earth to reflectors placed on the lunar surface by Apollo astronauts and carried on the Soviet LUNOKHOD automatic rovers. As a result, the Earth–Moon distance is now known to within a few centimeters.

Lassell, William (1799–1880)

British astronomer and instrument maker, discoverer of two satellites of Uranus and one of Neptune. Lassell improved construction techniques for large reflector telescopes, which then used mirrors of speculum metal. In 1846 he built a 24-inch (61-cm) metal-mirror reflector, using it that year to discover Triton, the largest satellite of Neptune. In 1851 he discovered the satellites Ariel and Umbriel of Uranus. Lassell also independently discovered the satellite Hyperion of Saturn, two days after W. C. BOND in 1848. He built a 48-inch (122-cm) reflector in Malta in 1860, with which he located 600 new nebulae.

latitude

A coordinate for determining positions on Earth north or south of the equator. An object's latitude is its angle north or south of the equator; all lines of latitude are parallel. The equivalent of latitude on the CELESTIAL SPHERE is DECLINATION; and it should not be confused with *celestial latitude,* which measures instead the angle north or south of the ecliptic (the Sun's apparent path around the sky). The ecliptic is inclined at $23\frac{1}{2}°$ to the celestial equator because the Earth's axis is tilted relative to the plane of its orbit.

Leavitt, Henrietta Swan (1868–1921)

American astronomer, who discovered that the period and brightness of CEPHEID VARIABLE stars is related (see PERIOD-LUMINOSITY RELATION). She made her discovery from a study of variable stars in the Magellanic Clouds, which are satellite galaxies of our own; because all the stars in each cloud are at roughly the same distance from us, their relative brightnesses can be directly compared. The period-luminosity

relationship, published by Miss Leavitt in 1912, was used by Harlow SHAPLEY to determine the size of our Galaxy. Miss Leavitt discovered a total of 2,400 variable stars. She also measured photographic magnitudes of stars to produce standard lists of star brightnesses, such as the famous *North Polar Sequence* published in 1917.

Lemaître, Georges Édouard (1894–1966)

Belgian astronomer who in 1927 formulated the BIG-BANG theory of cosmology, which holds that the Universe began in a giant explosion. Lemaître studied at the Massachusetts Institute of Technology in the mid-1920s, where he learned of Edwin HUBBLE's work showing that the Universe seems to be expanding. Lemaître proposed that all the matter in the Universe was once concentrated into what he termed the primeval atom, whose explosion scattered material into space to form galaxies, which have been flying outward ever since. Lemaître's idea has been developed by other astronomers, notably George GAMOW, and is now the most widely held cosmological theory.

lens

An optical glass component which either converges or diverges a beam of light passing through it. A converging lens is said to be *positive*; a diverging one is *negative*. A positive lens is capable of forming a real image of an object on a screen; a negative lens is not, and is used only for special purposes.

The usual task of a lens (or a system of lenses) is to form an image of an object. If the object is at infinity (as, in effect, all astronomical objects are), its image is formed at a distance from the lens called the FOCAL LENGTH. The focal length of a lens determines the size of the image it produces. If the focal length is 100 units, then the image of an object 1° across in the sky will measure 1.75 units. The scale is entirely independent of the aperture of the lens, which affects only the brightness of the image. If the lens is to be used for a powerful telescope, a long focal length, to give a large image-scale, is desirable, while a large aperture, giving a brighter image, will allow fainter objects to be seen.

The performance of a lens, by which is generally meant the sharpness of its images, is governed by the curvature of its faces and the type of glass employed. The image formed by any single lens will be colored, because of its tendency to disperse the light passing through it into a spectrum; this deformity is known as CHROMATIC ABERRATION. It can be minimized by combining two lenses made of different types of glass. Such a combination is said to be *achromatic*. The other principal fault to which a lens may be subject is SPHERICAL ABERRATION, brought about because different concentric regions have different focal lengths, and thus produce no single sharp focal point. Spherical aberration is strongly affected by the choice of curvature on the faces; by a careful choice of curves, and the correct combination of two or more lenses into a unit, spherical aberration can be practically eliminated. Astronomical OBJECT GLASSES rarely consist of more than two lenses, but camera lenses, which must work at relatively wide aperture and cover a large field of view in sharp focus, must contain a large number of individual lenses to give good results (see also REFRACTING TELESCOPE).

Leo (the lion)
A major constellation of the zodiac, best seen during the northern hemisphere spring; the Sun passes through Leo from mid-August to mid-September. The head of the lion is formed by a characteristic sickle-shaped line of stars, at the foot of which is the constellation's brightest star, REGULUS (Alpha Leonis). The second-brightest star is Denebola, magnitude 2.23 and 42 light-years distant. The LEONID METEORS appear to radiate from the constellation each November.

Leo Minor (the lesser lion)
A small and faint constellation in the northern hemisphere of the sky, north of Leo itself, introduced on the 1690 star map of Johannes HEVELIUS.

Leonid meteors
A regular meteor shower which appears for about two days around November 17 each year. The stream has a radiant 3° northwest of Gamma Leonis and reaches its greatest altitude at 6.30 A.M. Although activity from the stream is usually a modest 5 to 20 meteors each hour, major meteor storms occur at infrequent intervals. The last took place on the morning of November 17, 1966, when astonished observers in the central United States witnessed meteor rates peaking for 15 to 20 minutes at 100,000 meteors an hour. A similar storm on November 12, 1833, attracted the attention of the American astronomers Denison Olmsted (1791–1859) and Alexander Catlin Twining (1801–1844). They demonstrated that the storm phenomenon was caused by particles moving together in a solar system orbit, thereby founding the scientific study of meteors. An earlier storm had been recorded in 1799, and another took place in 1866. The stream had produced only mediocre displays in 1899 and 1933, and the great storm of 1966 came as a surprise. The stream is debris from comet Tempel-Tuttle 1866 I and moves in the same orbit. Storms occur when the Earth passes through the stream close to the position of the comet, and the next is timed to occur on November 18, 1999.

Leonov, Alexei Arkhipovich (b. 1934)
Soviet cosmonaut, the first man to walk in space. Leonov crawled through an airlock in space on March 18, 1965, during the flight of the two-man Voskhod 2 capsule, remaining outside for 10 minutes. He had a portable life-support system strapped to his back during the walk, but was tethered to the craft by a lifeline. Leonov was commander of the Soviet crew for the APOLLO-SOYUZ TEST PROJECT in 1975.

Lepus (the hare)
A constellation of the southern hemisphere of the sky at the foot of Orion, best seen during the northern hemisphere winter. The name of its brightest star, the 2.69-magnitude Arneb, is Arabic for hare. R Leporis, which lies near the borders of the constellation with Eridanus, is one of the reddest stars known. It is a variable star of long period, like MIRA, ranging between magnitudes 5.9 and 10.5 every $432\frac{1}{2}$ days.

Leverrier, Urbain Jean Joseph (1811–1877)
French mathematician and astronomer, who predicted the existence of the planet Neptune. In 1845 he began to study the motion of the planet Uranus, which was not following its predicted path in the sky. Leverrier calculated that it was being pulled out of place by an unknown body. In 1846 he told astronomers at the Berlin Observatory where he believed the new planet lay; it was immediately discovered by Johann GALLE, and eventually named Neptune. After the discovery came news that similar calculations had been made by the Englishman John Couch ADAMS. From 1847 until his death Leverrier was engaged on the monumental task of calculating accurate tables for all the planets, producing a set of standard references for astronomers into the 20th century. During this work, in 1859, Leverrier predicted the existence of an asteroid belt close to the Sun in order to explain the observed motion of the planet Mercury. The apparent acceleration of Mercury is now known to be an effect of RELATIVITY.

Libra (the scales)
A constellation of the zodiac, lying between Virgo and Scorpius, best seen during the northern hemisphere spring; the Sun passes through Libra during November. Alpha Librae is a double star, just resolvable to keen eyesight. Beta Librae, a SPECTROSCOPIC BINARY, is one of the few bright stars that appear green in color. Delta Librae is an ECLIPSING BINARY, similar in type to ALGOL, ranging from magnitudes 4.8 to 5.9 every 2.33 days.

libration
The slow east–west and north–south rocking motions of the visible face of the Moon about its mean position. The Moon's axial rotation period precisely equals its orbital revolution period, and so it should always in theory present the same face to the

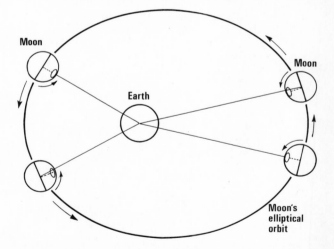

The movement of the Moon in its orbit when close to the Earth, **left**, and at its greatest distance, **right**. The rapid orbital revolution at left is faster than the steady rotation and the usual visible hemisphere drifts to one side, moving sideways normally central features such as the crater shown here. On the right, the steady rotation is faster than the slow orbital revolution, causing the visible hemisphere to drift in the opposite direction. The ellipticity of the lunar orbit and magnitude of the effect are greatly exaggerated here.

Earth. However, the Moon's orbit is elliptical, and it thus moves more quickly when closest to the Earth and more slowly when farthest away. The Moon's speed in orbit therefore changes rhythmically while its axial rotation remains uniform. When closest to the Earth, the Moon's axial rotation falls behind its orbital revolution so that it turns its face to the east; at its greatest distance, its axial rotation moves ahead of its orbital revolution, so that it turns its face to the west. This rocking effect is east–west libration. At the same time, the Moon's axis of rotation in inclined not at right angles but 83° to its orbital plane, producing a less pronounced north–south libration. The principal effect of libration is to allow features which would otherwise just be hidden to rock into view at the edge of the Moon's disk. In this way it is possible for Earth-bound observers to see 59 percent of the lunar surface.

Lick Observatory

The astronomical observatory of the University of California, situated at an elevation of 4,209 feet (1,283 m) on Mount Hamilton, 13 miles (21 km) east of San Jose, California. The observatory was established by a bequest from landowner James Lick (1796–1876). Its main telescopes are a 120-inch (304-cm) reflector opened in 1959, and a 36-inch (91-cm) refractor which, when the observatory was opened in 1888, was the largest in the world. Headquarters of the observatory moved in 1966 to Santa Cruz. In 1968 Lick astronomers selected the 5,862-foot (1,787-m) Junipero Serra Peak in Monterey County as a dark-sky outstation.

Life in the Universe

Many scientists now accept that life of some sort may exist on other bodies in space. Their view stems from recent advances in our understanding of the origin and early evolution of life, as well as evidence from astronomers that planetary systems may be very common in space (see PLANETS).

Attempts have been made to estimate the number of other civilizations that might exist in our own Galaxy, the Milky Way. These take into account many factors: the rate at which stars are formed; the number of stars that probably have planets; the number of planets on which life is likely to have arisen; the probability that such life might rise to a technologically advanced stage; and the length of time for which such civilizations might exist. These estimates—they are little more than educated guesses—suggest there may be a million advanced civilizations in the Galaxy—that is, one for every 100,000 stars. The average distance between two such civilizations (such as ourselves and another) would be a few hundred light-years.

Origin of life. Remains of algae, similar to those living today, are found in rocks 3,200,000,000 years old, showing that life has existed on Earth for at least three-quarters of our planet's existence. Traces of life may exist in even older rocks, and this suggests that simple forms of life arose rapidly once conditions on Earth became suitable.

The very first chemicals of life were probably created from the gases that surrounded the Earth at its birth—ammonia, methane, and water vapor. These same gases are found today in the atmosphere

James Lick, who founded the Lick Observatory on Mt Hamilton, Cal., lies buried beneath the pedestal of the 36-inch (91-cm) refractor, shown here. This telescope has been in use since 1888, and is still the second largest refractor in the world. The focal length of the lens is 57.8 feet (17.6 m). The weights on the right of the picture, on the end of the declination axis of the German-type mounting, counterbalance the telescope.

around Jupiter, which is thought to have changed little since the planets first formed. Experiments with mixtures of such gases have shown that they can combine to form the simple basic building blocks of life called amino acids. Amino acids form chains to make protein, the structural material of life. The gases of the Earth's early atmosphere would have been stuck together into amino acids with the energy from the Sun's ultraviolet radiation, or from lightning or shock waves. Raining into the oceans below, the amino acids and other chemicals would have formed an organic soup from which life could start to form. Although the origin of even the simplest organism from inert chemicals remains a mystery, biochemists are beginning to understand some of the pathways by which the very complex molecules of life may have built up on the early Earth, or any other planet, from simpler precursors.

Molecules between the stars. One of the most exciting developments in understanding the origin of life has been the discovery of organic molecules in space in regions such as the Orion nebula where stars are being born. In the dense gas-and-dust clouds of our Galaxy, radio astronomers have detected many molecules of varying complexity (see INTERSTELLAR MOLECULES). Similar molecules are also found in the gases of comets, which are believed to represent the left-over portions of the cloud around the ancient Sun from which the planets formed. The compounds now known in space include formic acid and methylamine, which can combine to give the amino acid called glycine. The predominance of carbon in all these compounds encourages biochemists in their belief that other life-forms will most likely be based on carbon, as life on Earth is.

Meteorites, lumps of rock from space that crash to Earth, have brought further evidence that complex organic chemistry once occurred in the cloud surrounding our growing Sun. Researchers have found a crop of amino acids in two stony meteorites that fell in recent years at Murchison, Australia, and in Kentucky. These meteorites have also revealed evidence of other important organic molecules, evidently formed nonbiologically in space.

Stars and life. Life could not form in the vicinity of all stars. Some stars flare up irregularly, emitting dangerous radiation; others change in brightness more regularly (see VARIABLE STARS). The very biggest stars burn out too quickly for life to begin around them. And the longest-lived stars are dim dwarfs that send out insufficient light and heat to warm up any planets they may have. The best sort of stars for life to form around are those which are average in age and brightness. Our Sun is just such a star, and there are many stars like it in the Galaxy.

For life like ours to form, there must be a planet in a region around the star that is neither so hot that all water evaporates, nor so cold that water freezes. Water is vital to life on Earth, and it so happens that the Earth lies in the very center of the region around our Sun where liquid water can exist. However, some biochemists think that life might form using liquid ammonia as a solvent instead of water, and this means that life could form in colder regions, perhaps among the clouds of a planet like Jupiter.

Mars is one planet in our solar system where Earth-type organisms might have arisen. The two American Viking spacecraft of 1976 were designed to search for life on Mars.

Communicating with the stars. No advanced creatures like ourselves exist on any of the planets in our solar system. They may, however, exist on planets of other stars, and our best chance of finding out is by radio contact. The first attempt to listen for radio messages from other civilizations was made in 1960 by Frank DRAKE in Project Ozma. He used the 85-foot (26-m) diameter radio telescope of the National Radio Astronomy Observatory at Green Bank, West Virginia, to listen to the stars Tau Ceti and Epsilon Eridani. These are both stars like the Sun, and are 12 and 11 light-years away from us, respectively.

Drake listened at a wavelength of 21 centimeters, at which hydrogen in the Galaxy naturally emits

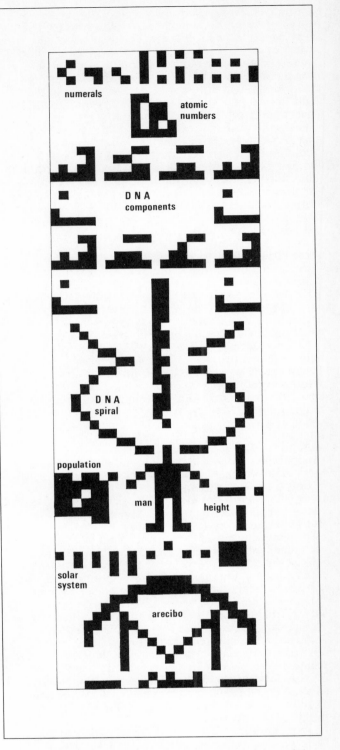

The test message sent from the Arecibo radio telescope on November 16, 1974, to the 300,000 stars of the globular cluster M13 in Hercules, 24,000 light-years distant. The message, sent at 12.6 centimeters wavelength, contained 1,679 on-off pulses which could be arranged into a pictogram, showing, first, the numbers 1 to 10 in the binary code used by computers, **top line, reading right to left**; then the atomic numbers of the basic elements that make up living cells; the formulae for the molecules of DNA, the genetic material of cells; a representation of the familiar double-helix spiral of DNA, with a "backbone" showing the number of DNA components; a figure of a human, flanked by numbers showing the approximate population of the Earth and the human's height; a sketch of the solar system indicating that planet three (Earth) is associated with the human; and a representation of the Arecibo dish transmitting the message.

radio waves. Anyone trying to attract the attention of other civilizations might choose this wavelength to signal on, because it is the one at which radio astronomers most frequently listen. Despite a total of 150 hours of listening to both stars, no signs of a message were detected. However, since only one civilization like our own would be expected for every 100,000 stars, the negative result was not at all unexpected.

Astronomers in the United States, the Soviet Union, and Canada have since extended the Project Ozma work to a total of nearly 1,000 stars. In 1975 Frank Drake and Carl SAGAN began to listen to nearby galaxies with the 1,000-foot (305-m) radio telescope at Arecibo, in search of signals from possible super-civilizations. All these searches have so far proved fruitless.

Stars farther and farther from the Sun will be searched, at various wavelengths, until success is achieved, or the possibilities are exhausted.

light
The part of the spectrum of electromagnetic radiation to which the human eye is sensitive; by extension, the term is often applied to all wavelengths of electromagnetic radiation.

Electromagnetic waves are regular periodic fluctuations in space of electric and magnetic fields. The distance between successive waves or troughs (the WAVELENGTH) determines the color of light. Blue light has a wavelength of about .0004 millimeter and red light about .0007 millimeter. Waves outside the visible region are of the same nature and travel with the same speed (see LIGHT, VELOCITY OF). The types of radiation, in order of increasing wavelength, are gamma rays, X rays, ultraviolet, visible and infrared light, microwaves, and radio waves.

Color, and other properties of light such as refraction, polarization, diffraction, and interference, can be explained by regarding light as a wave motion. But other properties can be described only if we assume that light is a stream of particles (photons), each a packet of energy of a definite size. Since as long ago as the time of Newton, there has been a conflict between the wave and particle theories of light. This has been resolved by the modern *quantum theory,* developed about 1900 by the German physicist Max Karl Ernst Ludwig Planck (1858–1947), which shows that the two views are compatible. The large-scale properties of the particles or photons approximate to those of waves, and light is therefore regarded as having a dual nature.

Early astronomers could exploit only visible light. Using modern techniques, we can observe the Universe at almost all wavelengths. Radio telescopes, in particular, have extended our perception of the Universe. Any limitations that are imposed are mainly those of the Earth's atmosphere. Infrared radiation is absorbed by water vapor in the atmosphere and ultraviolet light is cut off by the ozone layer. These problems are now being overcome by observing from rockets and satellites.

light, velocity of
The speed at which light travels in a vacuum; in denser mediums, such as glass or water, light travels more slowly. The speed of light was first estimated by the Danish astronomer Ole ROEMER in 1676. He found that when Jupiter was distant from the Earth, the eclipses of its satellites took place later than predicted, due to the greater traveling time of the light (called *light time*). He could not, however, obtain a good estimate for the velocity of light because the size of the solar system was not then accurately known.

The theoretical basis for assessing the velocity of light was established by the Scottish physicist James Clerk Maxwell (1831–1879). In Maxwell's equations of ELECTROMAGNETIC RADIATION, published in 1864, he showed that light always has the same velocity, regardless of the velocity of the object emitting it relative to the observer. This result was the starting point of Einstein's 1905 Special Theory of RELATIVITY.

The velocity of light (c) can be measured by determining independently the wavelength (λ) and frequency (v) of a particular SPECTRAL LINE. These are linked by the equation $c = \lambda v$. The modern value of the speed of light is 186,282.397 miles (299,792.458 km) per second. The speed of light in a vacuum is the same at all wavelengths.

light-year
The distance traveled by light in a vacuum during one year (strictly, a tropical year, see YEAR). It is equivalent to 5.8786 trillion miles (9.4607 trillion km), which represents 63,240 astronomical units, or 0.3066 parsec.

limb
The edge of a celestial body as seen from Earth.

Lindblad, Bertil (1895–1965)
Swedish astronomer, who in 1926 suggested that our Galaxy is rotating, to account for the phenomenon of star motion called STAR STREAMING, discovered by Jacobus KAPTEYN. With his son Per Olöf Lindblad (b. 1927), he made studies of the rotation of galaxies, suggesting that their spiral arms are caused by waves of density sweeping around galaxies, a concept developed in more detail by the Chinese-born American mathematician Chia-Chiao Lin (b. 1916). Lindblad developed a method of classifying spectra to determine the absolute magnitudes of stars. He also made surveys of faint stars in selected regions of the sky, improving knowledge of stellar statistics.

Local Group
A cluster of about 20 known galaxies including the Milky Way. The Local Group is believed to be a permanent cluster, held together by the gravity of its members. The Milky Way is near one edge of the group, and the ANDROMEDA GALAXY (M31), 2 million light-years away, marks the other edge. M31 is the largest and most massive galaxy in the Local Group, $1\frac{1}{2}$ times as large as the Milky Way. The third-largest galaxy, M33 in the constellation Triangulum, is only a tenth as massive as the Andromeda galaxy.

Apart from these three spiral galaxies, all the members of the Local Group are dwarf galaxies, with masses less than a hundredth that of the Milky Way. They contain between a million and a billion stars each, and some of the smallest are probably globular clusters that have escaped from the Milky Way. Some of these dwarf galaxies are in orbit around the spirals; the two MAGELLANIC CLOUDS are satellites of the

Milky Way, while the Andromeda galaxy has several small companions.

A large nearby galaxy on the edge of the Local Group was discovered in 1968. It cannot be seen optically because the view from Earth is obscured by dust clouds in the Milky Way, but it can be detected by infrared astronomy. This galaxy, Maffei 1, seems to be about as massive as the Milky Way and is of the elliptical type, containing no interstellar gas. Its distance is about 3 million light-years.

Lockyer, Sir Joseph Norman (1836–1920)

English astronomer, who in 1868 discovered the element helium, from lines observed by the French astronomer Pierre JANSSEN in the spectrum of the Sun's atmosphere during a total solar eclipse. Lockyer's name for the element came from the Greek *helios,* meaning Sun. At the Solar Physics Observatory in South Kensington, London, Lockyer made spectroscopic observations of sunspots in 1866, finding Doppler shifts in their spectral lines that indicated strong convective currents of gas in the Sun's outer layers. In the same year he devised a scheme to observe the bright prominences at the Sun's edge, normally visible only at an eclipse, by spreading out the Sun's light in a spectroscope, and observing the prominences at the wavelengths they emit most strongly. But he was unable to make his first observations until October 1868, by which time Pierre Janssen had already put the technique to work. Lockyer and Janssen found from their observations that prominences were eruptions from the outer layer of the Sun, which Lockyer called the chromosphere. Lockyer also pioneered the study of the astronomical alignments of stone circles and ancient temples. Perhaps his greatest contribution to science came in 1869, when he founded and became first editor of *Nature,* the world's leading scientific journal.

longitude

A coordinate for determining the position of an object east or west of the prime meridian of Greenwich, which is designated 0° longitude. Unlike lines of latitude, lines of longitude are not parallel; they originate in a common point at each pole, and are widest apart on the Earth at the equator. The Earth spins through 15° of longitude in an hour, and so the longitude of a location thus indicates its local time relative to Greenwich. The equivalent of longitude on the CELESTIAL SPHERE is RIGHT ASCENSION, usually measured in hours, minutes, and seconds. It is not to be confused with *celestial longitude,* a separate coordinate measured in degrees east of the vernal equinox along the ecliptic.

Lovell, Sir Alfred Charles Bernard (b. 1913)

British radio astronomy pioneer, founder of the famous Jodrell Bank Radio Observatory with its 250-foot (76.2-m) dish, for many years the world's largest fully steerable radio telescope. After World War II, Lovell used his wartime experience with radar to study radio echoes from cosmic ray showers. Many echoes turned out in fact to come from the trains of meteors, as Lovell's team conclusively proved in 1946. Lovell's radio studies located many daytime meteor showers that were previously unknown. In 1949 he began plans for the 250-foot

dish, which finally came into operation in 1957, in time to track the first Sputnik. The Jodrell Bank telescope was occasionally called upon for communications with space probes, but mostly Lovell pursued astronomical work, including the detection of radio emissions from outbursts on FLARE STARS.

Lovell, James Arthur (b. 1928)

American astronaut, commander of the Apollo 13 flight of April 1970, whose projected Moon landing was canceled because of an explosion in the service module. Lovell and his crew used the engine and electrical power of the attached lunar module to make an emergency return to Earth. Lovell's first flight was in the record-breaking Gemini 7 mission of December 1965, in which he and Frank BORMAN spent a total of 14 days in space. Lovell commanded the Gemini 12 flight in November 1966, during which Edwin ALDRIN made an extensive space walk. Lovell was also a crew member of Apollo 8 in December 1968, the first manned spacecraft to reach the vicinity of the Moon.

Lowell, Percival (1855–1916)

American astronomer, who instituted the search that led to the discovery of the planet Pluto. In 1894 Lowell, a wealthy amateur, founded the Lowell Observatory at Flagstaff, Arizona, the best observing conditions he could find, to observe Mars. Inspired by reports of "canals" on Mars by the Italian astronomer Giovanni SCHIAPARELLI, Lowell believed he detected an entire network of canals on the planet, and wrote works such as *Mars and Its Canals* (1906) and *Mars as the Abode of Life* (1908), in which he described a dying and arid planet whose inhabitants were forced to channel water from the poles to their crops nearer the equator. Despite its immense popular appeal, few astronomers agreed with Lowell's theory, nor could they confirm his canal observations. At the turn of the century Lowell began analyzing the motion of Uranus to determine if it revealed the gravitational effects of an unknown planet beyond Neptune. In 1905 he produced the first of a series of predictions of where a new planet might lie, which culminated in his famous 1915 memoir on the subject. A major search was instituted at Lowell Observatory, but only in 1930 was Pluto discovered. It is such a small body, however, that it could not have had any observable effects on Uranus, and most astronomers now assign Pluto's discovery to chance.

Lowell Observatory

An astronomical observatory founded by Percival LOWELL in 1894 at Flagstaff, Arizona, at an elevation of 7,250 feet (2,210 m). The observatory contains a 24-inch (61-cm) refractor installed by Lowell in 1896, together with 30-inch (76-cm), 24-inch, and 21-inch (53-cm) reflectors used mostly for photoelectric observations. The observatory also possesses the famous Pluto telescope, a 13-inch (33-cm) astrographic telescope used to discover Pluto. This is now situated at the dark-sky observing site opened in 1961 at Anderson Mesa, 12 miles (19 km) southeast of Flagstaff. The 72-inch (183-cm) Perkins reflector of the Ohio State and Ohio Wesleyan universities was moved to this site in 1961, and in 1968 a 42-inch

(107-cm) reflector was also installed. In 1965 a NASA-sponsored Planetary Research Center was opened in the grounds of Lowell Observatory; this is the headquarters of an international photographic survey of the planets, and contains the world's most extensive collection of planetary photographs.

luminosity

A measure of the total amount of radiation emitted by a star or other glowing object per second, usually expressed in relation to the luminosity of the Sun. A star's luminosity depends on its temperature and size. One way of expressing the luminosity of a star is its ABSOLUTE MAGNITUDE. Astronomers divide stars into six *luminosity classes*: Ia, Brightest supergiants; Ib, Less luminous supergiants; II, Bright giants; III, Giants; IV, Subgiants; V, Main-sequence stars (often termed dwarfs).

Lunar Module (LM)

The two-stage craft in which Apollo astronauts landed on the Moon. The Lunar Module was stored under the Apollo command and service modules at launch, and extracted by a docking maneuver once Apollo was on its way to the Moon. The Lunar Module had a total height with its four legs extended of 22 feet 11 inches (7 m), divided between the descent (lower) stage of 10 feet 7 inches (3.2 m) and the ascent (upper) stage of 12 feet 4 inches (3.8 m); its overall width was 14 feet 1 inch (4.3 m), and launch weight 32,000 lb. (14,500 kg). The two-man LM crew occupied the ascent stage, which was pressurized; the unpressurized descent stage contained the powerful braking engine, which could thrust from 1,050 lb. (476 kg) to 9,870 lb. (4,480 kg) for controlling the Moon landing. The LM had a roof hatch for docking with the command module, and a front hatch for exit onto the lunar surface; the astronauts, standing in the ascent stage, watched their descent through two triangular-shaped forward windows. The ascent stage contained room for lunar space suits and rock samples. To lift off from the Moon's surface the ascent stage explosively separated from the descent stage, and the 3,500-lb. (1,600-kg) thrust ascent stage engine boosted the astronauts into Moon orbit to meet the command module. Sixteen small reaction control system (RCS) thrusters, arranged in groups of four, were used for delicate maneuvers and attitude control of the LM.

Apollo 14 Lunar Module *Antares* stands deserted in the Moon's Fra Mauro crater in February 1971, while astronauts Mitchell and Shepard carry out EVA. On the foremost leg is the ladder which the astronauts used for access to the module. In this view the top of the module partly shields the Sun's glare.

A technician checks a Lunar Orbiter spacecraft before launch. In the center of the framework are the lenses of the two cameras; behind them is the self-contained darkroom and picture readout mechanism. Above the propellant tanks on the upper part of the craft can be seen a thruster nozzle; to the left is the directional antenna used to transmit the data to Earth.

Lunar Orbiter

A series of five American Moon-orbiting craft which photographed the entire lunar surface, front and back, revealing its features in detail and surveying possible landing sites for astronauts. Each Lunar Orbiter carried two cameras, one for detail shots and the other for wide-angle views. The photographs were recorded on film in the spacecraft and electronically transmitted to Earth. The first three probes entered orbits around the Moon's equatorial region, while Lunar Orbiters 4 and 5 were put into polar orbit and thus surveyed the whole Moon. Tracking the Lunar Orbiters revealed the existence of MASCONS, areas of dense rock on the Moon that cause a higher gravitational pull than normal. After their flights the Lunar Orbiters were crashed onto the Moon, so as not to interfere with later missions.

Probe	Launch date	Remarks
Lunar Orbiter 1	August 10, 1966	Entered lunar orbit August 14; impacted Moon October 29
Lunar Orbiter 2	November 6, 1966	Entered lunar orbit November 10; impacted Moon October 11, 1967
Lunar Orbiter 3	February 4, 1967	Entered lunar orbit February 8; impacted Moon October 9
Lunar Orbiter 4	May 4, 1967	Entered lunar orbit May 8; impacted Moon October 6
Lunar Orbiter 5	August 1, 1967	Entered lunar orbit August 5; impacted Moon January 31, 1968

Lunar Roving Vehicle (LRV)

The electrically powered Moon car used by Apollo astronauts for exploring the lunar surface. Lunar rovers were used on the flights of Apollo 15, 16, and 17. The wire-wheeled vehicle, 122 inches (310 cm) long and 45 inches (114 cm) high, allowed astronauts to drive at speeds up to 8 miles (13 km) per hour in search of rock and soil samples. Instruments on the control panel told the astronauts how far they had progressed from the Lunar Module, and in what direction. An umbrella-shaped antenna on the vehicle kept the astronauts in direct touch with Earth, and allowed mission controllers to point the on-board television camera by radio commands. The 480-lb. (218-kg) lunar rover, capable of negotiating foot-high obstacles, two-foot crevices, and slopes of up to 20°, was stored folded in the Lunar Module's descent stage; it was deployed by the astronauts once on the Moon, and left there after use.

Luna spacecraft

A series of Soviet Moon probes, originally called Lunik. Luna 1 missed its target but was the first object to leave Earth and enter an orbit around the Sun. Luna 2, in September 1959, became the first probe to hit the Moon and Luna 3 the following month flew behind the Moon, sending the first photographs of its far side. In 1966 Luna 9 became the first probe to make a successful soft landing, and Luna 10 became the first probe to orbit the Moon. The Luna program broadened in 1970 when Luna 16 automatically returned a small sample of Moon soil to Earth, and Luna 17 landed an automatic roving Moon car called LUNOKHOD.

Probe	Launch date	Remarks
Luna 1	January 2, 1959	Missed Moon by 3,728 miles (6,000 km); in solar orbit
Luna 2	September 12, 1959	Hit Moon September 13
Luna 3	October 4, 1959	Sent back first photographs of lunar far side
Luna 4	April 2, 1963	Missed Moon by 5,282 miles (8,500 km); possible soft-landing failure
Luna 5	May 9, 1965	Impacted Moon May 12; failed soft-lander
Luna 6	June 8, 1965	Missed Moon by 100,000 miles (160,000 km); failed soft-lander
Luna 7	October 4, 1965	Impacted Moon October 7; failed soft-lander
Luna 8	December 3, 1965	Impacted Moon December 6; failed soft-lander
Luna 9	January 31, 1966	Soft-landed on Moon February 3, in western Oceanus Procellarum. Returned photos for three days
Luna 10	March 31, 1966	Entered lunar orbit April 3; measured magnetic field, meteoroids
Luna 11	August 24, 1966	Entered lunar orbit August 28; successor to Luna 10

Probe	Launch date	Remarks
Luna 12	October 22, 1966	Entered lunar orbit October 25; took photographs, made measurements
Luna 13	December 21, 1966	Soft-landed on Moon December 24; returned photos, tested soil
Luna 14	April 7, 1968	Entered lunar orbit April 10; measured near-Moon conditions such as magnetic and gravitational field, solar wind particles
Luna 15	July 13, 1969	Impacted Moon July 21 in Mare Crisium; failed sample-return attempt
Luna 16	September 12, 1970	Landed in Mare Fecunditatis September 20; returned to Earth September 24 with 0.2 lb. (100 g) of Moon soil
Luna 17	November 10, 1970	Landed on Moon November 17, carrying Lunokhod 1 automatic Moon rover
Luna 18	September 2, 1971	Impacted Moon September 11; probably failed landing attempt
Luna 19	September 28, 1971	Entered lunar orbit October 3; studied lunar surface and near-lunar space
Luna 20	February 14, 1972	Landed near Mare Fecunditatis February 21; small soil sample returned on February 25
Luna 21	January 8, 1973	Landed in Mare Serenitatis on January 15, carrying Lunokhod 2 lunar rover
Luna 22	May 29, 1974	Entered lunar orbit June 2; studied Moon and near-lunar space
Luna 23	October 28, 1974	Landed on Moon November 6. Damaged drill prevented sample return

lunation
A complete cycle of lunar phases, the period from one new Moon, for example, to another. A lunation is therefore equivalent to a synodic MONTH, and lasts 29.53059 days.

Lunokhod
A Soviet automatic Moon car, driven by radio command from Earth. Lunokhod 1 was carried to the Moon aboard the Luna 17 soft-lander, which touched down on November 17, 1970. The eight-wheeled vehicle rolled off the lander, transmitting television pictures of its surroundings. Instruments detected cosmic rays and measured the chemical composition and physical nature of lunar rock; a small reflector panel bounced laser beams back to Earth, allowing scientists to determine the precise distance of the Moon. The 1,667-lb. (756-kg) craft had a length along its four wheels of 87 inches (222 cm) and a width of 63 inches (160 cm). By the end of its active life on October 4, 1971, Lunokhod 1 had covered a total distance of 34,588 feet (10,542 m), running in and out of craters and examining rocks as it moved over the lunar surface. An improved version, the 1,848-lb. (838-kg) Lunokhod 2, was delivered to the Moon's surface by Luna 21 on January 15, 1973. Lunokhod 2, traveling at twice the speed of its predecessor, completed its mission on June 4, having covered 23 miles (37 km) inside the crater Le Monnier. Among other features, it examined a large slab of Moon rock believed to have been ejected from a crater, and ran alongside a deep fissure.

Lupus (the wolf)
A constellation of the southern hemisphere of the sky, lying on the edge of the Milky Way between Scorpius and Centaurus. It contains no objects of particular importance.

Lynx (the lynx)
A faint constellation in the northern hemisphere of the sky near Ursa Major, introduced by Johannes HEVELIUS. Its brightest star is a red giant appearing of magnitude 3; there are no objects of major interest in the constellation.

Lyot, Bernard Ferdinand (1897–1952)
French astronomer, inventor of the CORONAGRAPH, an instrument that makes the Sun's corona visible without a solar eclipse. Lyot pioneered the study of the surface nature of planets by measuring the polarization of light reflected from them; he showed the dusty nature of the lunar topsoil and the existence of sandstorms on Mars in this way. To study the polarization of the Sun's corona, Lyot mounted his coronagraph at the high-altitude Pic du Midi Observatory in the French Pyrenees. He made the first spectrogram of the inner corona in 1930 and took the first corona photograph through his instrument in

Lunokhod 2, which landed on the Moon in January 1973 in the crater Le Monnier, at the edge of the Mare Serenitatis. Visible here are the directional antenna 1, TV cameras 2, solar cell array 3, omnidirectional antenna 4 and laser reflector array 5.

Lyra

1931. With a special monochromatic filter developed in 1933 to pass only narrow bands of light, Lyot in 1935 began to take the first motion picture films of prominences at the Sun's edge. The instruments he developed have revolutionized solar observations.

Lyra (the lyre)
A small but prominent constellation of the northern hemisphere of the sky, seen overhead in middle latitudes during summer. Lyra is particularly distinctive because its main star, VEGA, is one of the brightest in the sky: Beta Lyrae, also called Sheliak, is a remarkable ECLIPSING BINARY, varying between magnitudes 3.4 and 4.3 every 12.9 days. Matter is believed to be flowing from one star to the other, each of which is distorted in shape by the other's gravitational pull; hot gas is probably also spiraling away from them into space. Epsilon Lyrae is a famous multiple star called the "double double." Keen eyesight, or a pair of binoculars, shows two stars, of magnitude 4.68 and 4.5; in a telescope, each of these is also found to be double. Delta Lyrae and Zeta Lyrae are also double stars. Between the stars Gamma and Beta is the famous ring nebula in Lyra, cataloged as M57 or NGC 6720; it is actually a PLANETARY NEBULA, 4,100 light-years distant. Near the border of the constellation with Cygnus lies RR Lyrae, the prototype of a famous class of variable stars (see RR LYRAE VARIABLES).

Lyrid meteors
A regular meteor shower reaching a brief peak of about 8 meteors an hour on April 21 each year. The stream radiant, which moves eastward 1° a day, lies 8° southwest of Vega in Lyra at maximum, and reaches its greatest altitude at 4 A.M. Splendid displays about 2,000 years ago are recorded, but the stream is now very much in decline. It follows an orbit of very long period and is derived from periodic comet Thatcher 1861 I.

M

Magellanic Clouds
Two small galaxies that orbit our Milky Way Galaxy. They are easily visible to the naked eye from latitudes south of 20°N, and were first described by the Portuguese navigator Ferdinand Magellan (c.1480–1521) in 1521 during his voyage around the world. Both Clouds have the same fuzzy appearance as the Milky Way. The Large Magellanic Cloud (LMC), lying in the constellations Dorado and Mensa, is about 6° across. About 20° away in Tucana is the Small Magellanic Cloud (SMC), 2° in size. Long-exposure photographs reveal fainter outer regions of each extending to three times the naked-eye size.

Both Magellanic Clouds are 160,000 light-years distant, and are the nearest external galaxies to our own (twelve times closer than the Andromeda galaxy). They are comparatively small, with masses about 1/30 and 1/200 that of the Milky Way, containing both young and old stars. The LMC contains the largest-known gaseous nebula, 30 Doradus, called the Tarantula nebula, which has a mass of over a million Suns, and may be condensing to a globular cluster of

The Large Magellanic Cloud, **extending off the top of this photograph**, has a bright central bar, like some spiral galaxies. The huge bright nebula 30 Doradus is just above the left end of the bar. The superimposed squares are 1 degree across.

Many gaseous nebulae are visible in the Small Magellanic Cloud as small diffuse patches, and the stars can also be resolved by large telescopes. The brightest stars in this photograph and the two globular clusters, **right and top**, are foreground objects in our Milky Way Galaxy.

stars. Young globular clusters are comparatively common in the LMC, although unknown in the Milky Way.

Recent radio-astronomy observations have shown a streamer of hydrogen connecting the Clouds to the edge of the Milky Way. This hydrogen was apparently torn away from the Clouds as they passed through the disk of our Galaxy.

magnetic stars
Stars remarkable for their very strong magnetic fields. The nearest star to us, the Sun, has a generally weak magnetic field like that of the Earth, except locally in sunspots, where the fields can be thousands of times greater. In 1946 the American astronomer H. W. BABCOCK showed that some stars have general magnetic fields far stronger than the Earth's, the most powerful being over 50,000 times as strong. Both the magnetic field and the magnitude of many magnetic stars vary with periods generally between $\frac{1}{2}$ and 20 days. These variations represent the time the stars take to rotate on their axes. Magnetic stars may have enormous "starspots" (like sunspots) that appear and disappear as the star spins, causing the variations in light output. The magnetic field in the starspot is thought to be much stronger than elsewhere on the star's surface. Magnetic stars have a peculiar excess of certain rare metallic elements in their outer layers, presumably produced by nuclear reactions in giant flares, which are associated with the spots and magnetic fields.

magnetosphere

The outermost region of the Earth's atmosphere, extending beyond the IONOSPHERE, from an altitude of 300 miles (500 km) upward. In this region, also called the exosphere, ionized particles are controlled by the Earth's magnetic field. The magnetosphere forms the boundary region between the Earth's atmosphere and the SOLAR WIND in space. At the magnetosphere's edge, particles from the Earth's atmosphere escape into space. The outer edge of the magnetosphere (the *magnetopause*) extends about 10 Earth radii toward the Sun; beyond this is an unstable region called the *magnetosheath,* produced by shock waves as the solar wind encounters the magnetosphere. Away from the Sun, in the lee of the solar wind, the magnetosphere forms a long *magnetotail.*

Two zones of high density are centered above the equator; these are the VAN ALLEN radiation belts. The Van Allen belts were discovered in the late 1950s by early artificial satellites. In shielding the Earth from the solar wind, the magnetosphere deflects charged particles into the Van Allen belts, but some spill over into polar regions of the upper atmosphere where they excite atmospheric atoms to produce the colorful AURORAE.

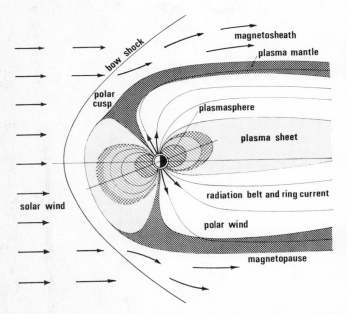

The Earth's invisible surroundings of magnetic zones. These regions have been detected by satellites and probes bearing magnetometers and particle counters; the arrows show the directions in which charged particles would move.

magnitude

A measure of an object's brightness. The Greek astronomer HIPPARCHUS first classified the naked-eye stars into six magnitude steps, with the brightest being first magnitude and the faintest he could see sixth magnitude. When stars were first measured with a PHOTOMETER, it was found that first-magnitude stars were roughly 100 times brighter than sixth-magnitude stars. In 1856 the English astronomer Norman Robert

Pogson (1829–1891) proposed the current classification system, in which a difference of five magnitudes is taken as representing exactly a factor of 100 in brightness. This means that one magnitude difference corresponds to a difference in actual brightness of 2.512 times (2.512 is the fifth root of 100). Objects more than 100 times brighter than sixth-magnitude stars are given negative (minus) magnitudes; the faintest objects are given increasingly large positive magnitudes. The term "magnitude" when used without further qualification usually means APPARENT MAGNITUDE, the brightness with which a star appears in the sky. Astronomers also use ABSOLUTE MAGNITUDE, which is a calibration of the actual light output of a star. (See also DISTANCE MODULUS; PHOTOGRAPHIC MAGNITUDE; PHOTOVISUAL MAGNITUDE.)

main sequence

The diagonal band ranging from bright blue stars to faint red stars on the HERTZSPRUNG-RUSSELL DIAGRAM, a plot of star temperatures against brightness. Main-sequence stars are termed DWARFS; consequently, the main sequence is also known as the *dwarf sequence.* The position of a star on the main sequence depends on its mass, the heaviest stars being the brightest. The majority of stars lie on the main sequence because the dwarf stage of a star's evolution, when it is in the prime of its life and producing energy by burning hydrogen in its core, lasts longer than any other. In theory, at the time they begin burning hydrogen in their centers, all dwarfs lie on a line in the H-R diagram called the *zero-age main sequence* (ZAMS). But in the actual observation of a young star cluster the line is broadened into a band, and the ZAMS is taken to be the lower edge of the observed main sequence.

major axis

The longest diameter of an ELLIPSE, one which passes through its two foci.

Maksutov telescope

A telescope system using both a lens and mirror, named for the Russian Dmitri Dmitrievich Maksutov (1896–1964), who published his design in 1944, although it was discovered independently by the Dutch optical manufacturer Albert Bouwers (1894–1972). The earlier SCHMIDT TELESCOPE corrects the blurred image, or SPHERICAL ABERRATION, of a spherical concave mirror by using a thin glass plate placed some distance in front of it. The cross-sectional shape or *figure* of this plate is difficult to achieve by mass-production methods. Maksutov substituted a thick, steeply-curved lens or *shell*, offering fewer production problems. Used as a wide-field camera, the performance of the Maksutov matches that of the Schmidt, with the added advantage that the tube is only about three-quarters the length. However, few large Maksutov-type cameras have been constructed because large shells are much thicker and heavier than the equivalent Schmidt plates. The Maksutov system has been exploited most widely in the production of relatively small-aperture telescopes for amateur use, and in telephoto *mirror-lens* systems for photography. In both cases, the central region of the convex face of the shell is coated with aluminum,

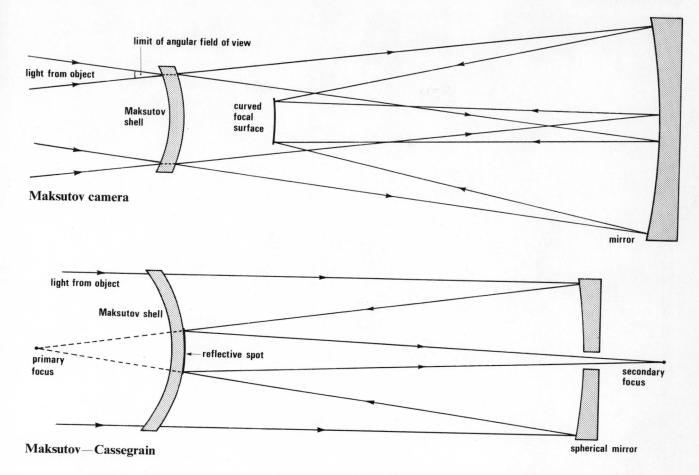

limit of angular field of view

light from object

Maksutov shell

curved focal surface

Maksutov camera

mirror

light from object

Maksutov shell

primary focus

reflective spot

secondary focus

Maksutov—Cassegrain

spherical mirror

the mirror so formed acting as a Cassegrain-type secondary (see CASSEGRAIN TELESCOPE). Such instruments are highly portable, and the coated reflecting surfaces, unlike those of ordinary reflecting systems, are protected from dust and corrosion in a sealed tube. Maksutov systems are relatively expensive, due to the care required in matching the curves and thickness of the shell to prevent false color in the image.

Mariner spacecraft

A series of American planetary probes. Mariner 2 was the first probe to reach another planet—Venus—successfully; Mariner 4 was the first probe to send results from Mars, revealing craters on the Martian surface. Mariner 9 became the first object to go into orbit around another planet when it reached Mars for a year-long photographic reconnaissance. Mariner 10 made history as the first double-planet mission, flying past Venus and then moving on to give astronomers their first detailed look at the surface of Mercury. Mariner 10 went into a close orbit around the Sun, re-encountering Mercury in September 1974 and March 1975. In 1977, Mariners 11 and 12 will be launched on double-planet missions to survey in detail the giant planets Jupiter and Saturn, arriving in 1979 and 1981.

Probe	Launch date	Remarks
Mariner 1	July 22, 1962	Launch failure; intended Venus probe
Mariner 2	August 26, 1962	Flew past Venus December 14 at a distance of 21,594 miles (34,752 km)
Mariner 3	November 5, 1964	Intended Mars probe; contact lost because spacecraft shroud failed to jettison
Mariner 4	November 28, 1964	Flew past Mars July 14, 1965, at a distance of 6,118 miles (9,846 km)
Mariner 5	June 14, 1967	Flew past Venus October 19, 1967, at a distance of 2,480 miles (3,990 km)
Mariner 6	February 25, 1969	Flew past Mars July 31 at a distance of 2,120 miles (3,412 km)
Mariner 7	March 27, 1969	Flew past Mars August 5 at a distance of 2,190 miles (3,534 km)
Mariner 8	May 8, 1971	Launch failure
Mariner 9	May 30, 1971	Went into Mars orbit November 13
Mariner 10	November 3, 1973	Flew past Venus February 5, 1974, at a distance of 3,585 miles (5,769 km); passed Mercury on March 29 at a distance of 431 miles (694 km)

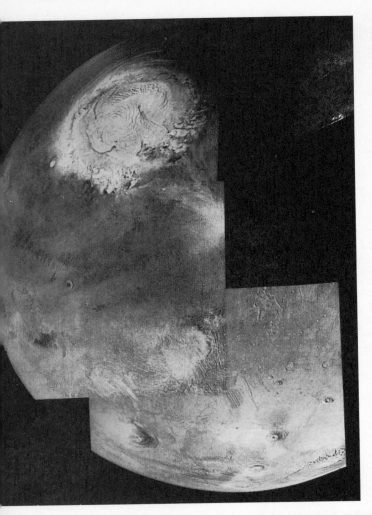

Mars

The fourth planet in order of distance from the Sun, commonly called the red planet because of its color. Mars is the planet that most resembles the Earth: it spins on its axis every 24 hours 37 minutes 23 seconds, and shows permanent surface markings, polar ice caps, and atmospheric clouds. Space-probe photographs have revealed canyons, shield volcanoes, ancient water courses, sand dunes, and heavily cratered landscapes resembling the highlands of the Moon. The climate is extremely cold and dry with violent dust storms; yet Mars is still the most likely extraterrestrial haven of life in our solar system.

Mars moves at 15 miles (24 km) per second in an elliptical path an average of 141,636,000 miles (227,941,000 km) from the Sun, taking 687 days to complete one orbit. The planet's equatorial diameter is 4,217 miles (6,787 km), about half that of the Earth, with a mass of 6.4×10^{20} tons. Mars has a mean density of 3.94 times that of water, compared with the Earth's 5.52. Its rotational axis is inclined at $25°$ (the Earth's tilt is $23\frac{1}{2}°$), and its surface gravity is 40 percent of Earth's.

Mars shows a wealth of fine detail—bright orange areas contrasting with dark bluish-gray markings, topped with brilliant white polar caps. Early theories held that the light areas were continents, the dark areas oceans, and the polar caps deep polar snow drifts. Eventually, however, it was realized that the presence of only minor cloud activity and thin polar caps indicated that water was very scarce. Scientists then postulated that Mars was dry and dusty, and that the dark areas were lowlands covered with primitive vegetation. This seemed to explain the darkening of the dusky areas in spring as the polar cap began to melt. As moisture was released the plants would burst into growth, breaking through the layers of dust deposited by the winter storms and darkening

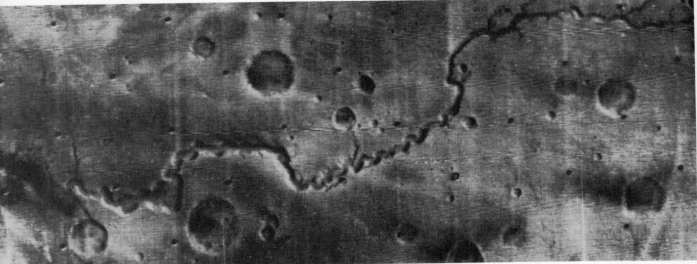

The northern hemisphere of Mars, in a mosaic of photographs from Mariner 9, taken at 8,500 miles range. At the top is the polar cap of frozen carbon dioxide; near the equator, **lower right**, lie the end of the huge Martian "Grand Canyon" and some Martian volcanoes, including the highest, Olympus Mons, **lower center**.

Nirgal Vallis is a winding channel 600 miles long in Mars' southern hemisphere. Its obvious tributaries suggest it is a dried-up river channel, formed when running water was abundant on Mars. The picture also shows some craters, which are common to the south of the equator, but rare further north.

the lowlands. Unfortunately again modern work has shown that Martian conditions are extremely bleak, making extensive vegetation of any kind unlikely.

Atmosphere. The Martian atmosphere has a surface pressure only 6 percent of the Earth's and is composed almost entirely of carbon dioxide with traces of water, oxygen, and carbon monoxide, with perhaps some nitrogen. Water cannot exist in a liquid state, although it could be frozen as a small part of the polar caps. Surface conditions are very severe. On the equator the surface temperature is $-120°C$ at dawn but rises to a peak of $25°C$ just after noon. It falls below $0°C$ again within two hours, reaching $-70°C$ by sunset and $-120°C$ by the end of the night. Noon temperatures range from $25°C$ at the equator to $12°C$ (at $20°$ latitude), $-15°C$ ($40°$), $-50°C$ ($60°$), $-90°C$ ($75°$), and $-120°C$ close to the pole. These are summer temperatures. In winter the temperature falls well below $-150°C$ at the poles and remains below freezing for the whole winter hemisphere.

Clouds appear on Mars, but they cover only about 5 percent of the surface even when most active. White clouds forming downwind from highland areas may be stationary, like those forming in the lee of mountains on Earth. Blue clouds, which form at an altitude of about 8 miles (13 km), appear to be similar to terrestrial NOCTILUCENT CLOUDS, formed by ice crystals around meteoritic dust high in the atmosphere.

Weather. Martian weather varies greatly with season and time of day. In winter, a massive temperature difference between equator and pole produces brisk westerly gales and creates intense low-pressure areas. The strong frontal activity lifts dust particles of all sizes to considerable heights. Great dust storms develop from hurricane-like disturbances over the central highlands, throwing up vast quantities of dust to form a layer at 12 to 18 miles (19–29 km) altitude. This can envelop the entire equatorial and temperate regions of the planet, as in 1909, 1911, 1956, and 1971. In winter, therefore, dust is being spread around the planet, deposited at ground level as winds abate in spring. As the polar cap begins to retreat, the atmosphere clears to reveal the Martian surface blanketed with dust and with indistinct light and dark areas. As temperatures rise, the gradient between the pole and equator is reduced, and the westerly winds drop. They are replaced by breezes toward the equator around dawn and dusk, with general easterly winds during the main part of the day.

The entire surface of Mars consists of a layer of dust either composed mainly of, or coated with, iron oxides, which are responsible for the planet's orange-red color. Keeping in mind that fine dust particles will always appear lighter in shade than larger dust particles of the same material, a meteorological explanation for the springtime wave of darkening is possible. The spring easterly winds can have little effect on heavy dust, but will scour exposed highland areas free of fine dust, carrying and depositing it on adjacent bright areas, increasing the contrast between dark highlands and light lowlands. Thus the dark areas on Mars are probably the exposed eastern slopes of highland regions.

Martian topography. The highlands of Mars, which rise to heights of 7 miles (11 km), are concentrated mainly in the southern hemisphere. In contrast, the northern half of the planet consists of dust-filled lowland areas with extensive fields of recent volcanic activity. The 7,000 photographs returned by Mariner 9 in orbit around Mars provide a detailed picture of the surface. The southern highlands have been heavily cratered by meteorites, indicating a great age—perhaps as much as 3.5 billion years. The largest craters are two immense impact basins, Hellas and Argyre. Hellas is a double-rimmed 3-mile-(5-km)-deep bowl some 1,000 miles (1,600 km) across resembling the Mare Imbrium basin on the Moon. The uplands also contain a number of degraded, weathered volcanic areas of great age; the whole area is similar to the highlands of the Moon. Part of the equatorial region is occupied by an extension of the southern uplands, dissected in spectacular fashion by a great 3,000-mile (4,800-km) rift zone. The major fault valley reaches a width of 45 miles (70 km) and a depth of 4 miles (6.4 km) dwarfing the Earth's Grand Canyon. There are many subsidiary faults of varying ages, some heavily eroded, others with young, clear outlines.

The northern hemisphere is strikingly different. It contains low-lying smooth basins and plains, with young, sparsely scattered impact craters. The lowlands form a wide belt around $65°$ north latitude and have many relatively young basalt flows. Close to the pole is an extensive volcanic plateau, 250 miles (400 km) across and 2 miles (3 km) high, with central calderas and extensive fault systems. But the most remarkable features are four huge shield volcanoes close to the equator. Reaching an astounding height of 18 miles (29 km), they consist of complex summit calderas atop cone-shaped basalt mantles. The largest, Nix Olympica, visible from Earth as a dusky spot, is some 320 miles (515 km) across at the base; all four volcanoes are quite clearly young features. Dust has collected in several sites at high temperate latitudes, forming fields of dunes resembling the Earth's sandy deserts. North of the equatorial plateau is a vast area of chaotic landscape, hundreds of square miles of jumbled surface blocks resembling the site of a major earthquake. Throughout the equatorial regions, and especially emerging from the chaotic zones, are numerous meandering valleys, many of them with extensive tributary systems. Hundreds of small channels complete a picture clearly indicating that once water flowed on Mars. Water in a liquid state must have existed at some distant epoch, when the atmosphere was much denser. The water courses appear of different ages, which suggests that Mars has experienced several short periods of warmth separated by lengthy glacial eras. The warm spells would periodically release water from the permafrost that must exist underground close to the poles. Remarkably, the close up photographs of Mars do not show any indication of the Martian "canals" drawn by such astronomers as Percival LOWELL. These observers apparently joined up quite disconnected scattered dusky markings and interpreted them as hard, dark lines.

The current polar caps are complex in nature. Their outer fringes are composed of solid carbon dioxide ("dry ice"), but the central cores clearly contain water ice. The caps re-form each winter on top of layered sediments as much as 4 miles (6 km) thick. Surrounding them is a pitted landscape where evaporating ice and wind-erosion have scoured depressions in which ice lingers as the cap melts. Beyond these are extensive dune fields, which become scarcer toward the equator. There seems to be a slow transport of dust from the windblown equatorial regions to build up the dune-fields and layered polar deposits.

The present geological structure of Mars, with ancient highlands in one hemisphere and youthful lowlands in the other, bears a certain resemblance to the Earth before continental drift began to break up the original giant continent Pangea. Internally, Mars does not seem to possess a liquid core; instead, it consists simply of mantle and crustal regions. Its internal composition may be uniform, with iron bound up with the bulk of Martian oxygen as oxides instead of being separated into a central iron core.

Evidence that warmer periods with abundant liquid water existed in Martian history greatly improves the chances that life evolved on the plant. Even with its present, inhospitable climate, it is just possible that life-forms may exist, perhaps some form of lichen, highly-adapted plant-life, or stress-adapted bacteria.

Moons. Mars has two tiny moons, Phobos and Deimos, which revolve in circular orbits in the planet's equatorial plane. These irregularly shaped, impact-scarred blocks of rock, may be captured asteroids (see DEIMOS; PHOBOS).

Mars probes

A series of Russian space probes to investigate the planet Mars (for American exploration of Mars see MARINER SPACECRAFT and VIKING). The first Soviet probes to reach their target successfully were Mars 2 and Mars 3 in 1971, both of which ejected lander capsules and then entered orbit around the planet. The Mars 2 lander, which crashed, was the first man-made object to reach the surface of Mars. The Mars 3 lander descended successfully; its transmissions failed 20 seconds after it had begun to send a television picture, which showed nothing. In 1974, Mars 4's braking engine failed and it swept past the planet; Mars 5 successfully entered orbit around Mars and sent back useful photographs. These orbiter craft were meant to act as communications relays for lander probes ejected from Mars 6 and 7. But contact was lost with the Mars 6 lander as it neared the surface, and the Mars 7 lander missed the planet entirely.

Probe	Launch date	Remarks
Mars 1	November 1, 1962	Radio contact lost after 66 million miles (106 million km) on March 21, 1963
Mars 2	May 19, 1971	Entered Mars orbit November 27; surveyed surface and atmosphere of planet. Lander capsule ejected but crashed
Mars 3	May 28, 1971	Entered Mars orbit December 2 and surveyed planet. Ejected lander, but transmissions ceased after 20 seconds
Mars 4	July 21, 1973	Passed Mars at a distance of 1,367 miles (2,200 km) on February 10, 1974, due to braking rocket failure

Probe	Launch date	Remarks
Mars 5	July 25, 1973	Entered Mars orbit February 12, 1974
Mars 6	August 5, 1973	Flew past Mars March 12, 1974, and ejected lander capsule which crashed
Mars 7	August 9, 1973	Flew past Mars March 9, 1974, and ejected lander capsule which missed planet

Marshall Space Flight Center (MSFC)

A NASA facility at Huntsville, Alabama, for the development of launch vehicles and spacecraft, named for General George C. Marshall. The center, established in 1960, grew out of missile work at the U.S. Army's Redstone Arsenal at Huntsville, which produced the Redstone, Jupiter, and Juno rockets. Under the directorship of Wernher VON BRAUN, the Marshall Space Flight Center developed the SATURN family of launch vehicles, in addition to the Skylab space station. MSFC provided the lunar roving vehicle used on the Apollo Moon missions, and it manages the High Energy Astronomical Observatory (HEAO) spacecraft project planned for the late 1970s. MSFC is developing the engines to be used for the Space Shuttle, and is also working on the Large Space Telescope and Space Tug assemblies to be flown in the Shuttle. MSFC has test stands for experimental firings of rockets under development. Rocket stages are manufactured at the Michoud Assembly Facility in New Orleans, and checked at the National Space Technology Laboratories in Bay St. Louis, which is now a separate NASA facility but until 1974 was part of MSFC, under the name of the Mississippi Test Facility.

mascon

An area where the Moon's gravitational field is increased; the name is a contraction of *mass concentration*. Mascons were first noted in 1968 from tracking the paths of space probes orbiting the Moon; minor irregularities in the space probes' orbits revealed the areas of increased gravity. A major mascon is associated with the Mare Imbrium lowland plain on the Moon, and further mascons are associated with a number of other lunar seas. Mascons are probably aggregations of denser rock under the Moon's surface where molten lava has collected in one area and solidified.

maser

A source of very intense microwave radiation (a few centimeters in wavelength); the name is an acronym for *Microwave Amplification by Stimulated Emission of Radiation*. The first maser device, invented in 1953 by the American physicist Charles Hard Townes (b. 1915), contained excited molecules of ammonia gas; these were stimulated to emit a very intense beam of microwaves when triggered by radiation of the same wavelength. Such maser amplifiers can be used in radio telescopes as very sensitive receivers.

In 1963, radio astronomers discovered an extremely powerful source of 6-centimeter wavelength radiation associated with clouds of dust and gas near our galactic center. It was identified as the hydroxyl radical (OH), but at an improbably high temperature. With the discovery of similar sources, including water at 1.35 centimeters, and weak emission from more complex molecules like methyl alcohol, astronomers concluded that they were observing cosmic masers. These small objects (comparable in size with the solar system) occur in larger dust clouds which radiate at infrared wavelengths. They are thought to be PROTOSTARS condensing from the interstellar medium.

At the other end of the stellar lifespan, red giant stars show maser action from a number of molecules, including hydroxyl, water, and silicon monoxide (SiO). This is believed to arise in a shell of dust and gas ejected from the star as it dies.

Maskelyne, Nevil (1732–1811)

Fifth British astronomer royal, who in 1766 began publication of the *Nautical Almanac,* a set of tables containing astronomical information for navigators. Maskelyne was sent in 1761 to the island of St. Helena to observe the transit of Venus across the Sun. Cloud prevented observations, but during the voyage Maskelyne tested a means of finding longitude at sea called the *method of lunar distances.* This involved measuring the Moon's position against the background of stars (the lunar distances were the stars' distances from the Moon). The Moon's position in its orbit acted as a standard clock in the sky, to give a universal time scale (like modern Greenwich time) that could be read off from tables. Comparing this standard time with the local time revealed the observer's longitude. But the success of the method depended on the accuracy of lunar tables. Maskelyne's trials showed the suitability of tables prepared by the German astronomer Johann Tobias Mayer (1723–1762), and in 1763 Maskelyne wrote *The British Mariner's Guide* to describe the technique. Maskelyne also helped test a rival "standard clock"—the chronometers of the English instrument maker John Harrison (1693–1776). Despite their greater accuracy the cost of chronometers meant that the method of lunar distances was long preferred, and the *Nautical Almanac* contained tables of lunar positions.

Eleven years later, Maskelyne began publication of the regular series of Greenwich Observations— previous astronomers royal had kept their observations to themselves. His catalog of 1790 showed the PROPER MOTIONS of 36 stars, from which William HERSCHEL determined the Sun's motion in space. Maskelyne successfully observed the 1769 transit of Venus, and from it calculated the Sun's distance to an accuracy of within one percent. In 1774 he performed a celebrated experiment to measure the density of the Earth. By measuring the deflection of a plumbline hung near the Schiehallion, a mountain in Perthshire, Scotland, he found a value for the Earth's density of 4.5 times that of water—lower than the true value of 5.5, but the best result of the time.

mass-luminosity relation

The relationship between a star's mass and its brightness, discovered by Sir Arthur EDDINGTON in 1924. The masses of stars are known reliably only from observations of the orbits of double-star systems. These masses bear a direct relation to the stars' intrinsic brightness or absolute magnitude. The relationship indicates that the biggest stars are also the hottest; a doubling of mass corresponds to an increase in brightness of 2.5 magnitudes. This mass-luminosity relation applies to most ordinary stars. The highly condensed white dwarfs, however, do not conform because of their exceptional nature; nor do the faint companions in certain types of double-star systems (known as the W Ursae Majoris stars). These latter may be the central cores of red giants, stripped of their outer layers by the gravitational attraction of their companion stars.

Mauna Kea Observatory

The world's highest astronomical observatory, operated by the University of Hawaii at an elevation of 13,800 feet (4,205 m) on Mauna Kea, a dormant volcano on the island of Hawaii. Opened in 1970, its main telescope is an 88-inch (224-cm) reflector; there are also two 24-inch (61-cm) reflectors. Mauna Kea Observatory has made valuable studies at infrared wavelengths; its altitude places it above most of the water vapor in the atmosphere that absorbs infrared radiation, and installation of a 127-inch (323-cm) infrared telescope is planned for 1977. France and Canada are cooperating to put a 144-inch (366-cm) telescope on Mauna Kea, scheduled for opening in 1978. Britain is building a 152-inch (386-cm) infrared telescope to be erected on the same site in the late 1970s.

McDivitt, James Alton (b. 1929)

Commander of the Gemini 4 mission in June 1965, during which astronaut Edward WHITE became the first American to walk in space. McDivitt, a qualified aeronautical engineer, also commanded the Apollo 9 mission in March 1969, which included the first manned flight of the Lunar Module; the success of the mission, which was confined to Earth orbit, assured eventual success of the Moon-landing program. In 1972 McDivitt left the space program for private industry.

mean Sun

A fictitious body representing the annual motion the true Sun would have around the CELESTIAL SPHERE if the Earth's orbit were circular instead of elliptical, and if its axis were not tilted but upright. The mean Sun therefore progresses along the celestial equator at a uniform rate; the true Sun appears to move along the ecliptic at a varying rate (see EQUATION OF TIME).

Mensa (the table mountain)

A faint constellation near the south pole of the sky, containing part of the Large Magellanic Cloud. It was introduced by Nicolas Louis de Lacaille.

Mercury

The closest planet to the Sun, Mercury is an arid, airless world with a meteorite-battered surface baked to above the melting point of lead at midday, and cooled to bitter temperatures at night. It can never have supported life and more closely resembles the Moon than any of the other planets.

Physical nature. Mercury is 3,032 miles (4,880 km) in diameter, intermediate in size between the Moon and Mars. Mercury is remarkably dense, with a mass of 3.30×10^{20} tons and a density of 5.45 times that of water, very nearly the density of the Earth. It rotates in 58.65 days about an axis perpendicular to the ecliptic. Mercury's orbit is unusually elliptical, taking it between 28.6 and 43.4 million miles (46 and 69.8 million km) from the Sun. The orbit is inclined at 7° to the ecliptic, and Mercury completes one circuit in 87.97 days. The position of Mercury's perihelion, the axis of the orbit, advances slowly around the Sun some 43 seconds of arc more per century than can be accounted for by planetary perturbations alone. This additional movement has been successfully accounted for by the theory of relativity, as a consequence of the CURVATURE OF SPACE near the Sun.

Visibility. Mercury shows a series of phases as it orbits the Sun. On the far side of the Sun, the tiny disk is seen nearly fully illuminated; as Mercury moves around into the evening sky it shows a decreasing gibbous phase until it reaches half phase at greatest elongation. As the planet starts to move between us and the Sun, its apparent size increases as it approaches us and its phase becomes an increasingly slender crescent. After passing between the Earth and Sun, the reverse sequence of phases follows as the planet emerges into the morning sky.

Occasionally, Mercury can be seen in TRANSIT across the solar disk, appearing as a slow-moving dark spot. This happens relatively rarely, because of the relative inclinations of the orbits of Mercury and the Earth. In transit Mercury is at its closest to the Earth, as little as 48 million miles (77 million km) distant. The last transit of Mercury took place on November 11, 1973; the next will occur on November 12, 1986, and November 14, 1999.

Mercury never strays far from the Sun in the sky. Its greatest apparent angle from the Sun varies from 28°, when the planet is at aphelion, to only 18° at perihelion. This makes the planet very difficult to observe, despite its maximum brightness of magnitude −1. Mercury can only be viewed against a bright, daytime sky, or low down after sunset or before sunrise.

Surface markings and conditions. Experienced observers detected faint dusky shadings on Mercury's disk, and careful study of these suggested that Mercury rotated on its axis every 88 days, with the same face permanently pointing toward the Sun—one hemisphere in perpetual sunlight, the other in perpetual night. This belief persisted for more than 80 years, until radar studies from 1965 onward showed that the correct rotation period was 58.65 days. This is exactly two-thirds of the 87.97-day orbital period and leads to a Mercurian day (from sunrise to sunrise) lasting 176 Earthdays, or Mercury's orbital periods. Another unusual effect is that at perihelion, the planet's rotation is slower than its motion around the Sun, so that the Sun ceases its slow westward movement, slipping back eastward for some hours before resuming its normal motion.

When Mercury is at perihelion, the Sun appears 50 percent larger than at aphelion, and shines with more than twice the intensity. The equatorial surface temperature reaches 415°C at perihelion, enough to melt lead and tin; even at aphelion it is 285°C. In sharp contrast, the dark-side temperature is as low as −175°C. This great temperature range between day and night is partly the result of an extremely thin atmosphere, a million millionth the density of the Earth's at sea level, or only ten times denser than the

This mosaic of Mariner 10 photographs of Mercury, taken from 124,000 miles, shows a heavily cratered terrain very similar to the Moon's highlands. The largest craters are about 120 miles in diameter. The brightest crater on Mercury, **center**, has been named after the late Gerard P. Kuiper, an eminent planetary scientist.

Moon's. The surface composition is very similar to that of the Moon, with a 7 percent reflectivity (ALBEDO), identical polarization, and a radar reflectivity like the Moon's. The faint dusky shadings and bright areas are the equivalents of the lunar maria and highlands.

Surface features. Previously sketchy information about Mercury was dramatically augmented by the three close passages of Mariner 10 to the planet in 1974 and 1975. Over 2,800 photographs were taken, including some with a resolution of only a hundred yards.

The photographs reveal a heavily cratered surface more closely resembling the Moon than Mars. The lack of erosion suggests that Mercury has been devoid of an appreciable atmosphere for practically its whole existence. Craters abound everywhere on the surface, from giant basin structures down to small craters only a few hundred yards across at the limit of the photographs. A study of about 25 percent of the surface has revealed no less than 17 impact basins more than 125 miles (200 km) across, and some larger than the mighty Mare Imbrium, the 420-mile (675-km) basin on the Moon. The largest, some 800 miles (1,300 km) across, has been named the Caloris Basin and contains a heavily fractured, lava-flooded floor.

Like the Moon, Mercury has two distinct hemispheres. One is covered with heavily cratered highlands, while the other contains smooth, lightly cratered plains similar to the lunar maria. The craters on Mercury and the Moon are similar, varying from sharply defined recent craters surrounded by bright halos and ray systems, to very ancient eroded craters distorted by later impacts. However, Mercurian craters tend to be shallower than their lunar counterparts. Central peaks caused by upward movement of underlying layers, or terracing of the crater walls caused by slumping, are present in most craters larger than about 9 miles (14 km) across,

A Mariner 10 closeup of Mercury's surface, from a distance of 54,000 miles, reveals craters down to a diameter of one mile; the brightest crater is 19 miles across. The large, old craters on the upper left seem to have been flooded by lava, like the lunar maria.

while they are not common in lunar craters less than 35 miles (56 km) across. The ejecta blanket and secondary craters surrounding large craters extend outward only half the distance normal on the Moon. All these effects are undoubtedly caused by the higher gravity on Mercury, more than twice that on the Moon.

The mare areas are very similar to those on the Moon, with similar amounts of cratering. A unique feature on Mercury, however, are the irregular scarps up to 2 miles (3.2 km) high, which can extend for several hundred miles. These are quite unlike anything on the Moon and cut through whatever lies in their path—basins, craters, or maria. Another highly unusual feature, a weird terrain of finely dissected hills, large rough valleys, heavily degraded craters, and great landslides, is found opposite the great Caloris Basin. Similar terrain on a much smaller scale has also been found opposite the Imbrium and Orientale basins on the Moon. Evidently, the basin impacts produced intense shock waves in the body of the planet which met at the antipodal point, causing great upheavals.

The detailed nature of the surface is probably identical with what we know about the Moon—a surface soil layer (REGOLITH) perhaps 50 to 100 feet (15–30 m) deep comprising rock debris of all sizes embedded in a glassy soil, with an underlying rock mass shattered by impacts. The high overall density of Mercury suggests a central iron-nickel core extending some 80 percent of the distance from the center toward the surface, with an overlying crust of silicate rocks. Mariner 10 detected a magnetic field around Mercury about one percent the intensity of the Earth's field. It is probably caused by the permanent magnetization of the surface rocks of the planet.

Evolution. The sequence of events in Mercurian history parallels the evolution of the Moon. After the planet formed, the surface was melted by the heat of intense meteoric bombardment. As this died away, the highland regions solidified. Further impacts scattered debris over the entire surface, leaving the craters and basins. Subsurface radioactive melting then filled some of the great basins with lavas to produce the mare areas. The surface was then further modified by smaller meteorite impacts. Internally, Mercury still possesses a partially molten core. The gradual cooling of the liquid core must have led to considerable contraction, and this may be the cause of the great scarps—the wrinkles that lace the planet's surface.

Mercury has no moon. Mariner 10 failed to detect any satellites as small as 3 miles (5 km) across.

Mercury project

The American program to launch a man into space. Its origins go back to experiments during the 1950s with high-altitude rocket planes, the X-1 and X-2 series that took men toward the edge of the Earth's atmosphere. From these developed the famous X-15 rocket plane, which during the 1960s established records for altitude—67 miles (108 km)—and speed—4,520 miles (7,274 km) per hour. Although proposals were made for boosting a winged vehicle like X-15 into orbit—similar to the concept of the SPACE SHUTTLE—it was decided in 1958 to extend the flight program to higher altitudes and greater speeds by launching an astronaut in a wingless capsule carried aloft by a rocket. By contrast, the Soviet manned space program was an ad hoc adaptation of large unmanned satellites, already well-advanced in design.

The Mercury project set itself the target of putting a man in orbit, testing his ability to function in spaceflight, and returning him safely to Earth. At the time the scheme was proposed, no one knew how well humans could withstand the rigors of launch, followed by weightlessness and the deceleration of reentry. Therefore the Mercury project was preceded by test firings carrying primates, whose successful flights on so-called suborbital paths, (dropping just short of orbit) and then into orbit itself, paved the way for the men who were to follow.

Mercury hardware. The Mercury capsule introduced the conical shape followed by Gemini, Apollo, and even the Soviet Soyuz spacecraft (the first Russian design, Vostok, was a sphere). Mercury was designed to hold one man. Its total length was 9 feet 7 inches (2.9 m) and its maximum diameter 6 feet 2 inches (1.9 m); it weighed nearly 3,000 lb. (1,360 kg). At its blunt end was the heat shield, to which was strapped a pack of retro-rockets which were jettisoned before reentry. Once the craft had entered the atmosphere and the parachutes began to deploy from the nose, the heat shield dropped down to extend an air bag; this cushioned the impact of splashdown, and then slowly filled with water to stabilize the Mercury capsule in the sea.

The astronaut sat in a contoured couch. Above him was an observation window. The astronaut wore a simple space suit, developed from high-altitude

The launch of a Mercury capsule by an Atlas booster. The capsule is at the top, just below the escape tower which could pull it clear of the main booster during the first few minutes of the flight in case of danger. The controlling vernier rockets can be clearly seen maintaining the rocket attitude.

aviation suits, to be inflated only in an emergency. He had a hand controller which fired gas jets to adjust the spacecraft's attitude in space, although orbital changes were not possible as they were with the later GEMINI. An escape tower above the spacecraft could pull the capsule free in the event of a launch rocket malfunction; the escape tower was jettisoned once the capsule was safely on its intended course.

The first two manned Mercury flights were suborbital lobs into the Atlantic Ocean. These flights followed a long, looping path into space and back again. The astronauts were able to control the orientation of the spacecraft, but they became weightless for only about five minutes. The two suborbital flights were launched by REDSTONE rockets, modified from their original role as intermediate-range missiles. Four other manned Mercury capsules were sent into orbit atop ATLAS ROCKETS, America's first long-range ballistic missiles.

The missions. Alan B. Shepard became the first American astronaut on May 5, 1961, when he was launched on a 304-mile (489-km) flight lasting 15 minutes 22 seconds and reaching a maximum altitude of 116.5 miles (187.5 km). On July 21, 1961, Virgil I. Grissom made a similar flight, lasting 15 minutes 37 seconds, reaching a maximum altitude of 118 miles (190 km). On splashdown, the capsule's hatch was ejected by accident, and the capsule itself sank, although Grissom was safely recovered. The first American in orbit was John H. Glenn, who circled the Earth three times on February 20, 1962. The flight was repeated three months later by M. Scott Carpenter. In October 1962 Walter M. Schirra made six orbits of the Earth, and in May 1963 L. Gordon Cooper made the last and by far the longest Mercury flight, 22 orbits (34.3 hours) in length.

Mercury flights

Mission	Launch date	Results
Mercury-Redstone 3 (*Freedom 7*)	May 5, 1961	Alan Shepard made suborbital flight
Mercury-Redstone 4 (*Liberty Bell 7*)	July 21, 1961	Virgil Grissom made suborbital flight
Mercury-Atlas 6 (*Friendship 7*)	February 20, 1962	John Glenn made 3-orbit flight
Mercury-Atlas 7 (*Aurora 7*)	May 24, 1962	Scott Carpenter made 3-orbit flight
Mercury-Atlas 8 (*Sigma 7*)	October 3, 1962	Walter Schirra made 6-orbit flight
Mercury-Atlas 9 (*Faith 7*)	May 15, 1963	Gordon Cooper made 22-orbit flight

meridian

The great circle passing overhead and through the observer's north and south horizon. Celestial objects attain their greatest altitude (*culmination*) on the meridian. The moment at which a celestial object crosses the meridian depends on the observer's longitude. One degree of longitude amounts to 4 minutes' difference in time. The timing of objects crossing the meridian, or in TRANSIT, therefore provides a basis for time measurement or determination of longitude.

Messier, Charles (1730–1817)

French astronomer best known for compiling a famous list of nebulae and star clusters. Messier became assistant to the French astronomer Joseph-Nicolas Delisle (1688–1768) in 1751; his interest was turned to comet hunting in 1758 with the predicted return of Halley's comet, which Messier was one of the first to see. In all, Messier discovered 15 new comets and claimed another six that were actually found to be first seen by others. In 1758 Messier was deceived by the comet-like appearance in his telescope of the Crab nebula, and made a special note to avoid it. From that point onward he began cataloging fuzzy-looking objects in the sky that might be mistaken for comets. Messier produced his first list, containing 45 objects, in 1771; a second list in 1780 added another 23 objects; and his final list of 1781 (published in 1784) had 103 objects. Not all were Messier's own discoveries; several had been seen first by other observers, notably the Swiss comet-hunter Jean Philippe Loys de Chéseaux (1718–1751). Another six objects were added to Messier's list by the French astronomer Pierre François André Méchain (1744–1804). Astronomers still refer to objects by their Messier, or M, numbers: for example, M1 (the Crab nebula) and M31 (the Andromeda galaxy).

meteor

The streak of light produced when a solid particle from space, known as a METEOROID, enters the Earth's atmosphere at high speed and burns up to produce fine dust. Meteors as bright as naked-eye stars are known popularly as *shooting stars*, while brighter meteors are termed FIREBALLS or bolides. Meteors occur between heights of about 55 and 65 miles (88 and 105 km). The majority of them have velocities of about 20 or 35 miles (32 or 56 km) per second, depending on whether they overtake the Earth or the Earth overtakes them. Significantly, no meteors have velocities clearly in excess of the escape velocity for the solar system, which indicates that meteors are solar system members.

Many meteors leave behind them a glowing train, which may persist for several seconds or even minutes. These occur at 55 to 60 miles (88–97 km) and are caused by the recombination of ionized atoms and molecules. Very bright daytime meteors or fireballs that penetrate to low altitudes can leave dark persistent trains behind them. These occur at 20 to 25 miles (32–40 km) and are suspended dust particles.

Sporadics and shower meteors. Most naked-eye meteors are lone travelers, termed *sporadic* meteors. The number of sporadics visible per hour rises from 6 after sunset to a peak of about 14 just before dawn, when the observer is facing in the direction of the Earth's motion. On several occasions each year, large numbers of meteors are seen moving outward from a RADIANT point in a meteor shower. These meteors orbit the Sun in parallel paths at the same velocity, forming a meteor stream. There are about 10 annual meteor streams, plus many minor streams of low

Characteristics of the Major Meteor Streams and their Orbits

Stream	Period of Visibility	Date of display peak	Radiant R.A. h m	Point Dec. o	Average peak rate meteors/ hour	Velocity miles/sec.	Perihelion distance a. u.	Stream orbit Inclination o	Period years	Associated comet
Quadrantids	January 2–5	January 4	15 28	+50	110	24.1	0.98	72.9	6.3	none
Lyrids	April 20–24	April 22	18 08	+32	8	29.2	0.92	79.5	200	Thatcher 1861 I
eta Aquarids	May 2–7	May 5	22 24	00	18	41.3	0.70	158.0	80	Halley
delta Aquarids	July 22–Aug. 10	July 31	22 36	−08	30	25.2	0.085	22.2	4.2	none
Perseids	July 27–Aug. 16	August 12	03 04	+58	65	37.0	0.95	113.0	140	Swift-Tuttle 1862 III
Orionids	October 17–25	October 21	06 24	+25	25	40.3	0.58	164.3	80	Halley
Taurids	Oct. 25–Nov. 25	November 8	03 44	+18	10	18.3	0.31	4.5	3.1	Encke 1786 I
Leonids	November 16–19	November 17	10 08	+22	15	43.6	0.98	162.1	33.3	Tempel 1866 I
Geminids	December 7–15	December 14	07 28	+32	55	21.7	0.129	24.9	1.63	none

activity almost indistinguishable from the background of sporadic meteors.

Meteors and comets. A meteor stream is formed by particles scattered by a periodic comet; the stream and comet occupy the same orbit. The stream is initially a compact, dense swarm of meteoroids, but slowly spreads out, eventually disintegrating completely to add to the stock of sporadic meteors. Very young streams give intense displays of up to 1,000 meteors per minute lasting for only a few hours at the longest. Such encounters, although rare, occurred in the Leonid meteor storms of 1799, 1833, 1866, and 1966, the Andromedid storms of 1872 and 1885, and the Giacobinid storms of 1933 and 1946. Meteor streams are named for the constellation in which their radiant lies, or for their parent comet. Meteor showers are best observed when their radiant is high in the sky. The ZENITHAL HOURLY RATE of a shower is the predicted rate with the radiant overhead; but because the radiant seldom is directly overhead, the actual observed rate is somewhat lower.

Observing meteors. The basis of modern meteor astronomy was laid down a century ago by naked-eye observers who recorded meteor paths, velocity, and brightness. An outstanding pioneer was the English amateur William Frederick Denning (1848–1931). Far more accurate photographic methods were later applied, culminating in the super-Schmidt cameras designed specifically for meteor work. With focal ratios of $f/0.65$ and a field of 56°, these can photograph meteors as faint as magnitude +4 at the rate of two every hour. Radio waves are reflected from the ionization path left by a meteor, and meteors as faint as magnitude +10 can be recorded by radar. This method has detected several meteor streams whose radiants are in the daylight sky and can never be observed visually.

By using all these techniques, meteors from magnitude −20 down to +10 can be observed, corresponding to bodies with masses from 10 tons down to 0.0001 gram. Most meteors brighter than magnitude −12 are sporadics, and a few of these are sufficiently massive and strong to withstand entry into the Earth's atmosphere to drop METEORITES. Meteor streams account for many fainter meteors—about

40 percent of meteors recorded photographically (mean magnitude −2) and 35 percent of naked-eye visual meteors (mean magnitude +2). However, shower meteors make up only a few percent of the faintest objects, as detected by radar and telescope (mean magnitude +8). Faint meteors are, of course, more numerous than bright ones. For ordinary naked-eye meteors, the numbers increase by 3.5 times for each magnitude step downward, while very faint meteors increase by a factor of 2.5.

Over a year, the Earth encounters about 10 trillion meteors brighter than magnitude +10, gaining some 2,000 tons of material. This is insignificant, however, when compared to the 200,000 tons added each year by micrometeorites and meteors fainter than magnitude +10.

The spectra of meteors reveal that their composition is generally similar to either stony or iron meteorites. Although for many years meteors were thought to be fragile dust-balls, many scientists now believe that most are composed of a very fragile form of gas-rich carbonaceous CHONDRITE, a type of stony meteorite. Ordinary stream meteors are clearly derived from the outer layers of young periodic comet nuclei, while fireballs may be scattered boulders from the disrupted nuclei of ancient periodic comets or debris from asteroid collisions.

meteorite

A lump of rock or metal from space that crashes to Earth. The study of such material is known as *meteoritics.* Accounts of pieces of stone or iron falling from the sky occur in the records of many early civilizations, but the earliest witnessed fall from which material still survives is the stone that fell near Ensisheim, Alsace, on November 16, 1492. The true nature of meteorites was first recognized by the German physicist Ernst Florens Friedrich Chladni (1756–1827) in 1794, but his views were only accepted after the fall of meteorites at L'Aigle, France, on April 26, 1803, documented by the French physicist Jean Baptiste Biot (1774–1862).

Frequency of falls and finds. A meteorite actually seen to hit the ground is known as a *fall;* a meteorite which is come across accidentally, having fallen at some undetermined time in the past, is known as a *find*. About 2,000 falls and finds have now been recovered, with roughly six falls and 10 finds being added each year. An estimated 500 meteorites fall each year over the whole surface of the Earth. Of these, 300 drop into the oceans and seas, while the bulk of the remainder fall either in uninhabited areas, or in places where they are not noticed. An increasing number have damaged buildings and other man-made structures, but there have been very few cases of injuries recorded.

Meteorites are broadly divided into three types: stones or aerolites, which are subdivided into CHONDRITES and achondrites; irons or SIDERITES, subdivided into hexahedrites, octahedrites, and ataxites; and stony-irons or SIDEROLITES, subdivided into pallasites and mesosiderites. The relative proportions of meteorites is shown below:

Distribution of Meteorite Types

Meteorite type	Finds %	Falls %
chondrites	38.6	84.8
achondrites	0.9	7.9
total aerolites (stones)	39.7	92.7
siderolites (stony-irons)	5.8	1.7
siderites (irons)	54.5	5.6

Although aerolites are the most common meteorites that fall to Earth, they weather rapidly and are difficult to trace long after a fall. By contrast, the rarer iron meteorites resist weathering, and can thus be found long after they have fallen.

Size of meteorites.   Small meteorites tend to be stones, but the proportion of irons increases with size, presumably because large irons more easily withstand collisions in space and entry into the Earth's atmosphere. The largest stone meteorite weighs just over 1 ton and was part of the Norton County, Kansas, achondrite fall of February 18, 1948. There are many irons much larger; the biggest is the ataxite Hoba, found in 1920 near Grootfontein in South West Africa (Namibia). This is the largest meteorite in the world, weighing over 60 tons, and has never been moved.

Fall phenomena.   Meteorite-producing bodies are heated by friction as they enter the Earth's atmosphere. The meteor produced is extremely bright, and fireballs with a peak brilliance of more than magnitude −22 have been recorded. During entry, the surface of the meteorite is heated to several thousand degrees; the stone or iron is vaporized and melted away more quickly than the heat penetrates into the meteorite, and the interior is unaffected. The meteorite slows to free fall at a height of about 10 to 15 miles (16–24 km) when the fireball fades; the meteorite then drops at about 150 to 200 miles (240–320 km) per hour, taking about $1\frac{1}{2}$ to 2 minutes to reach the ground. The meteorite arrives at ground level quite cold to the touch, and coated with a dull black fusion crust where the once-molten outer layer has cooled. Most meteorites break up into a number of large pieces during the fireball phase of descent. These fragments fall to the ground in an elliptical area, the heaviest pieces carrying the farthest; this is known as a

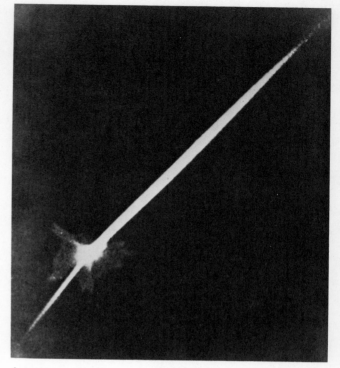

A meteor exploding as it ploughs through the Earth's atmosphere at a speed of about 30 miles a second (50 km/sec). These stony or metallic particles of interplanetary debris are usually only millimeters in size; this extremely bright example may have been caused by a chunk several centimeters across, weighing a few grams.

dispersion ellipse. Though most meteorites produce only a few fragments, in rare cases hundreds of individual meteorites are produced. The Allende meteorite fall of February 8, 1969, in Chihuahua, Mexico, scattered an estimated 5 tons of chondrite meteorite specimens up to 230 lb. (105 kg) in weight over an area some 30 by 4 miles (48 × 6.4 km).

Despite recent intensive efforts using networks of cameras to follow the passage of the bright fireballs that lead to meteorite falls, only two have so far been recorded and their orbits calculated. These are the chondrite falls of April 7, 1959, near Pribram, Czechoslovakia, and of January 3, 1970, near Lost City, Oklahoma.

The origins of meteorites.   Meteorite specimens recovered on Earth and preserved in museum collections are not representative of interplanetary debris. These meteorites represent only the 5 percent of debris which is strong enough to survive entry into the Earth's atmosphere. The other 95 percent disintegrate to produce flaring fireballs. These are thought to be composed of two different types of fragile material possibly similar to the carbonaceous chondrites. This material may come from comets, while most aerolites, siderolites, and siderites are believed to be fragments of asteroids. Meteorites contain a number of minerals not found on Earth, but these are always composed of known elements. Organic molecules have been found in some meteorites, presumably produced by chemical processes in space (see also LIFE IN THE UNIVERSE).

meteoroid

A solid object moving in interplanetary space, larger than a single molecule but smaller than an asteroid. Meteoroids smaller than about 5 microns (5 millionths of a meter) are decelerated in the Earth's atmosphere to fall as MICROMETEORITES. Meteoroids between 5 microns and about 6 inches (15 cm) in diameter are destroyed by ablation (frictional vaporization and melting) in the Earth's atmosphere, producing METEORS. Meteoroids larger than 6 inches, if they possess sufficient structural strength, can fall to earth as METEORITES.

meteorological satellites

See WEATHER SATELLITES

Metonic cycle

The period after which the Moon's phases recur on the same day of the year. The cycle was discovered in 432 B.C. by the Greek astronomer Meton (c.460 B.C.—?). The period lasts just over 6,939 days (exactly 19 years), during which there are 235 LUNATIONS. The Metonic cycle was used as a basis for lunar calendars.

Michelson, Albert Abraham (1852–1931)

Polish-born American physicist, who made a series of remarkably accurate measurements of the speed of light, and was the first to determine the diameter of a star other than the Sun. Michelson made his first measurement of the speed of light in 1878, and continued his experiments for the rest of his life. He obtained his best value, 186,285 miles (299,798 km) per second in 1926, over a 22-mile (35-km) path between Mount Wilson and Mount San Antonio in California. In 1881 Michelson began a series of tests culminating in the famous MICHELSON-MORLEY EXPERIMENT of 1887, which proved that light travels at a constant speed in space and refuted the ether hypothesis. In 1893 Michelson measured the length of the standard meter in terms of the wavelength of red light from heated cadmium, thus providing a precise standard of length that could be reproduced exactly. At Mount Wilson Observatory, Michelson set up an INTERFEROMETER to measure the diameters of several nearby giant stars in 1920. He used mirrors to feed two beams of light from a star onto the main mirror of the 100-inch (254-cm) telescope, which focused the light; the interference fringes produced by the two beams revealed the diameters of the star, the first time such a direct measurement had been made. In 1907 he was awarded the Nobel Prize in physics for his painstaking optical measurements. He was the first American scientist to receive the honor.

Michelson-Morley experiment

An attempt to measure the Earth's motion through the ETHER, carried out in 1887 by the American physicists Albert MICHELSON and Edward Williams Morley (1838–1923). According to the theory prevailing at that time, light moved at constant speed in an invisible substance called the ether. The Earth's motion through the ether would produce a kind of wind, so that the speed of light measured perpendicular to the Earth's direction of motion would be different from that measured in line with the Earth's direction of motion. Michelson and Morley compared the travel time of two light beams sent along optical paths at right angles to each other. Because of the predicted "ether wind" it was expected that the beam directed along the Earth's line of motion would return later than the perpendicular beam. In fact, no detectable difference was found, a result of stunning unexpectedness, which virtually invalidated the ether theory. The experiment has been repeated many times with increasing precision, but no difference in travel time has ever been detected. The result is consistent with the theory of RELATIVITY, which abandons the idea of any "fixed" medium in space and instead holds that the speed of light is constant relative to any observer.

micrometeorite

A particle from space which is small enough to be decelerated in the Earth's atmosphere without being vaporized. Smaller than 5 microns (5 millionths of a meter) across, micrometeorites fall gently to Earth at a daily rate of about 50 particles per square yard. Micrometeorites have been detected in ocean sediments, polar ice caps, and on clean exposed surfaces at ocean sites. They have also been collected by rockets and satellites above the Earth's atmosphere. Micrometeorites are probably derived from the periodic comets which release about 100,000 tons of fine dust a year into the solar system.

Microscopium (the microscope)

A small and faint constellation lying next to Sagittarius in the southern hemisphere of the sky, introduced by Nicolas Louis de Lacaille.

microwave background

See BACKGROUND RADIATION.

Milky Way

A spiral galaxy of some 100 trillion stars, of which the Sun is one. The Milky Way is a thin disk 100,000 light-years across and 2,000 light-years thick; the disk widens near its center into a flattened nuclear bulge. The Sun lies in the disk, about 30,000 light-years from the center. Surrounding the disk is a sparsely populated, spheroidal halo, up to 500,000 light-years in diameter. The three regions (halo, nucleus, and disk) contain very different types of objects.

Halo.   The halo represents the original extent of the gas cloud from which the Milky Way condensed 10 billion years ago. Rotation made it collapse into a disk, which now forms the most densely-populated part of the Galaxy; but some stars condensed out of the gas before the collapse occurred. Halo stars are therefore the oldest in the Galaxy. Many of them are found in the GLOBULAR CLUSTERS that are distributed symmetrically about the Galaxy, with a marked concentration toward the galactic center in Sagittarius. This distribution is shared by the RR LYRAE VARIABLE stars, found both in and out of globular clusters. Their short periods of variability (less than one day) and similar intrinsic brightness (ABSOLUTE MAGNITUDE +0.5) make them valuable probes for determining the structure of the halo. Non-cluster stars make up the greatest part of the halo's mass. Some areas of neutral hydrogen gas (H I REGIONS) still remain in the halo, distributed in

clouds which appear to be falling onto the disk at high velocities (up to 110 miles, or 177 km, per second); the reason for this is not known.

Nucleus. Apart from its outer margins, the nucleus can only be observed at radio or infrared wavelengths, which penetrate the dust clouds that obscure it. The nucleus measures 20,000 by 10,000 light-years, and contains most of the mass of the Galaxy, largely in the form of old stars (although younger than the halo population). Toward the galactic center these stars crowd together many thousands of times more densely than those near the Sun.

A thin, rotating disk of hydrogen extends 2,500 light-years from the center, containing a mass equivalent to 5 million Suns in gas, and probably 1,000 times more in stars. Many millions more stars lie within this zone, which also contains giant clouds of ionized hydrogen gas (H II REGIONS) nearer the center. The exact center of the Galaxy is marked by clouds of hot hydrogen and a point source of intense infrared radiation, less than one-third of a light-year across. This is possibly a supermassive object, 10,000 times heavier and 300,000 times brighter than the Sun; or it may be a BLACK HOLE.

Disk. The youngest stars in the Galaxy are concentrated in a thin plane surrounding the nucleus. The most recently-formed objects occupy the thinnest zone, while the older disk stars gradually merge with those of the halo, marking the progress of the Galaxy's collapse from a spherical gas cloud. Over 1 percent of the mass of the Galaxy is still in the form of cool hydrogen gas lying in the disk and not yet condensed into stars. Neutral hydrogen emits strong radio signals at a wavelength of 21 centimeters, enabling radio astronomers to map both its location and velocity. The gas forms a layer of uniform thickness close to the galactic plane in the inner regions, but becomes thicker and distorted at the rim, possibly due to tidal effects from the MAGELLANIC

Our Milky Way Galaxy, seen on a chart prepared by plotting the 7,000 brightest stars in the sky, and mapping in the distribution of nebulae from photographs. Gas and star clouds concentrate toward its nucleus, **center**, crossed by dark obscuring dust lanes, **left center**. The Magellanic Clouds, our companion galaxies, look like detached portions, **below right**.

CLOUDS. Within the uniform layer, the hydrogen concentrates into subclouds along two spiral arms winding outward from the nucleus. Three portions of these arms can be seen; the inner Sagittarius arm; the Orion arm, in which our Sun and the stars of Cygnus lie; and the Perseus arm, 6,000 light-years beyond the Sun. We see the stars of these arms superimposed on each other in the sky, forming the diffuse band of the Milky Way.

The Galaxy's spiral arms are rich in the gas from which stars form, and contain many hot, bright young stars. The clumpiness of the gas produces STELLAR ASSOCIATIONS, some of which are only a few million years old. These stars heat the hydrogen to temperatures above 10,000°K, forming regions like the ORION NEBULA. Interstellar clouds also contain quantities of dust particles, only a few millionths of a centimeter across. These consist of silica and graphite coated with ice, and are believed to provide surfaces for the formation of INTERSTELLAR MOLECULES. Dust grains are distributed throughout the disk, as well as in dark clouds like the COALSACK, absorbing and scattering starlight. As a result, optical telescopes can only penetrate about 6,000 light-years in the galactic plane, so that much of our information about the spiral arms comes from radio studies. A general background of galactic radio emission is also observed, arising from the acceleration of COSMIC RAY electrons in a large-scale magnetic field (SYNCHROTRON RADIATION) which follows the spiral arms.

The new Mills Cross has trough-like cylindrical arms each a mile long. The east—west arm, **right**, can be tilted to steer its beam to different declinations in the sky; the north—south arm, **far left**, is fixed, but its beam can be "steered" electronically.

The entire Galaxy rotates, with the inner parts moving faster than those farther out. This differential rotation indicates that the Galaxy's mass is mainly concentrated toward its center. The Sun, traveling at 170 miles (274 km) per second, takes between 200 and 250 million years to make one circuit.

Mills Cross
A type of radio telescope invented in 1953 by the Australian astronomer Bernard Yarnton Mills (b. 1920). The Mills Cross design uses two rows of aerials arranged at right angles to give a large cross-shaped array. The pattern of vision of each line of aerials is a narrow fan, but automatically switching between the two crossed aerial systems produces a narrow beam where the two fans overlap. The Mills Cross was an important forerunner of the APERTURE SYNTHESIS telescopes used to map fine details of radio sources.

Mimas
Saturn's second satellite in order of distance from the planet, with a diameter of about 300 miles (500 km). Mimas orbits Saturn every 22 hours 37 minutes at a distance of 115,320 miles (185,590 km). It was discovered in 1789 by William Herschel.

minor axis
The shortest diameter of an ELLIPSE, perpendicular to the MAJOR AXIS.

minor planet
See ASTEROID.

Mira
The first variable star to be discovered; its name means "wonderful." Mira is a red giant star in Cetus, 820 light-years away, and is the prototype of long-period variables. Mira was first noted by David FABRICIUS in 1596, but soon faded. It was bright again in 1603 when Johann BAYER labeled it Omicron Ceti on his star map. Mira has a period of roughly 332 days, expanding and contracting between 2.8 and 3.8 astronomical units in diameter. At its brightest, Mira is of the second magnitude, but it can fade to tenth magnitude. It has a blue dwarf companion orbiting it

every 14 years at a distance of some 40 a.u. During its expansion, Mira puffs off a cloud of gas which falls onto this companion, changing its brightness.

Miranda
The nearest satellite of Uranus, discovered in 1948 by Gerard Kuiper. It orbits Uranus every 33 hours 55$\frac{1}{2}$ minutes, at a distance of 81,060 miles (130,450 km). Miranda is about 350 miles (550 km) in diameter.

mirror
A reflecting component in an astronomical instrument. Unlike an ordinary mirror, those used for precise optical purposes have their reflective coating on the front surface rather than the back. This avoids the optical imperfections that would be caused if the light actually passed through the glass. The reflective coating is therefore relatively unprotected. The modern technique is to evaporate a very thin film of aluminum onto the mirror, and to coat it with a very thin layer of transparent silica; this can produce a surface lasting several years before it becomes tarnished and corroded. Mirrors have the advantage over lenses that they do not produce false color, or CHROMATIC ABERRATION. With the advent of photography this became particularly important, because most photographic emulsions were primarily sensitive to one special color, and a lens system could not be corrected to perform properly for all colors of the spectrum.

Furthermore, aluminum reflects very well in the ultraviolet part of the spectrum—an important region photographically—whereas normal glass transmits poorly at these wavelengths. Physically, mirrors have further advantages as light-collectors. The glass does not have to be perfectly clear, since only the front surface is used to form the image, and the disk can be supported across the back to reduce distortion. The difficulty of making large lenses and mounting them well has limited the refracting telescope to an aperture of 40 inches (102 cm); in contrast, there is no reason o suppose that the size limit for mirror disks has been reached even with the Soviet 236-inch (600-cm) reflector. Another advantage is that the FOCAL LENGTH (mirror-image distance) of a mirror can be much shorter than that of a lens, a very important consideration when a large-aperture instrument is being designed.

Although the material used for a mirror has no direct optical effect on its performance, glass is usually used because it can be polished to a smooth, accurate finish. Stainless steel, aluminum, and other materials have also been tried. Astronomical mirrors usually have a thickness about one-sixth their diameter for rigidity, making large disks extremely heavy. Mirrors were once made of plate glass cast into disks or cut from thick sheets, but the development of low-expansion, heat-resistant glass for ovenware led to its use in telescopes—the 200-inch (504-cm) Mount Palomar telescope mirror is made of Pyrex glass. Further development has produced glass with virtually zero expansion, the best-known being *Zerodur,* made by the West German firm of Schott and used in many very large modern instruments. The advantage of low-expansion glass is that temperature changes hardly affect the mirror's surface shape (see also REFLECTING TELESCOPE).

Mizar

The central star in the handle of the Big Dipper, also known as Zeta Ursae Majoris and 88 light-years from Earth. Mizar was the first double star to be discovered by telescopic observation, by G. B. Riccioli in 1650. It was also the first to be photographed, by G. P. BOND in 1857. The orbital period of the two stars is at least 20,000 years. The brighter star was the first SPECTROSCOPIC BINARY discovered, by E. C. PICKERING in 1889; its orbital period is 20.5386 days. The fainter is also a spectroscopic binary, with a period of 175.55 days. All four stars of Mizar are orbited every 10 million years by Alcor, which is visible to the naked eye nearby. The radial velocity of Alcor is variable, which suggests that it, too, may be a spectroscopic binary.

Molniya satellites

A series of Soviet communications satellites, the first of which was launched on April 23, 1965. They have orbits very different from those of the geostationary communications satellites situated over the equator, such as INTELSAT, which most other countries use. Satellites over the equator cannot adequately cover regions in high latitudes, which include much of the Soviet Union, and so the Molniya satellites have been put into orbits inclined at 65° to the equator. Their paths are so arranged that they spend as much time as possible over the Soviet Union. This is done by making them move in highly elliptical orbits, which rise to a maximum height of about 25,000 miles (40,000 km) over the northern hemisphere, but dip to 300 miles (500 km) on the opposite side of the Earth. This orbit, which it takes about 11 hours 50 minutes to complete, ensures that each satellite is visible from the Orbita network of Soviet ground stations for 8 to 10 hours at a time. A system of three Molniyas can therefore provide 24-hour coverage. The main body of each Molniya is a cylinder, roughly 11.3 feet (3.4 m) long by 5.2 feet (1.6 m) wide. The

satellites gain power from six solar-cell panels like the petals of a flower. The satellite, weighing about 1,800 lb. (800 kg) communicates with Earth via two 3-foot (0.9-m) dish antennae. Several Molniyas have also carried television cameras to photograph Earth's cloud patterns. On November 24, 1971, the first of an improved and heavier generation of satellites, called Molniya 2, was launched. These use frequencies similar to those of the Intelsat series; the Molniya 1 satellites use lower frequencies. On July 29, 1974, a Molniya satellite was launched into a geostationary orbit over the equator, but others have reverted to the usual highly elliptical orbit. On November 21, 1974, the first of a third generation of satellites called Molniya 3 was launched, apparently carrying improved transmission equipment.

Monoceros (the unicorn)

A constellation in the equatorial region of the sky, between Canis Major and Orion, best seen during winter, introduced in 1624 by Jakob Bartsch (1600–1633), German mathematician and son-in-law of Johannes Kepler. Monoceros lies in the Milky Way, and contains several interesting objects, including the famous Rosette nebula, NGC 2237–9. In front of the large cluster NGC 2264 is the irregular variable star S Mon (also known as 15 Mon), which varies between magnitudes 4.2 and 4.6 every few days. Nearby is a sixth-magnitude star named Plaskett's star, after the Canadian astronomer John Stanley Plaskett (1845–1941) who discovered in 1922 that it was a SPECTROSCOPIC BINARY with masses 76 and 63 times that of the Sun, the most massive pair known. The stars orbit each other every 14.4 days. Also nearby is NGC 2261 (otherwise known as R Mon), Hubble's variable nebula.

month

A unit of time based on the Moon's motion around the Earth. Astronomers refer to several types of months. The *synodic month* is the time taken for the Moon to go through a complete cycle of phases (a *lunation*); it lasts 29.530588 solar days (29d 12h 44m 2.9s). A *sidereal month* is the time taken for the Moon to return to the same position against the star background; it lasts 27.321661 days (27d 7h 43m 11.5s). The difference between the sidereal and synodic months is due to the Earth's motion in orbit around the Sun. The *draconic month* is the time between two passages of the Moon through the ascending node (from south to north across the plane of the Earth's orbit); it lasts 27.21222 days (27d 5h 5m 35.8s). The *tropical month* measures the passage of the Moon across the longitude of the equinox and back again; it lasts 27.321582 days (27d 7h 43m 4.7s). The *anomalistic month* is the time between successive close points (perigees) in the Moon's orbit; it takes 27.554551 days (27d 13h 18m 33.2s). The *solar month* is an artificial unit devised to fit into our calendar; it lasts one-twelfth of a solar year (30.43685 days).

Moon

The Earth's only natural satellite. The Moon is a stark, lifeless world whose surface, unprotected by substantial atmosphere, is alternately scorched by the Sun at lunar midday and frozen to bitter

A view of the rugged lunar far side, taken in 1969 from the orbiting Apollo 11 spacecraft. The large crater, center, with terraced slopes and central peaks is Daedalus, about 50 miles (80 km) in diameter.

Smaller craters are everywhere more numerous—halving the size leads to a fourfold increase in numbers. This simple rule holds roughly true for craters down to only a few yards across.

Most lunar craters are now believed to be the result of impacts by large meteorites. Eroded impact craters have been found on the Earth's surface (see CRATERS) and are also present on Mars and Mercury. However, some small craters—particularly those in chains—may be the result of eruptions, gas explosions, or collapse along lines of weakness. The lunar craters are almost perfectly circular, a feature not shared by volcanic craters. The volume of the central bowl is equal to that of the crater's raised rim, which spreads out to about $1\frac{1}{2}$ times the diameter of the central bowl. The ratio of crater depth to diameter decreases for larger craters, being 1:5 for 1-mile (1.6-km) craters, 1:10 at 6 miles (9.7 km), and only 1:50 for the largest lunar craters. These figures belie the impression given by photographs—that lunar craters are deep cavities with steep slopes. In reality, the craters are very shallow, with gentle 10° to 12° slopes. In the case of the 146-mile (235-km) crater Clavius, an observer standing at the center of the floor would see little of the rim because it would be below the horizon.

Craters are surrounded by an ejecta blanket, consisting of crushed rock blasted out of the crater by the meteorite impact. The ejecta appears bright around recent craters, but darkens with age. In some cases, the ejecta forms long rays.

Lunar craters show a wide range of ages. Some are clear and sharply defined, while others are weathered and partially filled in, often being obscured and distorted by later impacts. The youngest-appearing craters are usually quite small, suggesting that few large impacts have occurred in recent lunar history.

The maria.   The largest impact features are the Moon's dark circular plains called the maria. The Mare Imbrium, for example, is a giant crater 420 miles (676 km) in diameter, rimmed by mountain ranges. This giant basin was produced by the impact of an asteroid 80 miles (130 km) in diameter, an event whose violence is still evident in the form of gouges, valleys, and secondary craters visible for more than 1,000 miles (1,600 km). The basins were subsequently covered by later, smaller impact craters and eventually filled with lava when the Moon's interior had heated up through radioactive decay. Each mare was formed by many lava flows over a considerable period of time. The far side of the Moon contains several basins up to $3\frac{1}{2}$ miles (5.5 km) deep, but these have not been subsequently filled in with lava flows.

Mountains and valleys.   Instead of Earth-type mountain ranges, other, less spectacular features are present on the Moon. The dark lunar plains show wrinkle ridges a few hundred feet high formed by contraction of the centers of the maria. An unusual type of valley, known as a rille and akin to a terrestrial rift valley, has a flat floor and steep parallel walls. Rilles can be hundreds of miles long and are found close to the edges of the maria. Other rilles—the sinuous rilles—may be lava tubes under the maria surfaces that have collapsed. Finally, faults up to 100 miles (160 km) long occur, forming rock faces up to 1,200 feet (365 m) high.

The surface structure.   Lunar landings have

temperatures at night. The shattered surface is very ancient, showing the scars of more than 3 billion years of meteorite impacts. Having changed so little, the Moon can provide a vast amount of information on the early history of the solar system.

Physical characteristics.   The Moon is 2,160 miles (3,476 km) across, less than one-third the diameter of the Earth, and has a mass of 7.34×10^{19} tons. Its mean density is only 3.34 times that of water, in contrast to the Earth's 5.52, indicating a quite different internal constitution. The Moon moves in an elliptical path inclined at 5° to the ecliptic, taking 27.3 days to complete one orbit. The Moon also rotates on its axis in 27.3 days; it thus keeps the same face pointing Earthward. As the Moon orbits the Earth, we see different amounts of its sunlit side, leading to the familiar lunar phases, a complete set of which takes 29.5 days.

The Moon can come within 218,000 miles (351,000 km) of the Earth's surface; its average distance is 238,855 miles (384,400 km). It reaches magnitude −12.7 at full Moon, even though its dark surface reflects only some 7 percent of the incoming light. The Moon's proximity produces reactions in the Earth's surface, atmosphere, and seas, which most clearly show as the familiar ocean TIDES.

Other planets have satellites with no more than a few thousandths their own mass. By contrast, the Moon's mass is about 1/81 that of the Earth, and the Earth-Moon system can almost be regarded as a double planet.

Lunar craters.   The Moon's surface is divided into sharply contrasting bright rugged highland areas and dark smooth lowlands. The craters, scattered across the entire surface, are 16 times more abundant on the bright highlands than on the lowland plains.

Lunar Orbiter view of the Moon's far side from an altitude of 768 miles (1,200 km), showing the heavily cratered terrain. The large circular feature, left, is Mare Moscoviense, a walled plain 300 miles (480 km) across.

allowed detailed photography of the lunar surface, laboratory analysis of lunar rocks, and active investigation of the lunar surface and interior. The surface is clearly very ancient, a jumble of debris from more than 3 billion years of meteorite impacts. Large and small boulders and smaller rocks are interspersed with huge amounts of fine dust. Many of the fine particles are glass spheres caused by melting of the surface in the heat of meteorite impacts. This jumble of debris (regolith) is from 35 to 60 feet (11–18 m) deep and consists of 1 to 2½ percent of meteoritic material, the shattered remains of the original impacting meteorites. Below the regolith is a region of intensely shattered but more solid rock, which probably extends to depths of more than 10 miles (16 km). The continual impact of meteorites, which still continues, is constantly adding to and turning over the regolith. In addition, the exposed surface of rocks is being eroded by some 1 millimeter every million years through micrometeorite impacts. Ancient boulders would by now have lost some 6 feet (1.8 m) of material since the lunar surface was formed, thereby adding more dust to the lunar regolith.

The Moon's atmosphere is ten trillion times as rare as the Earth's atmosphere at sea level. It is formed from gases released by heating, impact melting, radioactive decay, or possibly volcanic venting during the lunar night, and is partially dispersed by the solar wind each lunar day. The lack of an appreciable atmosphere leads to enormous temperature ranges on the lunar surface, from 105°C at midday on the equator down to −155°C during the long lunar night.

The lunar rocks are unlike those of the Earth, being 6½ times richer in aluminum, calcium, and

titanium, but four times poorer in sodium, magnesium, and iron. The highland rocks also differ in many ways from the mare rocks, which accounts for their lighter color.

The interior. The network of seismometers left on the surface by the Apollo landings indicates that the Moon is seismically very quiet. Moonquakes are only a thousandth as strong as earthquakes, and instead of producing intense short-duration signals, as on Earth, they are dampened by the shattered surface to produce long-enduring ringing signals. About 40 major quake sources have now been identified at depths from 370 to 750 miles (595–1,200 km), in striking contrast to the Earth, where earthquakes only occur in the outer layers.

The interior of the Moon, like that of the Earth, consists of several distinct regions. The 40-mile(65-km)-thick outer crust has a density three times that of water; below this is the 740-mile(1,190-km)-deep lunar mantle, only slightly denser. Finally, there is a central core some 600 miles (1,000 km) across. Although crust and mantle are solid, the temperature of the core is perhaps 1,500°C, enough to melt its center. The Moon lacks a large nickel-iron core like the Earth's which accounts for its low density.

Lunar history. Dating of lunar samples has allowed scientists to piece together the history of the Moon. The Moon formed along with the rest of the objects in the solar system about 4.6 billion years ago. Almost immediately, its surface was melted by the heat of a heavy meteorite bombardment, which eventually died away. The lunar highlands formed about 3.95 billion years ago, and the declining meteorite bombardment scattered a thick blanket of impact debris across the lunar surface. Liquid magma then flowed out from under the Moon's nearside surface on various occasions from 3.9 to 3.3 billion years ago, forming the lunar maria. The surface has remained cold ever since.

The interior of the Moon has had a complex history.

135

Mare Imbrium, a vast, lava-flooded plain some 600 miles (960 km) across, was formed during a final burst of internal lunar activity over three billion years ago. It is encircled by the 18,000 ft (5,400 m) high Lunar Alps, **top right**, and Apennines, **bottom right**; the dark-floored crater, **top**, is Plato, and the bright ray crater Copernicus is at bottom of picture.

Initially the Moon was formed with a molten core that solidified with a strong magnetic field frozen into it. The molten crustal materials then solidified in this magnetic field, to become partially magnetized. Later, the core melted again from radioactive heating, losing its magnetism to produce the present low lunar magnetic field.

Lunar origins. A vast amount of information is now available, but it is still far from certain how the Moon originated. The Moon is currently slowly receding from the Earth; at one time it was as close as 40,000 miles (64,000 km). Theories of lunar origin therefore center on whether the Moon was formed in that position or was somehow captured. One theory suggests that the Moon was formed from material thrown out from the Earth's equator, while according to another, the Moon and Earth formed as a double planet from the same dust cloud. The strikingly different composition of the Earth and Moon is, however, a serious obstacle to these proposals. Yet another theory proposes that the Moon was formed as a small independent planet and was captured by the Earth. Although possible, this would involve an extremely unlikely series of events. At present, the most plausible theory traces the Moon's formation to material captured into Earth orbit. Planetesimals, low-density bodies like asteroids several tens of miles across, would have been plentiful in the early solar system; on passing very close to the Earth these

would have been disrupted, their lighter surface materials being captured and their heavier core materials continuing in solar orbit. This captured material in orbit around the Earth would eventually coalesce to form the Moon. This suggested origin satisfactorily explains the Moon's light density and its nonterrestrial composition.

Morgan, William Wilson (b. 1906)
American astronomer, who found the first evidence for the spiral structure of our Galaxy; previously, the existence of spiral arms in our Galaxy had only been assumed. Studying the distribution of blue giant stars in the Milky Way, Morgan in 1951 showed that they traced the outline of at least two spiral arms: one, toward the edge of the Galaxy, is called the Perseus arm, because it contains the stars of the double cluster in Perseus; the other, called the Orion arm, includes the Orion nebula and our Sun. Morgan also found evidence of another arm in the Sagittarius region, toward the center of the Galaxy. The spiral arms have since been mapped in detail by the radio-wavelength radiation given off by the hydrogen gas they contain. At Yerkes Observatory Morgan, together with Philip Childs Keenan (b. 1908) and Edith Kellman, developed a new classification system for stellar spectra, based on earlier work by Walter S. ADAMS and Bertil LINDBLAD. This was published in the 1943 Yale *Atlas of Stellar Spectra*; the system is usually referred to as the MORGAN-KEENAN CLASSIFICATION. In 1953, Morgan established a now-standard system for measuring the COLOR INDEX of a star, called the UBV SYSTEM. Morgan has also worked on the classification of galaxies, and in 1963 drew attention to supergiant elliptical galaxies that lie at the center of clusters of galaxies; these are the largest galaxies known.

Morgan-Keenan classification
A shorthand notation invented by W. W. MORGAN and Philip Childs Keenan (b. 1908) to describe the appearance of the spectrum of a star. The temperature is coded with one of a series of letters, called a spectral class, which, in order of decreasing temperature, run: OBAFGKM (mnemonic: "Oh, be a fine girl, kiss me"). Up to 10 subdivisions may be added as numbers after the letter: thus, B5 is four-tenths of a class cooler than B1. Roman numerals I, II, III, IV, and V, called the luminosity class, are used to distinguish supergiants, giants, and dwarfs (the smaller the numerals the brighter the star). The Sun's spectral classification, G2V, shows that it is a coolish dwarf, very similar to Alpha Centauri, also G2V. Betelgeuse, M2I, is a cool red supergiant, cooler and brighter than the yellow giant Aldebaran, K5III. Sometimes lower case letters are added to give extra detail in the spectral classification, such as B1Vpnne! meaning a hot (B1) dwarf (V) with some peculiarities (p) in its spectrum, very rapidly rotating (nn) and with strong emission (e!) lines. Some peculiar-looking spectra have special spectral classes.

morning star
Not a true star, but rather the planet Venus, seen shining brilliantly in the morning sky as it rises a few hours before the Sun. When it sets after the Sun in the evening sky, it becomes the brilliant evening star.

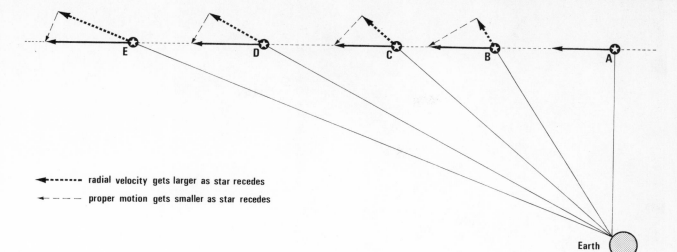

◄----------- radial velocity gets larger as star recedes

◄-- -- -- proper motion gets smaller as star recedes

Earth

Mount Palomar
The site of the 200-inch (508-cm) reflecting telescope of HALE OBSERVATORIES, more correctly called Palomar Mountain.

Mount Stromlo Observatory
An astronomical observatory operated by the Australian National University at an altitude of 2,520 feet (768 m) on Mount Stromlo, 7 miles (11 km) west of Canberra; it was founded in 1930. Its main telescope is a 74-inch (188-cm) reflector, completed in 1955; there are also 50-inch (127-cm) and 30-inch (76-cm) reflectors. A 26-inch (66-cm) refractor operated jointly by Yale and Columbia universities is situated on Mount Stromlo, as is the 20/26-inch (50/66-cm) Schmidt camera of the University of Uppsala, Sweden. Mount Stromlo Observatory operates a field station near Coonabarabran, New South Wales; this is called SIDING SPRING OBSERVATORY.

Mount Wilson
The location in California of the 100-inch (254-cm) reflecting telescope of the HALE OBSERVATORIES.

moving cluster method
A technique for finding the distance of a nearby star cluster whose stars are moving through space along parallel paths. If the cluster is receding, its stars appear to stream toward a single point in space, the *convergent point,* just as parallel railroad tracks seem to converge at a point on the horizon. The nearer a star to the convergent point, the more its speed is directed away from us (RADIAL VELOCITY); the closer a star is to the Earth, the greater its motion across the line of sight (PROPER MOTION). The relative amounts of radial velocity and proper motion for a star depend on its angular distance from the convergent point and its true distance from Earth. Once the convergent point has been identified, the actual distance of the star can be found.

The distance of the HYADES star cluster, determined in this way, forms the foundation for all distance-finding in astronomy (apart from the distances of a few nearby stars, which can be deduced from their parallaxes).

Five stars in a moving cluster belong to a star stream, with same space velocity. But the star A, nearest to Earth, has a zero radial velocity (speed along line of sight) and maximum proper motion (speed perpendicular to line of sight), while the more distant stars like E have a large radial velocity and small proper motion. The exact trade-off between proper motion and radial velocity depends on the angle between the convergent point and the line of sight. The measure of this angle, combined with the radial velocity and proper motion, gives the star's distance.

Mullard Radio Astronomy Observatory (MRAO)
The radio astronomy observatory of the University of Cambridge, situated at Lord's Bridge, near Cambridge, and founded in 1951. Cambridge radio astronomers, led by Sir Martin RYLE, have developed the APERTURE SYNTHESIS method of combining results from several small dishes to produce the view that would be obtained by one much larger dish. Their major instruments of this type are the One-Mile Telescope, opened in 1964, which consists of three 60-foot (18.3-m) dishes (two fixed, one movable) arranged in a line 1 mile (1.6 km) long; and the Five Kilometer Telescope opened in 1972, which has eight dishes (four fixed, four movable) of 42-foot (12.8-m) diameter arranged in a line 5 km (3 miles) long. The Half-Mile Telescope, using two 30-foot (9.1-m) dishes, is mounted on the ½-mile (0.8-km) length of railway track used by the movable dish of the One-Mile Telescope. Astronomers at MRAO have made several famous surveys of the radio sky, and many radio sources are referred to by their numbers in the third, fourth, or fifth Cambridge catalogs (3C, 4C, 5C). A fixed aerial array, intended to study scintillation of signals from quasars, produced the information that led to the 1967 discovery of PULSARS at MRAO by Antony HEWISH. (See picture p. 138.)

Musca (the fly)
A small constellation in the southern hemisphere of the sky, at the edge of the Coalsack nebula near Crux. It was introduced by Johann Bayer in 1603.

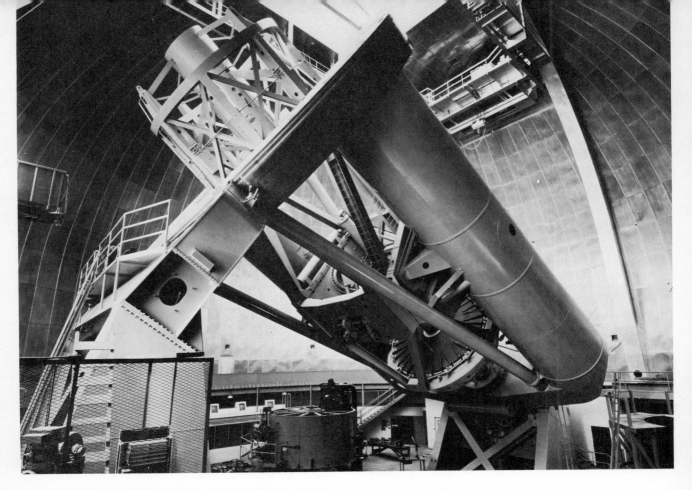

The 200-inch (5-m) Hale telescope at Mt Palomar, Cal. In this view, the telescope's open framework tube points north, along the polar axis. The mirror itself is hidden by the petals of a cover at the lower end of the tube. The scale of the instrument can be judged from the tables at ground level.

Aerial view of five of the eight dishes comprising the Five Kilometer radio telescope at the Mullard radio astronomy observatory, Cambridge. The furthest four can each be moved to eight positions on a ¾ mile long rail track. The other three dishes lie to the left, fixed at intervals of ¾ mile.

N

nadir
The point directly beneath an observer's feet, 90° below the true horizon and directly opposite the ZENITH. The nadir is the direction indicated by a freely suspended plumb bob.

NASA
The National Aeronautics and Space Administration, a federal agency established October 1, 1958, to administer all nonmilitary aeronautical research and space programs. It was successor to the National Advisory Committee for Aeronautics (NACA). NASA headquarters at Washington, D.C., coordinates the activities of the various NASA field stations: AMES

RESEARCH CENTER; Hugh L. Dryden Flight Center (formerly Edwards Flight Center), Edwards, California, concerned with manned flight in aircraft and spacecraft; GODDARD SPACE FLIGHT CENTER; JET PROPULSION LABORATORY; JOHNSON SPACE CENTER; KENNEDY SPACE CENTER; LANGLEY RESEARCH CENTER; Lewis Research Center, Cleveland, Ohio, concerned with aircraft and rocket propulsion and responsible for development of the Agena and Centaur rockets; MARSHALL SPACE FLIGHT CENTER; and WALLOPS FLIGHT CENTER.

National Radio Astronomy Observatory (NRAO)
The largest radio astronomy observatory in the United States, founded in 1957 at Green Bank, West Virginia. The first major telescope, 85 feet (26 m) in diameter, was completed in 1959. Two similar dishes, both movable, have since joined it to make an INTERFEROMETER with a maximum length of 1.65 miles (2.7 km). The interferometer is used in conjunction with a 45-foot (13.7-m) dish located 22 miles (35 km) away, which can receive wavelengths down to 3.7 centimeters. The observatory's largest telescope is a 300-foot (90.4-m) dish erected in 1962 and resurfaced in 1970; initially intended for observations at the 21-centimeter hydrogen line, it is now used down to a 6-centimeter wavelength. In 1965 an equatorially mounted 140-foot (42.7-m) telescope came into operation, working at wavelengths as low as 1.3 centimeters. Two years later a 36-foot (11-m) telescope designed for wavelengths of a few millimeters was installed by the observatory at Kitt Peak, Arizona; it has subsequently made important discoveries of INTERSTELLAR MOLECULES. The National Radio

The major instruments at the National radio astronomy observatory are the 140-foot fully steerable dish, **far left, in the distance**, the 300-foot transit telescope which tilts only north—south, **near left**, and three 85-foot dishes connected electronically to form an interferometer, **right**.

Astronomy Observatory contains a replica of the aerial with which Karl JANSKY detected the first radio waves from space, and it preserves the 31-foot (9.5-m) dish used by Grote REBER to continue Jansky's work. The NRAO is constructing the world's largest radio-astronomy instrument, the VERY LARGE ARRAY, near Socorro, New Mexico.

navigation satellites
Satellites that aid aircraft and ships in pinpointing their position. Initially satellite navigation was used mostly for naval vessels, but its use in merchant shipping and civil aviation is now increasing. A navigation satellite transmits information on its position and orbit. This can be picked up by small receivers, and the Doppler shift in the signals reveals the receiving station's location relative to the satellite, from which an extremely accurate position can be computed. The first navigation satellite was Transit 1B, launched on April 13, 1960 (see TRANSIT SATELLITES).

A series of satellites called Timation was begun in 1967. These carry atomic clocks and provide both navigational data and accurate time signals. The signals have been used to compare atomic clocks in the United States and Britain. The Timation system, initiated by the U.S. Navy, has been merged with an Air Force project to produce a coordinated navsat network known as the Global Positioning System. Timation satellites were launched on May 31, 1967; September 30, 1969; and July 14, 1974.

neap tide
See TIDES.

nebula
A region of gas and dust in the Galaxy, usually fuzzy in appearance. The name is derived from the Latin word for cloud. The obsolete terms "spiral nebulae" and "extragalactic nebulae" actually refer to GALAXIES, and they have nothing to do with genuine nebulae. Some nebulae, such as the famous COALSACK,

Clouds of hydrogen gas glow brightly from the radiation of hot, young stars within the emission nebula M16 in Serpens. Such nebulae are stellar birthplaces; the black globules, **below centre**, and the "elephant trunk", **center**, seen silhouetted against the bright nebula, are in the process of contracting and fragmenting to form stars.

are cold clouds of dust and gas that absorb light from stars behind them, thereby appearing as dark patches (see INTERSTELLAR ABSORPTION). Such *dark nebulae* look like "holes" in the Milky Way. *Bright nebulae* are similar clouds illuminated by stars.

Reflection nebulae are bright nebulae that shine by reflecting light from a nearby star; the light therefore exhibits the same spectrum as that of the star. The dust in the nebula tends to scatter the starlight, however, and because blue light is scattered more than red light, the reflection nebula appears bluer than the star.

The nebula around the star Merope in the PLEIADES cluster was the first reflection nebula to be recognized as such, by V. M. SLIPHER in 1913. A temporary reflection nebula was seen near Nova Persei, the year after its eruption in 1901. The burst of light produced by this NOVA spread into the surrounding space, illuminating a formerly invisible dark nebula that eventually disappeared as the nova faded back into insignificance. A curious reflection

nebula discovered in 1852 by the English amateur astronomer John Russell Hind (1823–1895) lies near the star T Tauri. Known now as Hind's variable nebula, it has fluctuated in brightness ever since. Changes in its brightness are not correlated with the brightness of T Tauri itself; instead, its variability is thought to be caused by clouds of dark dust sweeping between the nebula and T Tauri, thus temporarily blocking the star's light.

Emission nebulae are nebulae whose gas is excited to fluorescence by the ultraviolet light from a nearby hot star, thus producing emission of light at certain specific wavelengths. Such emission lines were first seen by William HUGGINS in 1864, when he examined the spectrum of the ORION NEBULA, the brightest of all emission nebulae. Some of the wavelengths of light given off by emission nebulae were immediately recognized as coming from such elements as hydrogen. Some however were unknown, and ascribed to a hypothetical element, "nebulium." This eventually proved to be oxygen which, in the rarefied conditions of space, emitted so-called FORBIDDEN LINES unknown in terrestrial laboratories. Such wavelengths of oxygen make emission nebulae appear green to the naked eye, although in photographs they often appear red because the photographic emulsion is more sensitive to the red wavelength of hydrogen called Hα. Emission nebulae are of three main kinds: PLANETARY NEBULAE, so called because they resemble the disk of a faint planet; *loops,* due to various causes including supernovae; and *diffuse nebulae* of irregular shapes. Inside diffuse nebulae, such as the Orion nebula, new stars are forming which will make the nebula shine. Diffuse nebulae and hot stars together are called H II REGIONS, areas composed principally of ionized hydrogen.

Neptune

The eighth planet in average distance from the Sun. The eccentric orbit of the outermost planet Pluto, however, brings it within the path of Neptune for relatively short periods. The existence of Neptune was calculated by mathematicians before the planet was discovered telescopically. Early in the 19th century, astronomers found that the newly discovered planet Uranus was not moving as expected. It appeared to be perturbed in its orbit by the gravitational pull of an unseen body. The perturbations of Uranus were analyzed by the English mathematician John Couch ADAMS, who in 1845 predicted the position of the unseen body. He sent his results to the Astronomer Royal, Sir George AIRY, but no search was begun until after similar results were announced the following year by the Frenchman U. J. J. LEVERRIER. Neptune was first recognized by J. G. GALLE at Berlin Observatory on September 23, 1846, in a search requested by Leverrier. Like Uranus, Neptune had been seen by previous astronomers, but was mistaken for a star.

Physical properties. Neptune's equatorial diameter is 30,750 miles (49,500 km), slightly smaller than Uranus. Neptune rotates once every 15 hours 48 minutes on an axis inclined at 28° 48′ to the vertical. The planet is slightly oblate (squashed) in shape, with a polar diameter about 600 miles (1,000 km) smaller than its equatorial diameter. Neptune orbits the Sun every 164.8 years at a mean distance of 30.06 astronomical units (2,794,100,000 miles; 4,496,700,000 km). The near-circular orbit, with an eccentricity of only 0.0086, is inclined at $1° 46'$ to the ecliptic. Neptune's mass is 17.2 times the Earth's, giving it an average density of 1.7 times that of water. Neptune appears blue-green in color, and is scarcely visible as more than a small disk in the largest telescopes.

Structure of Neptune. In size and structure, Uranus and Neptune are the most similar of all the planets in the solar system. Neptune's extensive atmosphere contains large amounts of methane, which are responsible for its greenish color (methane absorbs yellow and red light), although the major atmospheric constituents are believed to be hydrogen and helium. As on Uranus, ammonia probably forms clouds below the visible layers of Neptune's atmosphere. Observations of an occultation of a star by Neptune in 1968 suggests that Neptune has a warm stratosphere, with a temperature of $140°$ K, which is nearly $100°$ higher than what would be expected at the top of the atmosphere. The warming of the upper layers of Neptune's atmosphere is probably caused by absorption of solar heat by methane. The interior structure of Neptune is believed to be virtually identical to that of Uranus, with a central rocky core 10,000 miles (16,000 km) in diameter covered by a layer of ice 5,000 miles (8,000 km) thick; over three-fourths of the planet's mass is believed to be contained in this ice-coated core.

Satellites. Neptune has two known moons, TRITON and NEREID. Triton may be the largest satellite in the solar system. It moves around Neptune in a circular retrograde (east-to-west) orbit inclined $20°$ to the planet's equator. It has been suggested that Pluto was once a moon of Neptune, and that a close encounter of Pluto with Triton ejected Pluto into a separate orbit around the Sun and threw Triton into its retrograde orbit. Further evidence of some highly unusual event in Neptune's past is the orbital eccentricity of Nereid, which is the largest for any known satellite; the distance of Nereid from Neptune varies by over $5\frac{1}{4}$ million miles (8.5 million km).

Satellites of Neptune

	Discoverer	Diameter (km)	Orbit radius (10^3km)	Period (days)	Eccentricity
Triton	Lassell (1846)	6,000	355	5.877	0.00
Nereid	Kuiper (1949)	500	5562	359.881	0.75

Nereid

The more distant of Neptune's two satellites, discovered in 1949 by Gerard Kuiper. Nereid, with a diameter of about 300 miles (500 km), moves around Neptune every 359 days 21 hours 9 minutes in an eccentric orbit that takes it between 826,000 miles (1,330,000 km) and 6,100,000 miles (9,760,000 km) from the planet.

neutron star

A very small, dense star, so tightly packed that its protons and electrons have been compressed together to form the particles called neutrons. A neutron star's mass is roughly equal to that of the Sun, but its diameter is only a few miles. The resulting density is such that a pinhead of the star's material would weigh a million tons. In 1934 the astronomers Walter BAADE and Fritz ZWICKY first proposed that when a SUPERNOVA explosion blows off the outer layers of a star, its remaining dense core is compressed into a neutron star. The basic structure of neutron stars was calculated by the American physicist J. Robert Oppenheimer (1904–1967) in 1939, and although the details are still poorly understood, it is thought that the interior of a neutron star is liquid, with a relatively thin solid crust surrounded by an atmosphere of iron atoms less than a centimeter thick.

If the collapsing star is more than three times as massive as the Sun, it cannot become a neutron star because its gravity would be strong enough to crush even the neutrons, and it would continue to shrink until even light could not escape, thus becoming a BLACK HOLE.

The first neutron stars were discovered in 1967, as the regularly flashing objects called PULSARS. As expected, pulsars seem to be the remains of supernova explosions, as for instance in the CRAB NEBULA. Neutron stars also account for some of the X-ray sources in our Galaxy. Many X-ray sources are double stars, with the two components very different in size. As gas from a giant star's atmosphere falls onto the compact star, it is heated to a temperature of hundreds of millions of degrees and emits X rays.

In many sources the X-ray flux changes rapidly, indicating that the small star is only a few miles across, and must therefore be a neutron star.

new Moon

The Moon at the instant it lines up between the Earth and Sun. The illuminated side is therefore turned away from us, and the Moon is invisible. A solar eclipse would occur at each new Moon, were it not that the Moon's orbit around the Earth is inclined to the path of the Earth around the Sun.

Newcomb, Simon (1835–1909)

Canadian-born American astronomer, one of the foremost of all mathematical astronomers. He first made his mark in 1860, with a paper showing the orbits of the asteroids did not diverge from one point, a conclusion which undermined the existing belief that they were fragments of a single larger body; modern astronomers support Newcomb's view. In 1857 Newcomb had joined the American *Nautical Almanac* office, and at the U.S. Naval Observatory in Washington he began a series of observations to prepare more accurate tables of lunar and planetary motions. In 1877 he became head of the office, and began preparation of the most accurate tables ever made of the movements of objects in the solar system. The results, published in the *Nautical Almanac,* were used throughout the first half of the 20th century. Newcomb computed position tables for the Sun, Mercury, Venus, Mars, Uranus, and Neptune. His colleague George William Hill (1838–1914) undertook the calculations for Jupiter and Saturn. Newcomb studied in detail the complex problem of the Moon's motion, finding errors in the previously accepted lunar tables of the Danish astronomer Peter Andreas Hansen (1795–1874). From the work of Newcomb and Hill

on the Moon's motion, the English mathematician
Ernest William Brown (1866–1938) produced in 1919
new lunar tables for publication in the *Nautical
Almanac*.

Newcomb also produced a new and more accurate
set of astronomical constants, and in 1880–1882 he
remeasured the velocity of light. Newcomb called an
international conference in Paris in 1896, at which
astronomers from Europe and the United States
agreed to adopt a new, standardized set of
astronomical constants, based on Newcomb's figures.

Newton, Sir Isaac (1642–1727)

British mathematical physicist who developed the
concept of universal gravitation and made important
discoveries in optics. Newton's major discoveries began
in 1665 at his home in Woolsthorpe, Lincolnshire,
where he had retreated from Cambridge University
during the Great Plague. His work continued when he
returned to Cambridge and culminated in 1687 with
the publication of his *Philosophiae Naturalis Principia
Mathematica* (*Mathematical Principles of Natural
Philosophy*), usually known simply as the *Principia*.

Newton's great contribution to science was to
explain the physical Universe in mathematical terms,
thus bringing to completion the work of both Kepler
and Galileo. In the *Principia* he expounded in
mathematical form the laws of movement and reaction
of physical bodies on Earth and in space, enshrining
his results in his three now-famous laws of motion.
He also laid down the principle of universal
gravitation and gave this principle mathematical
precision by stating that every body attracts every
other body with a force that depends on how massive
the bodies are and the distance between them.

In the *Principia* Newton examined the actions of
bodies under gravitation, and his results solved the
problems of planetary motion that had faced
astronomers since ancient times. He showed how
KEPLER'S LAWS of planetary motion were a natural
consequence of universal gravitation, and he
demonstrated how the planets interacted with one
another in their orbits. Newton developed a new
mathematics—the calculus, later to be of great use
in science—so that he could obtain his results.
Newton's work was later pursued by his friend
Edmond Halley, who used Newton's principles to
compute the orbits of comets. In the late-18th
century William Herschel discovered pairs of
orbiting stars that obeyed Newton's law of gravity,
while in the 19th century the astronomers J. C. ADAMS
and Urbain LEVERRIER discovered the planet Neptune
by applying Newtonian theory to the perturbed

Sir Isaac Newton in 1702, when he was 60. His most
important work was done when he was in his early 20s,
though much of it was not published until later. In his
60s, however, Newton turned from academic work to
become an administrator at the Royal Mint—felt to be a
more suitable occupation for a great man.

orbital movements of the planet Uranus. During the
18th and 19th centuries, mathematicians worked out
other consequences of the Newtonian theory of
gravity as applied to planetary orbits, and investigated
the difficult problems of applying it to the Moon's
orbit.

Observational astronomy also owes an important
debt to Newton. His experiments on the dispersion of
white light into a colored spectrum showed that white
light was compounded of light of every color.
Newton believed that lenses would always form
images with colored fringes (CHROMATIC ABERRATION).
He therefore designed a reflecting telescope, using
mirrors instead of lenses, and in 1668 built such an
instrument himself. Over the following centuries the
reflecting telescope developed into one of the most
powerful of all astronomy's tools.

Newton's pioneering research in optics was
developed in the 19th century into the science of
spectroscopy which, in turn, gave rise to
ASTROPHYSICS—the study of the physical and
chemical nature of the stars. His theory of the nature
of light, eventually published in his *Opticks* of 1704,
proved invaluable to still later research.

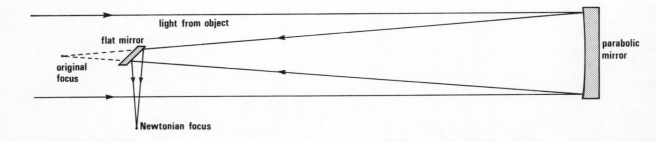

Newtonian Telescope

N-galaxy

A type of distant galaxy with a very small but bright central nucleus. N-galaxies were first discovered as the optical counterparts of some strong RADIO GALAXIES, and the name was devised by W. W. MORGAN to indicate the prominence of the nuclei (N) of these galaxies. They are probably enormously powerful SEYFERT GALAXIES, seen at larger distances than would otherwise be possible. Their nuclei vary in brightness over the years and can sometimes become so luminous that the rest of the galaxy cannot be seen. The star-like nuclei then resemble QUASARS, which are generally thought to be extremely powerful N-galaxy nuclei so distant that the surrounding galaxy is invisible. The first quasar to be identified, 3C 48, has in fact been reclassified as an N-galaxy because long-exposure photographs show the faint fuzz of a galaxy around it.

NGC

Abbreviation for *New General Catalogue of Nebulae and Clusters of Stars,* a now-standard list of deep-sky objects compiled by the Danish astronomer Johan Ludwig Emil DREYER. The NGC, published in 1888, contained nearly 8,000 objects, with another 5,000 added in two supplements, called the *Index Catalogues,* published in 1895 and 1908. Many nebulae and star clusters are referred to by their NGC numbers, although the M numbers given to the most prominent objects by the French astronomer Charles MESSIER also remain in use.

Nikolayev, Andrian Grigoryevich (b. 1929)

Soviet cosmonaut who in August 1962 piloted the Vostok 3 capsule during the joint flight with Pavel Popovich in Vostok 4. The following year Nikolayev married the Soviet female cosmonaut Valentina TERESHKOVA. He made a second flight in June 1970 on the Soyuz 9 mission.

Nimbus satellites

A series of six American weather satellites from which the Earth-sensing LANDSAT design was developed. Nimbus satellites tested new sensing devices which have been incorporated into other weather satellites, such as the NOAA series (see WEATHER SATELLITES). Nimbus was a butterfly-shaped satellite, 10 feet (3 m) long and 11 feet (3.4 m) wide across its two winglike solar panels. A circular tray 5 feet (1.5 m) wide carried the sensing devices, acting at wavelengths from infrared to ultraviolet. In addition to returning photographs of Earth's cloud cover, ice caps, and landforms, satellites of the Nimbus series also made temperature and humidity measurements at various depths in the atmosphere. The Nimbus satellites were launched into polar orbits so that they could scan the entire Earth. Nimbus I was sent into orbit in August 1964; the last satellite of the series was launched in June 1975.

noctilucent clouds

Clouds similar in appearance to cirrus or cirrostratus clouds, formed of ice crystals frozen around micrometeorites at heights of 50 to 55 miles (80–90 km) in the Earth's atmosphere. They appear as bright blue-white patches against the twilight sky, being visible only by reflected sunlight when the Sun is 10° to 15° below the horizon. Particles collected from the clouds by high-altitude rockets are found to be either true micrometeorites or dust derived from burnt-up meteors.

nodes

The points at which the orbit of a body intersects a reference plane, such as the plane of the Earth's orbit. Where the object is moving from south to north, the intersection is termed the *ascending node*; the *descending node* is the point at which the object moves across the reference plane from north to south. REGRESSION OF NODES is movement of the nodes caused by gravitational influence of other bodies.

Norma (the level)

A small, faint constellation in the northern hemisphere of the sky, lying on the edge of the Milky Way, introduced by Nicolas Louis de Lacaille.

northern lights

The popular name for the aurora borealis (see AURORA).

nova

A faint star that suddenly erupts in brightness, becoming visible where no star had been seen before; the term comes from the Latin word for "new." Two or three novae are discovered every year, often by amateur astronomers. Novae increase in brightness by thousands or tens of thousands of times, and can often be distinguished by the naked eye. They climb to their maximum magnitude very rapidly— sometimes in as little as a day or two—before fading again over several days, months, or even years. Novae are named after their constellation and the year of their maximum brightness. Nova Persei 1901, discovered by the Scottish amateur astronomer Thomas David Anderson (1853–1933), brightened by a factor of at least 10,000 in one day, temporarily becoming one of the brightest stars in the sky. Its spectrum showed that a shell of gas had been thrown off at a speed of 1,250 miles (2,000 km) per second; it is this rapid expansion of the surface area of a nova that is responsible for the dramatic increase in brightness. The total mass thrown off, however, is only .00001 of the star's original mass, so that a nova outburst does not greatly disrupt it. A SUPERNOVA explosion on the other hand virtually destroys a star. At maximum, the intrinsic brightness of a nova (its ABSOLUTE MAGNITUDE) reaches −7 or −8, about 10,000 times (10 magnitudes) less than that of a supernova.

Most novae, possibly all, are actually binary stars, as shown in 1962 by the American astronomer Robert Paul Kraft (b. 1927). One of the stars in each binary system is a WHITE DWARF, and the nova outburst is believed to be caused by gas flowing from the companion star onto the white dwarf, where it ignites and is ejected in a small nuclear explosion. Several novae have been seen to erupt more than once; Nova Pyxis holds the records for recurrent outbursts with four, in 1890, 1902, 1920, and 1944. Satellites have recently revealed the existence of X-ray novae (see X-RAY ASTRONOMY). These are probably also binary stars, with one of the components being a NEUTRON STAR or a BLACK HOLE.

nutation

A slight nodding of the Earth's axis in space, the result of a change in the angle of the Earth's axial tilt. Nutation is superimposed on the general motion of PRECESSION, which is caused by the gravitational attractions of the Sun and Moon on the Earth's equatorial bulge. Since the pulls of the Sun and Moon are not identical, the S-shaped wobble of nutation results. Each nutation cycle takes 18.6 years, during which the Moon's orbit shifts once around the Earth. This orbital shift has a similar cause to that of precession: the Moon's orbit is inclined at 5° to the Earth's own orbit in space, and the Sun exerts a force that tends to pull the two into the same plane, with the result that the whole orbit of the Moon precesses around the sky in 18.6 years. There are tiny extra cycles of a fortnight (half a lunar orbit) and six months (half the Earth's orbit) within the main nutational movement. The maximum amount of nutation is about 9 seconds of arc in each direction. Nutation was discovered in 1748 by James BRADLEY.

O

OAO satellites

The Orbiting Astronomical Observatory series of American satellites, which observed the Universe at ultraviolet and X-ray wavelengths. Since hot objects emit most of their radiation in the short-wavelength region of the spectrum, which is blocked by the Earth's atmosphere, they can only be studied accurately by orbiting observatories. Using clusters of telescopes, spectrometers, and X-ray and gamma ray detectors, the OAO satellites observed young, massive stars, nebulae believed to be the sites of star formation, and galaxies and quasars. They found that the most massive stars are even hotter than previously believed, and that galaxies emit considerably more energy than expected at ultraviolet wavelengths.

Satellite	Launch date	Remarks
OAO 1	April 8, 1966	Carried one 16-inch (41-cm) telescope and four 8-inch (20-cm) telescopes for ultraviolet studies, plus X-ray and gamma-ray detectors. Ceased operation after 2 days owing to power failure
OAO 2	December 7, 1968	Carried Celescope, a group of four 12¼-inch (32-cm) telescopes for ultraviolet observations of hot main-sequence stars; a 16-inch (41-cm) reflector for studies of nebulae; four 8-inch (20-cm) telescopes for stellar photometry; and two ultraviolet stellar spectrometers with 8-inch (20-cm) mirrors
OAO B	November 30, 1970	Failed to reach orbit because spacecraft fairing failed to jettison. Main telescope was 36-inch (91-cm) reflector
OAO 3 (Copernicus)	August 21, 1972	Carried 32-inch (81-cm) telescope for ultraviolet spectroscopy of hot stars, plus three X-ray detectors operated by British experimenters. Final OAO

Oberon

The most distant satellite of Uranus. Oberon's diameter is about 1,000 miles (1,600 km); it orbits Uranus at a distance of 364,270 miles (586,230 km) every 13 days 11 hours 7 minutes. It was discovered in 1787 by William Herschel.

Oberth, Hermann Julius (b. 1894)

German rocket pioneer, with the Russian Konstantin TSIOLKOVSKY and the American Robert GODDARD one of the three founding fathers of astronautics. Oberth was primarily a theorist and not an inventor. His writings, however, inspired the German interest in rocket research that led directly to the space age. As early as 1917 Oberth had drawn up designs for a long-range liquid-fueled rocket for military use. This design work was rejected for a Ph.D. thesis in 1922, but the following year he turned his research into a now-classic book, *The Rocket into Interplanetary Space*, which included the first detailed discussion of orbiting space stations. In 1929 Oberth published his major work, *The Road to Space Travel,* in which he foresaw the development of electric rockets and ion propulsion. The book won an award established by the French rocket pioneer Robert Esnault-Pelterie, the prize money supported Oberth's experiments with rocket motors in the German Society for Space Travel, one of his assistants being Wernher VON BRAUN. Oberth worked briefly at Peenemünde during World War II, and in 1955 spent three years with von Braun in the United States before returning to retirement in Germany.

object glass

The lens used in a REFRACTING TELESCOPE to form an image. A single lens, because of the defect known as CHROMATIC ABERRATION, forms a colored image; all precision object glasses therefore consist of two separate lenses to minimize this effect. The front lens is of crown glass (light in weight, consisting mainly of silica), and the rear lens is of flint glass, which is heavy because it contains lead oxide. Even with this modification, residual color still degrades the image if the FOCAL RATIO (lens-image distance divided by lens aperture) is less than about 12. Most object glasses therefore have focal ratios of between 15 and 20. In small sizes, up to about 6 inches (15 cm), the lenses are usually mounted together in a single cell, separated by three pieces of foil located around the edge; larger lenses may have separate cells with mutual adjustment.

observatories, history of

The earliest known observatories were constructed in Egypt about 2600 B.C., and in Babylonia some six centuries later. These buildings were as much religious as astronomical in purpose. In Britain and in Brittany (northwest France), astronomically aligned stone circles were being built around the same time as the Babylonian ziggurats. The earliest observatories were constructed to help make accurate calendars; stone circles such as STONEHENGE probably also produced eclipse predictions. Astronomers were mainly concerned with the stars, the Sun, and the Moon. Other astronomical observations were made, primarily of the five naked-eye planets, but there appear to have been few if any formal observatories as such. Ancient astronomers like HIPPARCHUS and PTOLEMY used observing instruments, but there is no indication they used permanent observing stations.

The first genuine astronomical observatories, in the sense of permanent places set aside and equipped with instruments, were privately owned. The Moslem astronomer Ulugh Beg (1394–1449) had an observatory with large stone instruments at Samarkand in the 15th century, where he produced a catalog of 994 stars; at almost the same time the German astronomer Regiomontanus (Johann Müller, 1436–1476) used an observatory in Nuremberg, at which he observed the comet of 1472, later identified as HALLEY'S COMET. But probably the most notable observatory of the pre-telescope era was established in 1576 by Tycho BRAHE on the island of Hven (modern Ven) in the sound between Copenhagen and Landskrona. Tycho constructed "Uraniborg," a large observing building with living accommodation, and "Stjerneborg," in which the instruments were housed in buildings largely underground and acting as windbreaks. Tycho made positional measurements using small instruments of the highest precision. This contrasted with Ulugh Beg, who used large stone instruments, including a quadrant with a vertical wall 180 feet (55 m) high, in the hope that large size would lessen the effect of errors in marking the scales; the same principle was followed in the 1720s by the Indian astronomer Jai Singh at Delhi. But Tycho had demonstrated that an observatory with small, precise instruments gave the best results.

The advent of the telescope in 1609 began a new era in astronomy, and in the 1670s national observatories were founded at Paris (see PARIS OBSERVATORY) and at Greenwich near London (see ROYAL GREENWICH OBSERVATORY). These were permanent institutions financed by the government and staffed by professional astronomers. Although amateur observatories proliferated, Paris and Greenwich set the pattern; other countries took many years, however, to establish their own national observatories.

Subsequent developments followed advances in observational techniques and in the teaching of science in the universities. In Britain, for example, Oxford and Cambridge had small observatories in the 17th century, yet not until 1778 was a full-sized university observatory—the Radcliffe at Oxford—in use. In the United States college observatories were not established until the following century, the HARVARD COLLEGE OBSERVATORY being the first, in 1839. Large telescopes were introduced by private astronomers; they included the 48-inch (122-cm) reflector of William HERSCHEL (1789) and the 72-inch (182-cm) reflector of Lord ROSSE (1845). Institutional observatories initially favored large refractors, notably the 36-inch (91-cm) refractor of LICK OBSERVATORY (1888) and the world's largest refractor, the 40-inch (102-cm) of YERKES OBSERVATORY (1897).

Leading astronomers such as George Ellery Hale realized early this century that the large telescopes of the future would be reflectors, and that large-aperture telescopes had to be sited in the best possible observing conditions. Southern California proved particularly suitable and in 1904 Mount Wilson Observatory was established near Los Angeles. It installed its famous 100-inch (254-cm) reflector in 1917, and it set the pattern for many later

observatories. The giant 200-inch (508-cm) reflector was erected on nearby Palomar Mountain after World War II (see HALE OBSERVATORIES). The famous French PIC DU MIDI OBSERVATORY is situated in the clear air of the Pyrenees, and a mountain-top site was chosen for the KITT PEAK NATIONAL OBSERVATORY near Tucson, Arizona. The need for clearer skies has forced even old-established observatories like the Royal Greenwich Observatory and the Paris Observatory to move to better sites.

In the southern hemisphere major observatories were relatively slow to be established. Not until 1813 was a permanent observatory established for southern hemisphere research, the CAPE OBSERVATORY in South Africa. Later in the century other southern observatories followed in South Africa, Australia, New Zealand, and in South America. Only since World War II, however, have really large observatories run by universities and research institutes spread widely across both hemispheres.

The observatory has changed markedly in appearance over the years. Originally it consisted of buildings housing measuring instruments; telescopes when they came were of very long focal length and had to be kept outside. Later, with improved technology, they were mounted in domes. Now, with the arrival of radio astronomy, astronomers have reverted to the outdoor observatory, for radio telescopes are far too large to keep in a dome (particularly APERTURE SYNTHESIS arrays). This also applies to the intensity INTERFEROMETER, for which two or more optical telescopes, mounted on rails, are used in the open. The outdoor observatory has returned as a feature of 20th-century astronomy.

The latest development of the observatory is the orbiting observatory or space station. In 1962 the satellite ARIEL 1 began radio-astronomy studies, and soon a series of radio-controlled Orbiting Solar Observatory (OSO) and Orbiting Astronomical Observatory (OAO) satellites were in space, observing in wavelengths that the Earth's atmosphere blocks from the ground. The LARGE SPACE TELESCOPE, which will be put into orbit in the 1980s, will achieve a level of performance far higher than that of the finest instruments on Earth. The SKYLAB space station has already presaged a totally new kind of astronomical observatory, which will come to fruition with the Spacelab module to be launched by the SPACE SHUTTLE, in which astronomers will operate their own instruments in orbit.

occultation

The obscuring of one astronomical body by another. The Moon, for example, occults numerous stars as it moves across the sky. A star disappears behind the Moon almost instantaneously; by timing the moment of occultation precisely, astronomers can compute the Moon's exact position in its orbit. Such observations have improved knowledge of the Moon's motion considerably. When a star appears to slide along the Moon's limb near the poles (a grazing occultation), its light flashes as it passes behind lunar mountains and reappears in valleys. The occultation of planetary satellites, asteroids, and radio sources by the Moon gives precise information about their sizes and positions. The slow fading of stars as they pass behind planets with dense atmospheres, such as

Venus or Jupiter, reveals much about the nature of these planetary atmospheres. Objects of uncertain size, such as Pluto and certain asteroids, have had their diameters more accurately determined by observing their occultations of stars.

Octans (the octant)

A faint constellation at the south pole of the sky, introduced by Nicolas Louis de Lacaille. Its brightest stars are only of the fourth magnitude.

o.g.

Abbreviation for OBJECT GLASS, the front lens of a refracting telescope.

OGO satellites

The Orbiting Geophysical Observatory series of American satellites, which studied the Earth and its surroundings in space. OGO satellites were box-shaped with extendable booms carrying sensors. The satellites measured cosmic ray particles and solar emissions, the atomic processes in the Earth's upper atmosphere that cause the AURORAE, and the Earth's magnetic shell in space called the MAGNETOSPHERE. OGO results have included better mapping of the Earth's magnetic field and radiation belts. OGO satellites were launched into near-polar or very eccentric orbits that took them a third of the way to the Moon at their farthest. OGO 1 was launched in September 1964; the final satellite, OGO 6, went into orbit in June 1969.

Olbers, Heinrich Wilhelm Matthäus (1758–1840)

German astronomer, discoverer of two asteroids and five comets, and author of a famous cosmological paradox. In 1796 Olbers invented a new and simplified way of calculating comet orbits called *Olbers' method,* which was soon adopted by many astronomers. On January 1, 1802, Olbers relocated the first asteroid, Ceres, in the position calculated by Carl Friedrich GAUSS, after it had been lost by its initial discoverer, Giuseppe PIAZZI. While following Ceres, Olbers on March 28, 1802, discovered another asteroid, Pallas, the second to be discovered. Olbers became convinced that the asteroids were the shattered remains of a larger body; he searched for more fragments, and on March 29, 1807, discovered Vesta. He proposed in 1811 that comet tails always point away from the Sun because material is blown away from the head by solar radiation; this was before the discovery of RADIATION PRESSURE. The comet he found in 1815 is now called Olbers comet. Olbers is possibly best known for the so-called OLBERS PARADOX, stated in 1826. According to this paradox, light from all the stars in space should make the night sky bright. Olbers himself believed that the sky appeared dark at night because dust clouds blocked off light from the most distant stars. Astronomers now believe that the expansion of the Universe weakens the light from other galaxies that reaches Earth.

Olbers' paradox

The darkness of the night sky has long been thought to be a cosmological paradox, because in an unchanging Universe filled with an infinite number of stars the whole sky should shine with the brightness of a typical star like the Sun. In 1826 Heinrich Wilhelm OLBERS suggested that wherever one looks in the sky there must be a star, even if it is at a very great distance. Every tiny patch of the sky should thus be shining like the surface of a typical star, and the whole sky should be bright.

The discovery that all stars are concentrated into galaxies does not alter the paradox, but it is now known that the Universe is neither infinite in extent nor unchanging with time. The general expansion of the Universe was discovered a century after Olbers had formulated his paradox. Light from the stars in distant galaxies is shifted to longer wavelengths by this cosmological expansion, and in the process its energy is reduced. The most distant stars are thus progressively dimmer than they would have been in a nonexpanding Universe, and Olbers' argument breaks down.

Oort, Jan Hendrik (b. 1900)

Dutch astronomer, who has made major studies of the size and structure of our Galaxy. Oort's mentor was Jacobus KAPTEYN, who had proposed that the Sun was near the center of a lens-shaped Galaxy of stars. But in 1917 the American astronomer Harlow SHAPLEY showed that the Galaxy was much larger than Kapteyn had envisaged, and that the Sun lay toward the edge. Oort analyzed the systematic star motions discovered by Kapteyn and termed STAR STREAMING. Like the Swede Bertil LINDBLAD, Oort realized the streaming was due to the Galaxy's rotation about a center far from the Sun, with stars at different distances orbiting at different rates. From a detailed analysis of star motions and distribution Oort in 1927 accurately calculated the mass of our Galaxy (about 200 billion times the Sun's mass), its dimensions (100,000 light-years in diameter), and the Sun's distance from the center (about 30,000 light-years), thereby confirming and extending Shapley's discovery. During World War II Oort realized that radio astronomy, as pioneered by Grote REBER, provided a new technique for observing the Galaxy's structure. Oort's student Hendrik van de Hulst calculated in 1944 that hydrogen would radiate at a wavelength of 21 centimeters, and in 1954 Oort's team published a famous radio map of the spiral arms of our Galaxy, as revealed by hydrogen emission. Motions in this gas confirmed Oort's calculation that the Galaxy rotates once every 225 million years at the Sun's distance. Other studies by Oort showed that clouds of gas are being ejected from the nucleus of our Galaxy as if from an explosion; he also drew attention to gas clouds from space that the Galaxy appears to be sweeping up. Observations of the Crab nebula in 1956 showed that radio waves from it are strongly polarized, and are thus emitted by so-called SYNCHROTRON RADIATION (electrons spiraling in magnetic fields). In 1950 Oort proposed that comets exist in a vast cloud at the edge of the solar system, about 1 light-year from the Sun; they are perturbed toward the Sun by the gravitational pulls of nearby stars. This "comet cloud" hypothesis is now widely accepted.

open cluster

A shapeless, loosely packed cluster of stars. Open clusters occur in the spiral arms of the Galaxy, and are thus also called GALACTIC CLUSTERS. The HYADES and PLEIADES are typical open star clusters, showing a loose concentration of bright, blue stars and up to 400 known fainter members, in a volume typically less than 100 cubic parsecs. Stars in some open clusters are so sparsely scattered that it is difficult to decide where the cluster ends. Star-formation in a cluster occurs in a short burst until the birth of massive stars, which radiate so much energy they prevent new PROTOSTARS from condensing. Because the stars in open clusters formed virtually simultaneously, astronomers can see the different rates at which stars of different masses evolve. Open clusters have thus provided valuable data on the evolution of stars. The distances to open clusters can be found by comparing the brightness of their MAIN SEQUENCE stars with similar stars of known distance (*main-sequence fitting*), or by the MOVING CLUSTER METHOD.

Ophiuchus (the serpent holder)

A constellation in the equatorial region of the sky, visible in the northern summer; it is depicted with a serpent (represented by the constellation SERPENS) twined around it. Its brightest star, called Ras Alhague, is a hot white star of magnitude 2.04. The star 70 Ophiuchi, 16.7 light-years away, is a pair of yellow dwarf stars cooler than the Sun, orbiting each other every 87.8 years; their magnitudes are 4.27 and 6.0. The brighter of the two is also a SPECTROSCOPIC BINARY, with a period of 18 years. Another nearby star is 36 Ophiuchi, 18.3 light-years distant, also composed of two yellow dwarfs, both of magnitude 5.3, orbiting every 549 years. The star RS Ophiuchi is a well-known recurrent NOVA. Near the star 66 Ophiuchi is BARNARD'S STAR, the second-closest star to the Sun. Ophiuchus lies in the Milky Way, and includes several interesting star clusters. Although Ophiuchus is not a constellation of the zodiac, the Sun spends 20 days within its boundaries, from November 27 to December 17.

opposition

The instant at which a planet farther from the Sun than Earth appears opposite the Sun in the sky. (Planets between the Earth and Sun cannot come to opposition.) At opposition, a planet appears on the meridian (north–south line in the sky) around midnight. Opposition is the best time to observe a planet, for it is then at its closest to Earth.

orbit

The path followed by one astronomical body as it moves around another. In reality, both bodies move around their common center of gravity. For example, the center of gravity of the Earth-Moon system is 2,900 miles (4,700 km) from the Earth's center, roughly 1,000 miles (1,600 km) under the Earth's surface. The Moon follows an elliptical orbit about this point while the Earth wobbles slightly each month. For a very tiny body orbiting around a massive one, however, the movement of the large body is insignificant.

The shape of an orbit is governed by the laws of gravity and motion formulated by Isaac NEWTON and Johannes KEPLER. An orbit is precisely defined by a set of factors known as the *elements of the orbit*. These can be shown by considering a comet

following an elliptical path around the Sun (see diagram accompanying the entry COMET). The orientation of the orbital plane is defined by three angles: i is the inclination of the orbital plane; Ω defines the point where the orbit plane crosses the Earth's orbit; and ω fixes the comet orbit axis. The semimajor axis a defines the length of the ellipse, and q defines the perihelion distance, the closest point on the orbit to the Sun. The position of the comet at a precise moment is also required, normally, the time of perihelion passage, T, is given. There are thus six elements defining a comet's orbit.

The eccentricity of the ellipse, e, is often quoted and equals $1 - q/a$. For a circle, e is exactly zero; for a parabola e is exactly 1. Circular and parabolic orbits are ideal cases and are never found in reality because the slightest perturbation turns them into either ellipses or hyperbolas. An elliptical orbit is closed; an object in a hyperbolic orbit makes only one close passage to the central body and then escapes along an open orbit.

Orbits are computed by the careful analysis of observed positions, taking into account the gravitational perturbations caused by other bodies. The study of orbits is not confined to the solar system. Astronomers also calculate the relative motions of binary stars and can follow the orbit of star clusters and individual stars around our Galaxy.

Orbiting Astronomical Observatory, Orbiting Geophysical Observatory, Orbiting Solar Observatory

See OAO, OGO, OSO.

Orion

One of the major constellations of the sky, representing the hunter of Greek legend. Orion lies astride the celestial equator; it is visible in the south during the northern hemisphere winter. Its main stars are BETELGEUSE and RIGEL. Gamma Orionis, called Bellatrix, is the constellation's third-brightest star, of magnitude 1.64 and 360 light-years distant; it is 8.2 times the Sun's diameter. The central star in Orion's belt, Epsilon Orionis or Alnilam, is a blue-white star of magnitude 1.70, 1,800 light-years away and 54 times the Sun's diameter. Zeta Orionis, called Alnitak, is a double star of total magnitude 1.74, 1,200 light-years away; just south of it lies the famous *horsehead nebula* of dark gas. The third star in the belt, Delta Orionis or Mintaka, is a triple star of combined magnitude 2.46. The two visual components have individual magnitudes of 2.48 and 6.87; the brighter component is in fact a slightly variable eclipsing binary of period 5.7 days. Probably the constellation's most famous feature is the ORION NEBULA, which makes up the sword of Orion hanging from his belt; at its center is the quadruple star Theta Orionis. The entire constellation is bathed in a tenuous nebulosity; it is the center of a large field of new-formed stars in our own spiral arm of the Galaxy.

Orion nebula

A giant cloud of gas and dust in the constellation Orion, also called M42 and NGC 1976. The nebula, which makes up the sword of Orion, is about 1,500 light-years away and 15 light-years in diameter. New stars are being formed inside it. Among them are four which make up the *Trapezium,* collectively

Visible to the unaided eye as a hazy spot under Orion's belt, the Orion nebula is revealed here as a delicate fan of glowing gas, penetrated by lanes of obscuring dust. Its central regions contain luminous, young blue stars, some of which are known to have been born in the last fifty years.

termed Theta Orionis. The brightest of these is chiefly responsible for making the nebula glow (see NEBULA). The Orion nebula is the brightest nebula in the sky, visible easily in binoculars. It was discovered in 1610 by the French astronomer Nicolas Claude Fabri de Peiresc (1580–1637). Behind the bright part of the nebula is a much larger nonluminous cloud in which INTERSTELLAR MOLECULES are found. The Orion nebula is the core of a much larger STELLAR ASSOCIATION. Encircling the entire constellation is Barnard's Loop nebula, a "bubble" being blown in the interstellar gas and dust by the combined output of so many bright stars in one region.

OSCAR satellites
A series of amateur-built satellites for use by radio hams, carried into space in the spare payload capacity of American rockets. The name OSCAR stands for *O*rbiting *S*atellite *C*arrying *A*mateur *R*adio. The first two Oscar satellites carried simple beacon transmitters in the 2-meter amateur band for tracking. Oscar 3 had a repeater in the 2-meter band, allowing the first satellite contacts between radio hams; Oscar 4 worked at both 2 meters and 70 centimeters. Oscar 5, the first to be built by the Amsat amateur radio satellite corporation, broadcast telemetry at 2 meters and 10 meters, wavelengths used on Oscar 6 for amateur communications. Oscar 7 relayed communications at 70 centimeters, 2 meters, and 10 meters. The Oscar satellites have been used for educational demonstrations of space techniques. Future Oscars may include geostationary satellites. The first Oscar was sent in orbit on December 12, 1961; later launches occurred in 1962, 1965 (Oscars 3 and 4), 1970, 1972, and 1974.

oscillating Universe
A theory of cosmology which holds that the Universe expands and contracts in cycles. If its density of matter is sufficiently great, the present Universe will start to contract, slowly at first, and then catastrophically, to a final fireball similar to the BIG BANG. In this holocaust, space, time, and perhaps all laws of nature must break down completely. It is therefore impossible to know what, if anything, comes after the end of the collapse. Some astronomers, however, have suggested that the Universe "bounces" out on another cycle of expansion and contraction. This process could continue indefinitely. Such a Universe would then have no determinable beginning or end. The fireball phase might destroy all physical structure, so that each cycle would begin with new matter, and possibly even new laws of physics. The time direction of all physical processes might even be reversed, so that the next cycle would run "backwards." Unless some such phenomenon occurs, the oscillating Universe would become progressively hotter with each cycle. Current observations suggest, however, that the expansion of the Universe will not be halted by gravity, but will continue indefinitely (see HUBBLE'S CONSTANT).

OSO satellites
Orbiting Solar Observatory series of American satellites, which studied the Sun and its terrestrial effects. Experiments measured radiation from the Sun and space at ultraviolet, X-ray, and gamma-ray wavelengths, and studied interplanetary dust and emissions from the Earth's upper atmosphere. Numerous solar flares, which emit most of their intense energy at short wavelengths, were monitored by the OSOs. Mapping the Sun at different wavelengths also produced pictures of the Sun's atmosphere at different levels. Each OSO consisted of a spinning platform for all-sky scanning, surmounted by a sail containing solar cells and pointed directly at the Sun. The OSO series monitored solar activity through a complete 11-year cycle.

Satellite	Launch date	Remarks
OSO 1	March 7, 1962	Carried 13 experiments
OSO 2	February 3, 1965	Carried 8 experiments
OSO C	August 25, 1965	Launch failure
OSO 3	March 8, 1967	Carried 9 experiments
OSO 4	October 18, 1967	Carried 9 experiments
OSO 5	January 22, 1969	Carried 8 experiments
OSO 6	August 9, 1969	Carried 7 experiments
OSO 7	September 29, 1971	Carried 6 experiments
OSO 8	June 21, 1975	Carried 8 experiments. Last OSO

P

Pallas
The second asteroid to be discovered, also known as minor planet 2, found by Wilhelm OLBERS on March 28, 1802. Pallas orbits the Sun every 4.61 years

in an elliptical path between 2.11 a.u. and 3.43 a.u.; its orbit is inclined at 43°. At its brightest Pallas reaches magnitude 6, at which it is just visible to the naked eye in clear dark skies. It has a diameter of 347 miles (558 km) and a mass of about 4×10^{17} tons. Pallas has a reflectivity of only 8.7 percent; its dark surface is probably similar to the carbonaceous meteorites in composition.

pallasite
A type of stony-iron meteorite (see SIDEROLITE).

parabola
The curve produced by cutting a cone parallel to one of its sides, identical to the trajectory of a ball thrown in the air. A parabola can be thought of as an ELLIPSE so elongated it has an infinite distance between its foci, and an ECCENTRICITY of 1. It thus has one focus. The path of an object falling from an infinite distance toward the Sun would be a parabola.

paraboloid
The solid surface formed by rotating a PARABOLA about its axis of symmetry. A concave paraboloidal telescope mirror reflects light rays traveling in a line parallel to the axis to one single point, the focus of the parabola, thus forming an image.

parallax
The change in position of an object when viewed from two different positions. A finger held at arm's length jumps from right to left when viewed alternately with each eye. The jump from side to side is termed a *parallactic shift*. In astronomy, the nearer stars show a parallactic shift against the backdrop of more distant stars when viewed from opposite sides of the Earth's orbit. The change from the star's mean position is called its parallax, and the amount of parallax is a measure of the star's distance: the nearest stars show the largest parallax. The reciprocal (inverse) of a star's parallax gives its distance in PARSECS. The first star to have its parallax measured was 61 Cygni, by Friedrich BESSEL in 1838. Almost simultaneously, the parallax of Alpha Centauri was measured at the Cape of Good Hope by the Scottish astronomer Thomas Henderson (1798–1844), and that of Vega was measured by Friedrich STRUVE. The parallax of the nearest star, PROXIMA CENTAURI, is 0.762 arc seconds. The measurement of these tiny parallax shifts is a prime concern of ASTROMETRY.

The Moon, Sun, and planets also reveal a parallactic shift when seen from different points on Earth. This was the method first used to determine their distance. The parallactic shift of the Moon is almost 1° when seen from opposite sides of the Earth. The Earth's rotation carries an observer from one side to the other each day; the daily shift shown by nearby objects is termed a *diurnal parallax*.

Paris Observatory
Observatory founded in 1667 at the suggestion of the French astronomer Jean Picard (1620–1682), noted for his accurate determination in 1671 of the size of the Earth. The observatory's first directors were the distinguished CASSINI family; they were succeeded by Dominique François Jean Arago (1786–1853), and then by Urbain LEVERRIER, whom Arago had

encouraged to analyze planetary motions, leading to the discovery of Neptune. In 1876 an observatory for solar-system studies was established at Meudon, $2\frac{1}{2}$ miles (4 km) south of Paris; it was taken over by the main Paris Observatory in 1926. The Paris Observatory is now headquarters of the International Time Bureau, which standardizes the time systems of world observatories. The main telescope at Paris is a 12-inch (30-cm) refractor. Meudon has a 39-inch (100-cm) and a 24-inch (61-cm) reflector, and a 33-inch (83-cm) refractor. The Paris Observatory operates a major radio-astronomy station at Nançay, 100 miles (160 km) south of Paris; this has a MILLS CROSS radio telescope with arms 1 mile (1.6 km) and $\frac{1}{2}$ mile (0.8 km) long, and a meridian radio telescope with a fixed aerial 1,000 feet (305 m) long and 110 feet (33.5 m) high, and a movable antenna 700 by 130 feet (213×40 m).

Parkes Observatory
The National Radio Astronomy Observatory of Australia, situated 15 miles (24 km) north of Parkes, New South Wales, at an altitude of 1,230 feet (375 m), run by the Commonwealth Scientific and Industrial Research Organization (CSIRO). Its main telescope is a 210-foot (64-m) dish, opened in 1961 to work at wavelengths of 75, 21, and 10 centimeters; it has since been resurfaced to work at wavelengths down to 1.35 centimeters. This telescope was responsible for several important discoveries of INTERSTELLAR MOLECULES. A 60-foot (18-m) dish runs on rails nearby to provide interferometer spacings with the main radio telescope.

The 210-foot diameter radio telescope at Parkes, soon after completion. The height of the supporting building, which contains the control room, allows the dish to be tipped into an almost vertical position.

parsec

A measure of astronomical distance. A parsec is the distance at which the Earth and Sun would appear to be 1 second of arc apart. A star at this distance would therefore show a shift in position (PARALLAX) of 1 arc second in the sky as observed from opposite sides of the Earth's orbit. (Actually, no star is quite this close.) The origin of the term parsec—a contraction of *par*allax of one *sec*ond—is attributed to the English astronomer Herbert Hall Turner (1861–1930), an expert on measuring star positions. A thousand parsecs is termed a *kiloparsec*; a million parsecs is a *megaparsec*. The distance of a star in parsecs is the reciprocal (inverse) of its parallax in seconds of arc. One parsec is 3.2616 light-years, 206,265 astronomical units, 19.174 trillion miles, or 30.857 trillion km.

Pavo (the peacock)

A constellation of the southern hemisphere of the sky, introduced on the 1603 star map of Johann BAYER. It contains the bright globular star cluster NGC 6752, about 20,500 light-years away and 100 light-years in diameter. None of the stars of Pavo is of particular note.

Pegasus

A prominent constellation of the northern sky, best seen during the northern hemisphere fall; it represents the winged horse of Greek mythology. Its most famous feature is the so-called square of Pegasus, one corner of which is actually marked by the star Sirrah of the neighboring constellation Andromeda. The star Alpha Pegasi, of magnitude 2.57, is called Markab, an Arabic name meaning saddle. Beta Pegasi, called Scheat, is a red supergiant about 90 times the Sun's diameter; it varies irregularly between magnitudes 2.1 and 3.0 about every month.

Pegasus satellites

Three American satellites designed to monitor micrometeorites around the Earth. Each Pegasus satellite had long wings covered with aluminum foil that registered micrometeorite impacts. Their results showed that micrometeorites were not a significant hazard to men in space. The Pegasus satellites were launched on test flights of the SATURN 1 rocket: Pegasus 1 on February 16, 1965; Pegasus 2 on May 25, 1965; and Pegasus 3 on July 30, 1965. They were switched off on August 29, 1968.

penumbra

The partially shaded area in an eclipse. Areas in the Moon's penumbra at a solar eclipse see the Sun only partially covered. The term penumbra is also used to denote the outer, light-shaded portion of a SUNSPOT.

perfect cosmological principle

The principle that the Universe is always basically the same, wherever and whenever it is observed. Cosmologists have always assumed that the appearance of the Universe is independent of the observer's position in space (the cosmological principle), but that it must change with time as the stars age, and as the galaxies move apart in what is now believed to be an expanding Universe. In 1948 Hermann BONDI and Thomas GOLD suggested that the Universe as a whole is the same at all times. For this to be so, new galaxies must be forming continuously to replace dying galaxies and to preserve the average "density" of the Universe as the older galaxies move away (see CONTINUOUS CREATION). This principle formed the foundation of the STEADY-STATE THEORY of the Universe.

periastron

Point at which an object in orbit around a star comes closest to that star. The term is usually used to describe the point at which the two stars of a binary system are closest together, although it can also indicate a planet's closest approach to a star.

perigee

Closest approach to Earth by an object in the solar system. When the Earth is at PERIHELION, the Sun is at perigee. The farthest point from Earth is termed the *apogee*.

perihelion

Closest point to the Sun of a body in orbit around it. The Earth is at perihelion on about January 2 each year at a distance of 91,397,000 miles (147,090,000 km). The farthest point from the Sun is termed the *aphelion*.

period-luminosity relation

The very precise way in which CEPHEID VARIABLE stars of longer period are systematically more luminous than shorter period Cepheids. Henrietta LEAVITT discovered this relation in 1912, when investigating Cepheid variables in the Small MAGELLANIC CLOUD. All these Cepheids are at effectively the same distance from the Sun, and so the relation between period and luminosity appeared as a correlation between period and apparent magnitude.

Harlow SHAPLEY first determined the ABSOLUTE MAGNITUDE (intrinsic luminosity) of Cepheids from a statistical study of these stars in our Galaxy. By using the results of Leavitt's and Shapley's work, a Cepheid's absolute magnitude can be deduced from its period, and a comparison with its apparent

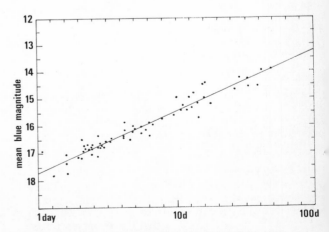

The relation between the apparent magnitude of the cepheids in the Small Magellanic Cloud and their period, plotted logarithmically.

magnitude allows its distance to be determined
(see DISTANCE MODULUS). In 1929 Edwin HUBBLE
used the apparent magnitudes of the Cepheid variables
in the ANDROMEDA GALAXY to prove that it is outside
our own Galaxy, the Milky Way. The Cepheid
period-luminosity relation is still the most accurate
method for determining the distances of nearby
galaxies.

Perseid meteors

A major annual meteor stream, first detectable about
July 25, and reaching a peak of about 65 meteors an
hour on August 12 to 13, then declining over a period
of five days. The stream radiant, which moves eastward
$1\frac{1}{4}°$ a day, is 8° north of Alpha Persei at maximum
and is at its greatest altitude at 6 A.M. Perseid meteors
enter the Earth's atmosphere at $37\frac{1}{2}$ miles (60 km) per
second, and are noted for their brilliant terminal
flares and persistent trains. The stream follows an
elliptical 140-year orbit, moving out beyond the orbit
of Uranus, and is associated with periodic comet
Swift-Tuttle 1862 III.

Perseus

A prominent constellation of the northern hemisphere
of the sky, lying in a rich part of the Milky Way;
it is best seen during the northern hemisphere fall.
The constellation represents the figure Perseus of
Greek mythology. Alpha Persei, called Mirfak, is a
supergiant yellow-white star of magnitude 1.90.
The most famous star in Perseus is the variable-
brightness ALGOL, or Beta Persei. The star Rho
Persei is a red giant that varies irregularly between
magnitudes 3.3 and 4.0 every one to two months.
In 1901, the brilliant Nova Persei flared up at the
center of the constellation, throwing off a shell
of gas still visible today. A prominent feature of
the constellation is the so-called *double cluster*, a
twin star group cataloged as NGC 869 and 884
but better known as h and Chi Persei. Both clusters
have 300 to 400 stars and are 7,350 light-years away,
with diameters of 75 to 80 light-years. M34
(NGC 1039) is a loose cluster of about 100 stars,
1,400 light-years away. The strong radio source
Perseus A is associated with galaxy NGC 1275
near the center of Perseus. Perseus is the radiant
of the PERSEID METEORS.

perturbations

Slight disturbances in the motion of a body caused
by the gravitational pull of another object.

phase

The proportion of an illuminated body that is
visible to an observer. Objects such as the Moon,
Venus, and Mercury go through a complete cycle
of phases from new (illuminated side not visible)
through a crescent, half phase (dichotomy), gibbous
phase, to full, and back again. Planets farther from
the Sun than Earth do not show such a cycle of
phases.

Phobos

The larger of the two satellites of Mars, discovered in
1877 by Asaph HALL. Phobos orbits a mere
3,700 miles (5,955 km) above the surface of Mars
(5,810 miles, 9,350 km from the planet's center). Its

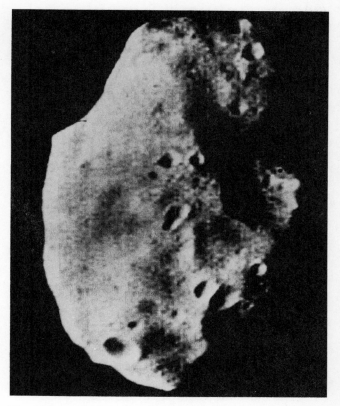

The larger of Mars' two satellites, Phobos, is shown to be
heavily cratered in this photograph taken by Mariner
9 from a distance of 3,400 miles. The indentation in the
outline at top left is a large crater seen in profile; another
large crater is visible at center right.

orbital period is 7 hours 39 minutes 14 seconds, and
it could thus be seen rising, moving across the sky,
and setting at least twice a day. Phobos rotates on its
own axis in the same time as it takes to orbit, thereby
keeping one face permanently turned toward Mars.
As photographed by the Mariner 9 space probe in
1971, Phobos appears a rocky, potato-shaped body of
dimensions 17 by $13\frac{1}{2}$ by 12 miles ($27 \times 21.5 \times 19$ km).
Like its companion moon DEIMOS, Phobos is believed
to be a captured asteroid.

Phoebe

The most distant satellite of Saturn, discovered in
1898 by W. H. PICKERING. Phoebe orbits Saturn every
550 days 8 hours 5 minutes at an average distance of
8,047,650 miles (12,951,440 km). Its diameter is about
120 miles (200 km). Phoebe orbits Saturn in a
retrograde direction (east to west).

Phoenix (the phoenix)

An inconspicuous constellation in the southern
hemisphere of the sky near the foot of Eridanus,
introduced on the 1603 star map of Johann BAYER.
There are no objects of particular interest.

photoelectric cell

Device which produces an electrical output
corresponding to the amount of light striking a
sensitive surface (the photoelectric effect). Simple
photocells, such as those used in light meters or

cameras, are rarely sensitive enough for use in astronomy. Instead, the most widely used device is the *photomultiplier tube,* which amplifies the energy of electrons liberated by the impact of light on the sensitive surface.

The tube is made of glass, with the sensitive layer at one end. Metal rings along the interior of the tube carry increasing electrical potentials which accelerate the electrons given off. The electron flow constitutes an electric current proportional to the light's intensity, and can be measured on a suitable meter. Such devices are known as photoelectric photometers.

photographic magnitude

A star's apparent brightness as measured on a photographic plate sensitive to the blue end of the spectrum. Until the 1920s, black-and-white photographic plates used for astronomy were sensitive only to blue and ultraviolet light, and brightnesses measured from them did not compare with eye estimates. A blue star, for example, would appear bright on a photograph while a red star, which might appear equally bright to the eye, would show up only faintly on the photograph. Photographic magnitudes (symbol: m_{pg}) are still occasionally used, but they have been largely replaced by the B magnitude of the UBV SYSTEM.

photometer

A device for measuring the brightness of objects. Early photometers required a human observer to make visual comparisons between the object under study and a reference of known brightness, such as a standard star. Today, electronic devices incorporating PHOTOELECTRIC CELLS (or, more usually, *photomultiplier tubes*) are used; these are known as photoelectric photometers. Filters can also help measure the brightness of the object in different wavelengths (see UBV SYSTEM).

photometry

The measurement of an object's brightness; the devices used are called *photometers.* Astronomers employ the MAGNITUDE scale for recording brightnesses. The simplest photometry is based on visual estimates of star magnitudes, as first used by the Greek HIPPARCHUS. Many amateur astronomers still make accurate visual estimates of star brightnesses to chart the fluctuations of VARIABLE STARS. However, most modern photometry is carried out either by measuring the brightnesses of individual stars with a telescope and photometer, or by photographing a field of stars and measuring the sizes of the images on the developed plate.

Photoelectric photometry—using photoelectric cells—can record brightness to within a hundredth of a magnitude and gives immediate results, but it can use too much valuable telescope time if a large number of stars are to be measured. Photographic photometry requires simpler apparatus and less telescope time, but is generally not as accurate.

The light from a bright star will spread out more in the photographic emulsion than would light from a faint one; it "burns" a larger image, and the diameter of a star's photographic image thus reveals its brightness. Because of the nature of photographic emulsions, however, a doubling in star brightness does not produce a doubling in image area—the relationship varies with brightness. Consequently, stars of known magnitude must be used as references. A photoelectric photometer, by contrast, has more regular characteristics, and fewer reference stars over the entire range are needed.

In either system, filters may be placed in the light path in order to match the color sensitivity of the detector to the UBV SYSTEM. This is known as *multicolor photometry*.

photosphere

The visible surface of the Sun. The photosphere is not an actual solid surface, but, instead, a layer of relatively dense gas about 200 miles (300 km) thick, whose temperature ranges from 9,000° at the bottom to 4,300° where it merges with the CHROMOSPHERE. Almost all the light we receive from the Sun comes from the photosphere, although the energy source lies far deeper inside. It is believed that energy travels by radiation through the Sun's core and is transported to the surface by great convection currents, the tops of which form GRANULATION cells making up the fine structure of the photosphere. Each hexagonal granule is only about 600 miles (1,000 km) across, and so the network is extremely difficult to see through the Earth's turbulent atmosphere. The centers of granules consist of rising gas about 100° hotter than the surroundings, while cooler gas flows downward at their dark boundaries; their lifetime is only 10 minutes. In addition to the granules, the photosphere also shows larger-scale turbulent motions. Other features, such as SUNSPOTS and FACULAE, are visible on the surface; but these only appear when strong magnetic fields prevent the normal convective motions.

The spectrum of the photosphere is that of a hot body at 6,000° (continuous spectrum) crossed by dark absorption lines (the FRAUNHOFER LINES). These lines reveal the relative abundances of chemical elements making up the Sun's outer layers, found to be 90 percent hydrogen and 8 percent helium, with small amounts of oxygen, carbon, nitrogen, magnesium, silicon, and iron. The gases of the photosphere are thought to be representative of the original material from which the Sun and planets formed; only the innermost parts of the Sun have been altered by nuclear reactions.

photovisual magnitude

The magnitude of brightness of a star measured by using a photographic system sensitive to the same range of colors as the eye. The color sensitivity of photographic emulsions does not quite match that of the eye; some colors are slightly emphasized and some diminished on a photograph. Filters used in combination with panchromatic emulsions do, however, provide an overall color sensitivity very similar to that of the eye. Magnitudes measured from a photograph taken this way are photovisual (symbol: m_{pv}).

Piazzi, Giuseppe (1746–1826)

Italian astronomer, who on January 1, 1801, discovered the first asteroid, Ceres, while making a star catalog at the Observatory of Palermo, of which he was founder and was appointed first director in 1790. Although Ceres looked like an ordinary star, Piazzi was

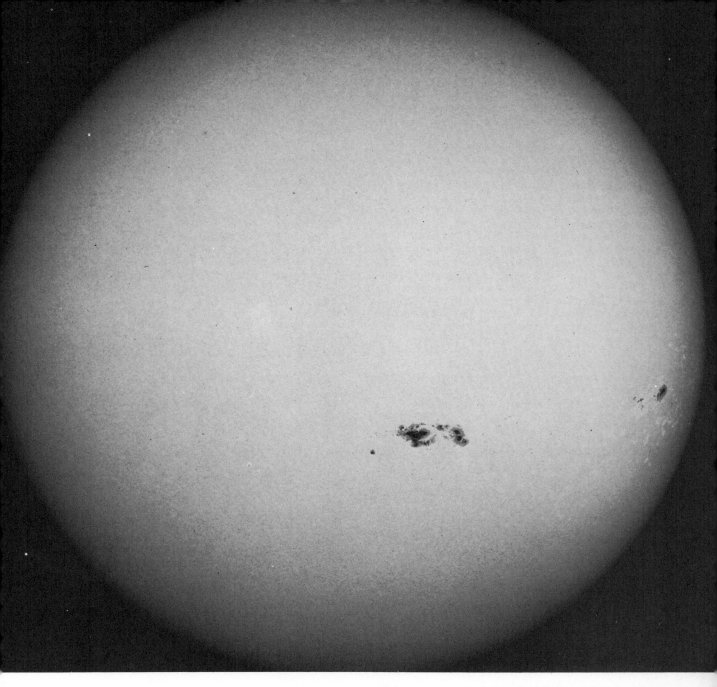

The Sun's photosphere, showing typical surface features; a large sunspot group, **center**, and a smaller foreshortened group, **right**, which is about to be carried around the limb by the Sun's rotation. The pronounced limb-darkening allows the bright facules surrounding the right-hand group to be clearly seen.

suspicious because it was not marked on the star list of Nicolas Louis de LACAILLE. He found that it moved from night to night like a small planet or comet; he lost it on February 11, 1801, as it moved into the Sun's glare. Calculations by the German mathematician Carl Friedrich GAUSS showed that the object was moving in the large gap between the orbits of Mars and Jupiter, where a group of German astronomers were searching for an unknown planet;

one of these astronomers, Heinrich OLBERS, relocated Ceres in 1802 from Gauss' predictions. In 1814 Piazzi published his catalog, containing 7,646 stars and showing the proper motions of many of them. One faint star, 61 Cygni, showed a particularly large proper motion, indicating its closeness to the Sun, a fact verified by the German Friedrich BESSEL. The thousandth asteroid, discovered in 1923, was named Piazzia in his honor.

Pic du Midi Observatory
An astronomical observatory operated by the University of Toulouse, at an altitude of 9,390 feet (2,862 m) in the French Pyrenees. Founded in 1882, it was one of the world's first high-altitude observatories, designed to exploit the advantages of improved seeing high above the turbulent layers of the Earth's

atmosphere. It was here that Bernard LYOT installed his coronagraph in 1930. A 24-inch (61-cm) refractor, installed in 1943, was used for renowned observations of the Moon and planets; it was removed in 1972. In 1964 a 43-inch (110-cm) reflector came into operation, and a 78-inch (200-cm) reflector is under construction.

Pickering, Edward Charles (1846–1919)

American astrophysicist, renowned for his major catalogs of star brightness and spectra. Trained as a physicist, Pickering became director of Harvard College Observatory in 1877 and inaugurated astrophysical studies there. He developed the meridian photometer, a device for visually comparing the brightness of a given star with Polaris, chosen as a magnitude standard. The first great star-brightness catalog, containing 4,260 stars, was published in 1884; it is known as the *Harvard Photometry*. The *Revised Harvard Photometry* later extended the work to a total of over 50,000 stars. (Unfortunately, Polaris has since been found to be slightly variable in brightness and therefore provides an unreliable comparison.) Another major project of the observatory was the *Henry Draper Catalogue*, a classification of stellar spectra; most of this work was performed by Annie J. CANNON. Pickering believed in the value of astrophotography, and set about producing a PHOTOGRAPHIC MAGNITUDE scale for stars. In 1903 he issued the first-ever *Photographic Map of the Entire Sky*, consisting of 55 plates down to magnitude 12 taken at Harvard and its southern station at Arequipa, Peru. Additionally, Pickering photographed large areas of sky on each clear night, building up a 300,000-plate Harvard photographic library, invaluable to astronomers searching for changes in the brightness and position of objects. In 1889 Pickering found the first SPECTROSCOPIC BINARY star, Mizar; he worked with Antonia Caetana de Paiva Pereira Maury (1866–1952), a niece of Henry Draper, who also developed a classification system for stellar spectra. Pickering, with his international contacts, made Harvard the center for dissemination of astronomical news and information, a role it has maintained in conjunction with the SMITHSONIAN ASTROPHYSICAL OBSERVATORY.

Pickering, William Henry (1858–1938)

American astronomer, brother of E. C. PICKERING. In 1891 he helped found Harvard's southern station at Arequipa, Peru, where in 1898 he discovered Saturn's ninth satellite, Phoebe. Pickering also reported a tenth satellite, which he called Themis, in 1905; it has never been seen again. Probably he mistook a faint star, passing asteroid, or blemish on the photographic plate for a new satellite. In 1903 Pickering published a famous photographic atlas of the Moon, which he had taken at Harvard's station in Jamaica; this became his private observatory after retirement in 1924. Pickering claimed he saw evidence of vegetation and frost on the Moon, but in fact he misinterpreted small spots on the surface whose contrast changes under varying angles of illumination. Pickering helped set up Percival LOWELL's observatory at Flagstaff, Arizona. Pickering, like Lowell, was interested in Mars, on which he discovered the dark spots termed "oases;" also like Lowell, he tried to predict the position of a planet beyond Neptune. Photographic plates taken

for Pickering at Mount Wilson Observatory actually contained images of Pluto, but one was masked by a blemish and the other was superimposed on a star; consequently they were not noticed until after Pluto had been officially discovered by Clyde TOMBAUGH in 1930.

Pictor (the painter)

A faint constellation of the southern hemisphere of the sky, introduced by Nicolas Louis de Lacaille; it contains no important objects.

Pioneer spacecraft

A continuing series of American space probes to explore the solar system. The first Pioneers were intended as Moon probes, though none succeeded. Pioneers 5 through 9 were interplanetary monitors, measuring radiation from solar storms and changes in the interplanetary magnetic field. Their results helped evaluate the danger to astronauts of solar flares, and by monitoring the side of the Sun turned away from Earth gave notice of solar outbursts that the Sun's rotation would bring into view. Pioneers 10 and 11 were the first space probes to reach Jupiter. Pioneers 12 and 13, scheduled for 1978, will orbit Venus and eject sub-probes to examine its atmosphere.

Craft	Launch date	Remarks
Pioneer	August 17, 1958	Launch failure
Pioneer 1	October 11, 1958	Intended lunar orbiter, to send back TV pictures. Fell short due to insufficient thrust, but reached 70,717 miles (113,800 km) from Earth, mapping extent of Van Allen radiation belts
Pioneer 2	November 8, 1958	As Pioneer 1. Launch failure
Pioneer 3	December 6, 1958	Intended lunar flyby; insufficient launch thrust. Reached 63,580 miles (102,300 km) from Earth, mapping intensity variations of Van Allen radiation belts
Pioneer 4	March 3, 1959	Passed 37,300 miles (60,000 km) from Moon
Pioneer	November 26, 1959	Intended lunar orbiter; launch failure
Pioneer 5	March 11, 1960	Interplanetary probe, orbiting Sun between Earth and Venus; sent data on solar flares and particles until June 26, 1960
Pioneer	September 25, 1960	Intended lunar orbiter; launch failure
Pioneer	December 15, 1960	Intended lunar orbiter; launch failure
Pioneer 6	December 16, 1965	Interplanetary probe, orbiting Sun between Earth and Venus
Pioneer 7	August 17, 1966	Interplanetary probe, orbiting Sun between Earth and Mars; with Pioneer 6 monitored solar activity
Pioneer 8	December 13, 1967	Interplanetary probe; orbiting Sun slightly farther than Earth

Craft	Launch date	Remarks
Pioneer 9	November 8, 1968	Interplanetary probe; orbiting Sun between Earth and Venus
Pioneer E	August 27, 1969	Intended interplanetary monitor; launch failure
Pioneer 10	March 3, 1972	Bypassed Jupiter at 81,000 miles (130,000 km) on December 3, 1973. Now on a trajectory that will eventually take it out of the solar system
Pioneer 11	April 5, 1973	Bypassed Jupiter at 26,725 miles (43,000 km) on December 3, 1974. Now on path that will take it to Saturn in mid-1979

Pisces (the fishes)

A constellation of the zodiac, best seen during the northern hemisphere autumn. The Sun passes through Pisces from mid-March to mid-April. None of its stars is particularly prominent; it is best located by its proximity to the famous square of Pegasus. Alpha Piscium, or Alrisha, is a multiple star of combined magnitude 3.94; the components are of magnitudes 4.33 and 5.23, revolving with a period of 720 years. Each star is also a spectroscopic binary. The most important feature of Pisces is that it contains the VERNAL EQUINOX—the point where the Sun's path around the sky cuts the celestial equator. This is the point from which the celestial coordinate of RIGHT ASCENSION is measured.

Piscis Austrinus (the southern fish)

A constellation of the southern hemisphere of the sky below Cetus, best seen during the northern hemisphere summer; its name is also written Piscis Australis. Its brightest star is FOMALHAUT.

plage

A brighter, denser region found at all levels of the solar atmosphere. The term is often reserved for a bright, dense region in the CHROMOSPHERE, where it can be studied with a spectrohelioscope. A similar region in the upper PHOTOSPHERE is often called a FACULA. X-ray and radio plages can be observed in the CORONA.

planet

A nonluminous body that shines by reflecting sunlight. Planets can be made of rock and metal, like our Earth, or of gas, like Jupiter. Any such object orbiting a star can be termed a planet; some planets may also drift dark and unknown in space, free from any star. In our own solar system there are nine major planets orbiting a glowing star, the Sun. Smaller bodies, which range from about the size of bricks to several hundred miles across, are termed minor planets or ASTEROIDS; sometimes the name *planetesimal* or *planetoid* is used. However, no real distinction can be drawn between a major planet and what is termed an asteroid. The upper limit to the size of a planet is about 1 percent the mass of our Sun (roughly 10 times the mass of Jupiter); above this the temperature and pressure at the object's center are enough for nuclear reactions to begin, so that it becomes a small star in its own right.

Formation of planetary systems. Planetary systems are believed to form as natural byproducts of the origin of stars. A star is born when a cloud of dust and gas in the Galaxy begins to callapse under its own gravitational pull (see STAR). The embryo star forms at the cloud's center, surrounded by a rotating disk of left-over gas and dust. The dust grains stick together to form a carpet of solid particles in the plane of the doughnut-shaped disk, growing bigger by collisions until their gravity is enough to pull more material toward them, eventually forming a planet. The heat of the young star evaporates the most volatile substances near to it, leaving an inner band of rocky planets. Farther away, the forming planets can acquire a dense cloak of gas from the cloud around the young star to produce gaseous giants like Jupiter; while in the farthest reaches of the planetary system the gases freeze to ice, producing a cloud of comets.

The remaining gas is driven away from the planetary system forever by the SOLAR WIND of atomic particles from the star, and left-over rock and dust particles are gradually swept up by the fully-formed planets, producing a bombardment such as scarred the surfaces of the Moon, Mars, and Mercury. Our Sun and planets are believed to have formed in this way 4.6 billion years ago. Computer models of planetary formation show that other planetary systems would look very similar.

Planets of other stars. About half the stars in the sky have one or more visible companions, forming double or multiple star systems. Given enough material in the disk around a young star, a second star may be formed, producing a close binary system. Twin systems of wider separation, with orbital periods over 100 years, may arise when the initial collapsing cloud of gas and dust fragments into several parts. The remaining stars, which seem to have no companions, may instead have planets. If each planetary system contains as many planets as our own solar system, there may even be more planets than stars in the Galaxy. Planets of other stars are too faint to be visible with Earth-based telescopes, although sensitive telescopes in space may eventually detect them. Instead, astronomers must look for the wobble in a star's PROPER MOTION across the sky, caused as a star and its planetary system rotate around their common center of gravity, like a dumbbell. Such a wobble has been detected by Peter VAN DE KAMP for BARNARD'S STAR, a red dwarf that is the second-closest star to the Sun (the closest star is actually the triple system of ALPHA CENTAURI). According to Van De Kamp, two giant planets similar to Jupiter may orbit Barnard's star; smaller planets the size of Earth might also exist, but their effects would be too small to notice. Other nearby stars believed to have planetary companions are Epsilon Eridani, Lalande 21185 (in the constellation Ursa Major), and possibly Tau Ceti (see also LIFE IN THE UNIVERSE).

planetary nebula

A gaseous shell surrounding a hot central star. Their resemblance to planets (especially URANUS) when viewed through a small telescope led Sir William HERSCHEL to give them their name in 1785. They exhibit a variety of forms, as shown by such names as

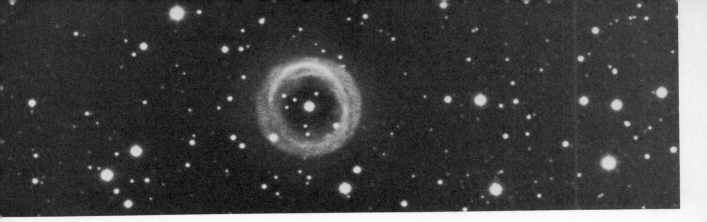

The planetary nebula Shapley I, showing the faint central star surrounded by its huge ejected shell of gas. Although the gas cloud is made to glow brightly by the radiation from the star, it is very tenuous and has barely one-tenth of the Sun's mass.

the "Dumb-bell" (M27), the "Ring" (M57), and the "Owl" (M97), although others have a more formless structure. Some 1,000 are known (it is estimated that perhaps 60,000 exist), concentrated toward the center of our Galaxy. This location demonstrates that they are old, POPULATION II objects.

The central star of a planetary nebula is very hot, with a surface temperature between 30,000°C and 150,000°C. At these temperatures, most of the radiation is emitted at ultraviolet wavelengths, so that the star appears faint optically; in some cases it is completely invisible. Although most of these stars are about the same mass as the Sun, a few are more massive WOLF-RAYET STARS. The nebula itself is a thin shell of gas that glows by absorbing ultraviolet radiation from the central star and reemitting it as visible light. Most of the gas is hot hydrogen, only 1,000 times denser than the hydrogen in space. The nebulae have diameters of the order of 40,000 a.u.; yet the amount of mass in this volume is estimated to be only 10 percent that of the Sun. Spectroscopic observations show planetary nebulae to be in a state of turbulent internal motion due to intense heating from the central star; they are also expanding at 12 to 30 miles (20–50 km) per second. This indicates they have been ejected from the central star, and that they disperse into space over a period of only about 10,000 years.

It has been suggested that planetary nebulae explosions are one way in which a massive star can throw off enough material to come within the Chandrasekhar limit (1.5 times the mass of the Sun), below which a star must fall in order to become a WHITE DWARF. Rough calculations involving the numbers of white dwarfs and planetary nebulae in the Galaxy tentatively support this theory, indicating that the planetary nebula phase may be a rapid one in the evolution of many stars.

plasma
A very high-temperature gas consisting of negatively charged electrons and positively charged atomic nuclei, or IONS. The atoms of the gas are broken up into these constituent parts either by collisions occurring between the atoms in a hot gas, or as a result of being struck by high-energy ultraviolet radiation from a nearby hot star. As a whole, a plasma cloud has no electric charge, but unlike an ordinary gas it is an electrical conductor and can thus be strongly affected by magnetic fields in space. Most of the matter in the Universe is in a plasma state (at temperatures from 10,000 to 100 million degrees), including the interiors of stars, interplanetary gas (like the SOLAR WIND), much of the interstellar gas (H II REGIONS), and the gas that fills giant clusters of galaxies.

Pleiades
A cluster of stars in the constellation Taurus, also known as M45. The Pleiades contain six or seven naked-eye stars, together with some 200 fainter members within a radius of 1°. The Pleiades are young stars, born within about the last 50 million years, and are still surrounded by remnants of the nebula from which they formed. The densest part of the nebula surrounds the magnitude-4.25 star Merope. The Pleiades are about 415 light-years away, three

This long-exposure photograph reveals clouds of dusty gas which reflect the light from the young, bright stars of the Pleiades cluster. Only six or seven stars are seen by the unaided eye, but the telescope reveals about 200 members, formed only about 60 million years ago. *The Kitt Peak National Observatory*

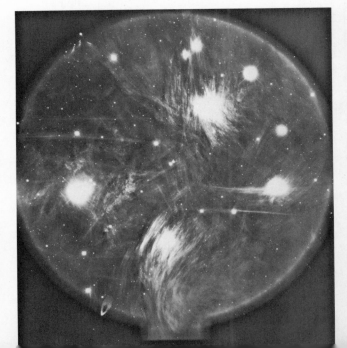

times farther than the HYADES cluster which lies nearby in the sky. The brightest Pleiades stars are BLUE GIANTS, about one-twelfth the age of the fainter and less massive stars of the Hyades. One of the bright stars, Pleione, is a so-called *shell star,* which is rotating rapidly and occasionally throws off a shell of gas in a small NOVA-like explosion. The brightest of the Pleiades is Alcyone (Eta Tauri), of magnitude 2.96.

Plesetsk

Soviet space launch site 105 miles (170 km) south of Arkhangel in northern Russia. The first space launch from Plesetsk, a former missile site, was Cosmos 112, on March 17, 1966; it has since become the most important Soviet launch site for unmanned satellites. Satellites from Plesetsk are sent into polar or high-inclination orbits, for communications, weather monitoring, and military reconnaissance.

Plow

See URSA MAJOR.

Pluto

The farthest known planet from the Sun, so distant that to any observer on Pluto the Sun would appear as little more than a bright star. Pluto's diameter is about 3,700 miles (5,900 km), and its mass is about 0.08 that of the Earth, giving a density of about 4.5 times that of water. Its surface temperature is about −230°C. The planet rotates in 6.39 days about an axis inclined at 50° to the orbital plane and has a reflectivity of 15 percent, twice that of the Moon. Pluto follows a highly unusual elliptical path inclined at more than 17° to the solar system plane, moving between 2.750 and 4.582 billion miles (4.425 and 7.375 billion km) from the Sun; its average distance is 3.666 billion miles (5.90 billion km). The planet takes 247.7 years to complete one orbit and will not return to its discovery position until the year 2177. Because of its highly elliptical orbit, Pluto can come closer to the Sun than Neptune; this will happen between January 1979 and March 1999.

Discovery. Pluto was the second planet to be discovered as the result of a deliberate search, and the first to be found by photography. After the discovery of Neptune in 1846 through calculations of its predicted gravitational effect on the motion of Uranus, Uranus still did not seem to be moving as expected. Astronomers therefore suspected the perturbing effect of yet another planet. Various astronomers, notably Percival LOWELL and William H. PICKERING, issued predictions of the unknown planet's position. Nothing was discovered, however, and so in 1929 the Lowell Observatory in Arizona began a concerted search using a 13-inch (33-cm) photographic telescope. The planet Pluto was discovered on February 18, 1930, by Clyde TOMBAUGH, from plates taken on January 21 and 29.

Pluto's nature. Unfortunately, the newcomer was smaller and fainter than expected, for it showed no rounded disk like Uranus or Neptune. Long-exposure photography failed to reveal any satellites, and it was thus impossible to make an independent determination of the planet's mass. Estimates of Pluto's diameter in 1950 by Gerard KUIPER, using the 200-inch (508-cm) Hale telescope, gave a figure of under 4,000 miles (6,400 km), later confirmed by occultation

Pluto, marked by the arrow, appears no different from the background stars in this photograph from the 200-inch Mount Palomar telescope. This most distant planet is not shown as a disk by even the largest telescopes, and its planetary nature was only revealed by its motion.

observations. This ruled out previous suggestions that Pluto was a large planet with a shiny surface that reflected light from one central spot, like a ball bearing. Subsequent studies have confirmed that Pluto is a terrestrial-type planet, intermediate in size between Mars and Mercury. The planet is therefore far too small to have had a significant effect on the paths of Uranus and Neptune; it was in fact found quite by accident. The apparent perturbations in the motion of the two outer giant planets must have been caused by inaccurate early observations, together with a degree of uncertainty about their true orbits.

It is conceivable that Pluto is not a true planet at all, but an escaped satellite of Neptune. Another possibility is that Neptune is the most distant of the family of giant planets, and that Pluto is the brightest member of a swarm of small planets moving in inclined, elliptical orbits in the outer regions of the solar system.

polar distance

The angle between an object and the celestial pole.

Polaris

The north pole star, lying in the constellation of Ursa Minor, the lesser bear, and also known as Alpha Ursae Minoris; an alternative name is Cynosura. Polaris is a double star. The main star, a yellow supergiant, is actually a Cepheid variable, of the so-called W Virginis type, varying between magnitudes 2.1 and 2.2 every 3.97 days (the average value is 2.12). It has a spectroscopic companion orbiting it every 29.6 years. Polaris also has an optical companion (not physically connected) of magnitude 9. Polaris does not lie exactly at the celestial pole, but about 1° from it. Because of PRECESSION, the celestial pole will pass closest to Polaris about the year 2012, and then move away again.

polarization

Radiation whose waves oscillate only in certain planes along the direction of travel is said to be polarized. Ordinary unpolarized light can be thought of as consisting of waves which oscillate at right angles to the direction of travel—not only up and down but in all planes. Polarization occurs if the oscillations are restricted so that only those waves in a particular plane are transmitted; this is known as *linear polarization. Circular polarization* occurs when the plane of polarization rotates with time. Radiation can be polarized when it is reflected from interstellar dust or from the dust in a nebula such as that around the Pleiades (see NEBULA). Radiation emitted by electrons in a magnetic field (SYNCHROTRON RADIATION) is also polarized. The polarization of light waves can be analyzed by a polarizing filter. Polarized radio waves are detected by a dipole antenna (a simple rod shape). The rod will receive only those waves that are polarized at right angles to it. Turning the dipole at different angles will vary the signal strength if the source is polarized.

pole

The end of an axis of rotation. The Earth's geographical poles are at the north and south ends of its rotation axis. Projected onto the CELESTIAL SPHERE, these points form the north and south celestial poles, about which the sky seems to rotate (actually an effect of the Earth's spin). The *galactic pole* is at 90° to the plane of the Galaxy, and the *ecliptic pole* lies at 90° to the plane of the Earth's orbit round the Sun (the ecliptic).

pole star

The bright star nearest the celestial pole. The north pole star is currently POLARIS. Because of the drifting effect of the Earth's axis called PRECESSION, the position of the celestial pole against the stars changes with time, so that other stars can become the pole star. In about 5,000 years, for example, the star Alpha Cephei will be the pole star, and in 12,000 years Vega. There is currently no bright star near the south celestial pole.

Pollux

The brightest star in the constellation Gemini, the twins; it is also known as Beta Geminorum. Pollux is an orange giant star, 35 light-years away, and of magnitude 1.15. It is the nearest giant star to the Sun.

Pond, John (1767–1836)

Sixth British astronomer royal, appointed in 1811. Pond's private observations, made before he had become astronomer royal, had revealed errors in star positions measured at Greenwich, caused by warping of the observatory's quadrant instrument with old age. As astronomer royal Pond therefore began to reequip the Greenwich Observatory for modern astronomy; he also established new observing methods and increased the staff from one to six. The result was an 1833 catalog listing 1,113 stars to hitherto unattained standards of accuracy. In the same year Pond introduced the first daily public time signal at Greenwich, by dropping a time ball at one o'clock down a long pole. In 1820 he recommended the founding of the CAPE OBSERVATORY in South Africa. Pond's predecessor, Nevil MASKELYNE, had begun publication of tables for seamen, called the *Nautical Almanac*. During Pond's time a separate *Nautical Almanac* Office was set up to supervise the publication; it continues today. Pond retired in 1835.

populations, stellar

The division of stars and star systems into two categories on the basis of age, population I being young stars, and population II old stars. Walter BAADE was the first to make this distinction, after observing that the bright stars in the center of the ANDROMEDA GALAXY are red, while those in its spiral arms are blue. He called the former region population II, and noted the similarity with GLOBULAR CLUSTERS in our own Galaxy. The population I spiral arms resemble our younger OPEN CLUSTERS. Baade's classification is still useful, although it is now recognized that there is a continuous range of intermediate types.

Population II stars formed some 10 billion years ago, when the galaxies first condensed. Since that time, all the massive, hot, blue stars have evolved to become red giants, and only red stars (both giants and dwarfs) remain. These stars condensed before SPIRAL GALAXIES like our own flattened to a disk, and they consequently still form a spherical halo around the Galaxy. (In elliptical galaxies all the gas condensed at the same time, and they contain only population II stars.)

Population II stars near the Sun are often found to have high velocities, because they are in fact relatively stationary while the Sun orbits the galactic center at 150 miles (250 km) per second. Because the galaxies condensed from gas containing only hydrogen and helium, population II stars contain very little of the so-called HEAVY ELEMENTS.

Many of the stars in the Galaxy's disk are only slightly younger than the population II stars, and are known as *intermediate population stars*.

Concentrated in a thin layer in the disk (the galactic plane) are the young population I stars, the youngest of which (extreme population I) are still forming from the interstellar gas and dust clouds. These nebulae and their associated young, hot, blue stars are also observed in small irregular galaxies (like the MAGELLANIC CLOUDS), and in the disks of other spiral galaxies.

Population I stars near the Sun share its motion around the Galaxy, and therefore have low measured velocities. Their composition is similar to the Sun's (a population I star). They contain heavy elements, which were created within previous generations of stars and returned to the nebulae by SUPERNOVA explosions.

position angle

Relative positions of two objects, such as the two components of a double star, measured as an angle from north through east.

Poynting-Robertson effect

The slowing in the motion of small meteor particles through their collision with solar radiation. This causes their orbits to become smaller, so that they spiral into the Sun. The effect was first predicted in

1903 by the British physicist John Henry Poynting (1852–1914), and calculated in detail in 1937 by the American physicist Howard Percy Robertson (1903–1961). Very large bodies, such as the planets, are unaffected by the collisions, but millimeter-sized particles in the asteroid belt are forced into the Sun in about 60 million years. However, the very smallest specks of dust and gas molecules in the solar system are actually blown away from the Sun by RADIATION PRESSURE.

precession
A slow wobbling of the Earth on its axis, like the wobble of a spinning top whose axis is not upright, but far slower. The net effect is to change the part of the sky at which the Earth's axis points. The Greek astronomer HIPPARCHUS discovered this effect 2,000 years ago, although its explanation had to wait for Isaac Newton's work on gravitation.

Precession is caused by the gravitational pulls of the Moon and Sun on the Earth's slight equatorial bulge. This bulge arises because of the Earth's rotation; the planet's equatorial diameter is about 26 miles (42 km) greater than its polar diameter. The Earth's equator, and hence also the bulge, is inclined at about $23\frac{1}{2}°$ to the plane of the Earth's orbit (the *ecliptic*); the Sun and Moon pull on the bulge, as if to tilt our planet back to the vertical. But instead of tilting upright, the Earth's axis swings in a cone-shaped motion, still at $23\frac{1}{2}°$ to the vertical; neither the tilt of the Earth's axis, nor the position of the poles on the globe, is changed by precession. The Earth's axis takes about 26,000 years to swing around once; this is called a *cycle of precession*.

During each cycle of precession the Earth's poles trace out a circle on the sky. Precession therefore slowly changes the position of the celestial poles. Although POLARIS is the pole star today, in 12,000 years the pole will have drifted near to Vega. The changing orientation of the Earth with respect to the stars affects the positioning against the star background of the *equinoxes,* the points at which the Earth's equator intersects the ecliptic (the plane of the Earth's orbit). The equinoxes slide once around the sky every 26,000 years, in what is termed the *precession of the equinoxes*; the equinoxes move about 50 seconds of arc westward against the star background each year. Precession also shifts the star coordinates known as right ascension and declination; this is why star positions are always given for a certain *epoch,* or reference date (currently 1950 or 2000). A winter constellation such as Orion will be seen in the summer skies after half a cycle of precession. The seasons will not be affected, however, because our calendar is based on the movement of the Sun, and the first day of northern spring will always fall around March 21 (the spring equinox). The spring equinox lay in the constellation of Aries 2,000 years ago, and is still referred to as the First Point of Aries. However, precession has now moved it into the constellation of Pisces, and it will reach the constellation of Aquarius in about 600 years. So the much-heralded age of Aquarius will not be with us for some time yet.

prime focus
The point at which the primary mirror in a reflecting telescope focuses an image. The advantages of using the prime focus for photographic work include greater efficiency, since only one reflection occurs; smaller image scale because of the shorter focal length (mirror-image distance), and hence faster registration of extended objects such as nebulae; and a wider field of view. Very large instruments, such as the 158-inch (400-cm) reflector at Kitt Peak, Arizona, and the 200-inch (508-cm) reflector at Mount Palomar, California, have a "cage" inside the tube so that the observer can sit and guide from the prime focus while taking a photograph.

Procyon
The eighth-brightest star in the sky, and the brightest star in the constellation Canis Minor; it is also called Alpha Canis Minoris. Procyon is a brilliant yellow star of magnitude 0.34, 11.4 light-years away. Its diameter is 2.17 times that of the Sun, and its mass 1.74 Suns. Procyon is actually a double star, with a faint white dwarf companion of magnitude 10.8, mass 0.63 times the Sun, and 0.01 the Sun's diameter, orbiting every 41 years.

Prognoz satellites
A series of Soviet scientific satellites to monitor solar activity and the interaction of solar atomic particles with the Earth's surroundings. The name Prognoz means "forecast." The Prognoz satellites travel in elliptical orbits, 124,000 miles (200,000 km) from Earth at their farthest, almost halfway to the Moon. They are therefore similar in nature to the American IMP (Interplanetary Monitoring Platform) series. Prognoz 1 was launched in April 1972; the most recent (December 1975) was Prognoz 4.

prograde motion
Movement of an object from west to east, the usual direction in the solar system. An alternative name is *direct motion*. Movement in the opposite direction (east to west) is called RETROGRADE MOTION.

prominence
A hot, bright cloud of gas projecting from the Sun's CHROMOSPHERE into the CORONA. There are two main types: quiescent prominences, which may last for months; and eruptive prominences, which have lifetimes of a few hours and are often associated with FLARES.

Quiescent prominences are also called "filaments," because they appear as dark, ribbonlike structures projected against the Sun's disk in spectroheliograms. They can be thousands of miles long. About a third are associated with sunspot groups, in which the prominence material, mainly hydrogen gas at a temperature of 10,000°, is observed to stream downward into the spots. Such filaments may suddenly become activated and their material blown away from the Sun at speeds of hundreds of miles per second. Most, however, are long-lived and unchanging; they are supported in the corona by the magnetic field in a manner still not understood.

Eruptive prominences may be composed either of material ejected from the Sun after a flare (surge prominences), or of condensing material falling back to the surface long after a flare has taken place (loop prominences). Their trajectories also seem to be

A quiescent prominence arching 38,000 miles (60,000 km) above the solar limb, following the loops of magnetic field between two sunspots on the photosphere below. The material flows downward, toward the points where the field comes out of the Sun; but if activated, it may suddenly blow off into space.

determined by the local magnetic field, although their velocities are higher than those of quiescent prominences—up to 600 miles (1,000 km) per second. The gas of which they are composed is also at a higher temperature, and may reach 30,000°.

Both types of prominences owe their origin to loops of magnetic field breaking through the solar photosphere, a phenomenon which also gives rise to SUNSPOTS, PLAGES, and flares. Like these, the frequency of prominences is governed by the SOLAR CYCLE.

proper motion

The motion of a star across the sky, expressed in arc seconds per year. The absolute proper motion of bright stars can be determined from precise position measurements over a period of years. The relative proper motion of a fainter star is found by comparing its position relative to other stars on photographs taken with the same telescope, some 5 to 50 years apart. Its absolute proper motion can then be calculated from the known proper motions of the brighter comparison stars. The average proper motion of all naked-eye stars is about 0.1 arc second per year; nearby stars generally show larger proper motions. The largest known proper motion, 10.27 arc seconds a year, belongs to BARNARD'S STAR.

Proton launcher

A Soviet rocket first used in 1965 to launch the series of PROTON SATELLITES, and subsequently employed for Soviet lunar and planetary probes and the SALYUT space station. In its performance it is midway between the American Saturn 1B and Saturn V rockets. The Proton launcher has a first stage made of a central core surrounded by six strap-on boosters, with upper stages added as necessary. It can place about 50,000 lb. (23,000 kg) in Earth orbit, send 12,000 lb. (5,500 kg) to the Moon, or 10,000 lb. (4,500 kg) to the planets. The diameter of its central body is about 13 feet (4 m), and its overall length is up to 225 feet (68.5 m), including the Salyut space station. Its first stage has a thrust of about 3,300,000 lb. (1,500,000 kg). The Proton launcher is sometimes known in the West as the D type.

proton-proton chain

A nuclear process, similar to that which occurs in the hydrogen bomb, by which energy is produced in stars like the Sun. At the beginning of the proton-proton chain, two protons (hydrogen nuclei) combine to form a deuterium nucleus or *deuteron*. A positron (positively charged electron) and a neutrino are also produced in this reaction. Because the neutrino is electrically neutral, its passage through the star is unimpeded. The deuteron is short-lived; within about five seconds, it interacts with further protons to form helium, releasing electromagnetic radiation in the form of gamma rays. This initial form of helium, helium-3, is unstable, and the final step in the chain is the reaction between two colliding helium-3 nuclei. A stable helium-4 nucleus is formed with the release of two protons. The net effect of the chain is to transform four hydrogen atoms into one helium atom. In the Sun, about 600 million tons of hydrogen are converted to helium each second; in the course of this process, roughly 4 million tons of matter are turned into energy. The proton-proton reaction predominates at temperatures below about 15 million degrees K. At the hotter temperatures inside large stars, the CARBON-NITROGEN-OXYGEN CYCLE prevails.

Proton satellites

A series of Soviet Earth satellites whose stated intention was to study high-energy cosmic rays. However, their great weight—12 tons for the first three and 17 tons for the last—together with the fact that they were launched with a new high-performance rocket (the PROTON LAUNCHER) suggested that they

The preparation of a Viking spacecraft destined for Mars. The solar panels of the orbiting craft are folded prior to the launch; the lander itself is nuclear powered. After assembly, the lander was sterilized to prevent contamination of the Martian surface by Earth organisms.

Left During Apollo Moon missions, the Lunar Module was stored inside the third booster stage during launch, as seen here during the Apollo 9 practice flight. The Command Module had to dock with the Lunar Module in order to extract it from the booster.

Below left An Apollo Moon mission takes off using the giant Saturn 5 booster. Nine Apollo craft went to the Moon: three of them (8, 10 and 13) did not land, however. The Apollo 9 mission was restricted to Earth orbit. The total cost of the Apollo program was 25 billion dollars, of which one quarter was spent on developing Saturn 5.

Below The Soviet Soyuz 9 flight being prepared for launch from the Tyuratam Cosmodrome in 1970—the first Soviet launch to be televised. This same pad on the barren steppes of Kazakhstan was used for the launches of Sputnik 1 in 1957 and Yuri Gagarin—the world's first spaceman—in 1961.

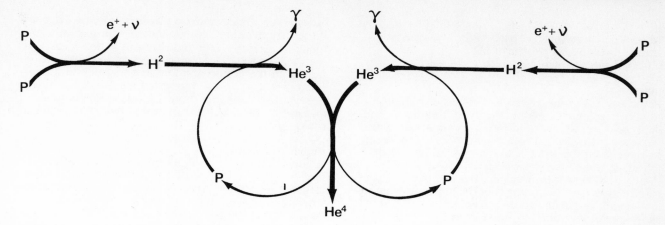

Four protons (p) are converted to a helium nucleus with release of radiation (r) neutrinos (v) and protons (e⁺) in a chain involving deuterium (H^2) and the lighter isotope of helium (He^3).

were actually associated with development of future space projects such as the SALYUT manned space station. Proton 1 was launched on July 16, 1965; Proton 2 on November 2, 1965; Proton 3 on July 6, 1966; and Proton 4 on November 16, 1968.

protoplanet
A cloud of gas, dust particles, and rocks in the process of forming into a planet.

protostar
The early stage in a star's formation, between the collapse of a gas cloud and the onset of nuclear burning at the star's core. Some astronomers think that small, spherical interstellar clouds called GLOBULES are such contracting clouds. As the cloud contracts it becomes hot, radiating energy at infrared wavelengths. Such radiation has been detected from protostars in the Orion nebula. T TAURI STARS are believed to be protostars at a late stage, but still surrounded by remnants of the cloud from which they formed.

Proxima Centauri
The nearest star to the Sun, 4.3 light-years away. It is a faint red dwarf, of magnitude 10.7, the third member of the ALPHA CENTAURI star system. It is called Proxima because it is about 0.1 light-year nearer to us than the other two stars of Alpha Centauri. Proxima Centauri was discovered in 1915 at Johannesburg by Robert Thorburn Aytoun Innes (1861–1933), Union Astronomer of South Africa and a double-star specialist who in 1927 published the *Southern Double Star Catalogue*. Proxima Centauri appears over 2° away from the other two stars of Alpha Centauri; it must take about 1 million years to orbit them. Proxima is a FLARE STAR.

Ptolemy (Claudius Ptolemaeus; c.100–c.178)
Alexandrian astronomer, geographer, and mathematician, the last great astronomer of the ancient world. Ptolemy wrote many works: the *Tetrabiblos* on astrology, treatises on planetary

hypotheses, on mechanics, on music, on map projections, and one on optics in which he discussed the effect of refraction in the Earth's atmosphere on the apparent positions of celestial bodies. His most important works were his *Guide to Geography* and *Almagest*.

The *Guide to Geography* included details about mapmaking, and lists of places with their longitudes and latitudes. Although Ptolemy placed the equator in too high a latitude and took a value for the Earth's size nearly one third less than that determined by ERATOSTHENES, the book was regarded as authoritative until the great age of discovery began.

Ptolemy's greatest work was *The Mathematical Collection*, later known to the Arabs (through whom the book came to us) by the Greek superlative "megistee" (the greatest), corrupted into the name *Almagest*. In this encyclopedic work Ptolemy discussed the motion of the planets and the layout of the Universe. He extended the star catalog of the great Greek astronomer HIPPARCHUS with his own observations; and he discussed in detail whether or not the Earth is fixed in space, concluding that all evidence indicated it must be stationary. Ptolemy explained the movements of the planets using the basic concept of uniform motion in a circle, together with mathematical devices proposed in the third century B.C. by APOLLONIUS OF PERGA. These made use of large and small circles (*deferents* and *epicycles*) and a movable eccentric (a large eccentrically mounted circle), and with them Ptolemy was able to describe all planetary motions in detail.

A persuasive synthesis of all Greek astronomical knowledge, Ptolemy's *Almagest* exerted a profound and lasting influence on all subsequent generations of astronomers. It would not be seriously challenged for 1,300 years.

Pulkovo Observatory
The Central Observatory of the U.S.S.R. Academy of Sciences, located at an elevation of 250 feet (75 m) near Leningrad. It was founded in 1839 by F. G. W. STRUVE, its first director, with a 15-inch (38-cm) refractor that was then the largest in the world. In 1885 a 30-inch (76-cm) refractor by Alvan CLARK was installed, the world's largest until the opening of the LICK OBSERVATORY in 1888. Destroyed during World War II, the observatory was reopened in 1954 and now has a $27\frac{1}{2}$-inch (70-cm) reflector, $25\frac{1}{2}$-inch

(65-cm) refractor, and a 20/27½-inch (50/70-cm) Maksutov camera. The observatory also operates a transit radio telescope measuring 345 by 10 feet (105 × 3 m).

pulsar

A radio source that emits short pulses of radiation at very regular intervals, typically of about a second. The radiation is produced by a very small and dense NEUTRON STAR, which rotates and "flashes" a beam of radio waves like the beam from a lighthouse.

Pulsars were discovered accidentally in 1967 by Antony HEWISH, who was investigating the scintillation (or "twinkling") of distant radio sources. The regularly pulsing sources seemed so artificial they were first dubbed "Little Green Men" signals; but Thomas GOLD soon showed that a rotating neutron star could in theory produce the pulses. This hypothesis was confirmed in 1968, when a pulsar was discovered in the center of the CRAB NEBULA, the debris of a star which was seen to explode in the year 1054. Walter BAADE and Fritz ZWICKY had predicted that neutron stars are formed in such SUPERNOVA explosions, and the radio pulsar coincided with a star long thought to be the remains of the supernova. In 1969 the light from this star was found to be flashing in time with the radio pulsar, which has a period of only 1/30 second. In the same year the pulsar was found to emit X rays with the same periodicity. The radiation at all wavelengths is produced by electrons moving in the neutron star's magnetic field, which is a trillion times as strong as the Earth's field. This SYNCHROTRON RADIATION travels outward in a beam, causing the pulsar to flash as it rotates.

Over 100 pulsars have now been discovered, and their periods range from the 1/30-second of the Crab pulsar to about 3 seconds. Apart from the Crab pulsar, they can only be detected at fairly long radio wavelengths of about 1 meter.

The electrons in space between the pulsar and the Earth slow down the radiation, so that different wavelengths from each pulse arrive at different times. The measured delay between wavelengths reveals the distance of the pulsar. Most pulsars lie in the disk of our Galaxy, where the stars and gas are concentrated, and are at distances of 300 to 30,000 light-years from the Sun.

Although their timekeeping is almost as good as that of atomic clocks, many pulsars are slowing down at a rate of about a billionth of a second a day as they radiate away their energy. This provides a key to a pulsar's age. The Crab pulsar is found to be about 1,000 years old, in good agreement with the date of the supernova outburst. Most pulsars, however, are millions of years old, and the comparative youth of the Crab pulsar probably accounts for its easy visibility at optical and X-ray, as well as at radio wavelengths. The second-fastest pulsar is about 10,000 years old; it lies in the constellation Vela amid a filamentary nebula, which is the gaseous remnant of the supernova that produced it. Occasionally, both the Crab and Vela pulsars suddenly speed up during their otherwise steady slow-down in spin. This is probably the result of an internal change in the neutron star (a "starquake").

After 10 million years a pulsar's radio emission fades away. Older neutron stars can only be detected if they are orbiting another star, so that gas falls onto them, heating them up and emitting X rays (see X-RAY ASTRONOMY). Some sources emit X rays in pulses, with a period of a few seconds; in one star (Hercules X-1) optical pulses are also seen. In these X-ray pulsars the radiation is produced by hot gas, and not by the synchrotron process which operates in radio pulsars.

Puppis (the stern)

A constellation of the southern hemisphere of the sky next to Canis Major; it originally formed part of the larger constellation Argo Navis, the Argonauts' ship, which was broken into smaller parts by Nicolas Louis de LACAILLE. Puppis lies in the Milky Way and contains several star clusters; the most prominent is M47 (NGC 2422), 3,700 light-years away and about 27 light-years in diameter. L_1 and L_2 Puppis are two stars that appear close to each other in the sky but are not physically connected (they are an *optical double*). L_1 is of magnitude 5.04 and 540 light-years away. L_2 is a red giant, 180 light-years away, which varies irregularly between magnitudes 2.6 and 6.0 about every 140 days.

Pythagoras (c.580 B.C.–c.500 B.C.)

Greek philosopher and mathematician, who believed that the Earth was spherical in shape and that it lay at the center of a spherical Universe. He realized that the Sun, Moon, and planets follow their own paths around the sky. He introduced the idea that each is carried around the Earth on a crystalline sphere; each sphere produced a musical note (the music of the spheres) as it turned. The concept of the heavenly spheres was developed into an extremely intricate system by EUDOXUS and others.

Pyxis (the compass)

A faint and insignificant constellation of the southern hemisphere of the sky, below Hydra, once part of the larger constellation of Argo Navis; it was made a separate constellation by Nicolas Louis de LACAILLE. Pyxis contains no objects of interest.

Q

QSO

Abbreviation for quasi-stellar object, popularly known as a QUASAR.

Quadrantid meteors

One of the principal meteor streams, reaching a sharp peak of from 45 to 250 (average about 110) meteors an hour on January 3 to 4 each year; the display lasts only a few hours. The stream is named for the obsolete constellation of Quadrans Muralis. At maximum the radiant is 25° east of Eta Ursae Majoris. Currently the strongest annual shower, the Quadrantids are bluish meteors with fine persistent trains; they have no known associated comet.

quartz-crystal clock

A highly accurate clock based on the rapid oscillations of a crystal of the mineral quartz. When a quartz crystal is compressed, a small voltage forms across it.

Similarly, a voltage applied across a quartz crystal distorts its shape. (Technically, this is termed the *piezoelectric effect*.) Feeding back the voltages formed by the distortion of the quartz crystal will set it oscillating about 100,000 times a second. These oscillations act like a very fast but highly accurate pendulum, which can be used to control an electric circuit. The oscillations are reduced by frequency dividers to drive an electric clock. Quartz-crystal clocks superseded pendulum clocks for keeping Greenwich Mean Time in 1942, and they have since been widely adopted as accurate timekeepers throughout the world. At their best they are accurate to about .000001 second a day. The quartz crystal is kept in an evacuated bulb at constant temperature to insulate it from outside effects. However, the frequency of a quartz crystal will slowly change with age, and to maintain its accuracy a quartz crystal clock must therefore be regularly adjusted. This is done by calibrating it against the standard frequency produced by an ATOMIC CLOCK.

quasar

An object which appears as a star-like point of light, but which emits far more energy than an entire galaxy. The name quasar is a contraction of quasi-stellar object (QSO). Many quasars also emit radio waves, and these are sometimes called quasi-stellar radio sources (QSS).

Quasars were discovered in 1963, when the first accurate measurements of radio-source positions showed that some coincided with star-like objects with very unusual optical spectra. Maarten SCHMIDT interpreted these spectra as having very large RED SHIFTS. Over 200 quasars are now known, showing red shifts ranging from 0.036 to 4.53. This means that, in the case of the latter quasar, OQ 172, the wavelength of its radiation has been lengthened 4.53 times, so that the visible light received from the quasar was actually emitted at ultraviolet wavelengths, 4.53 times shorter.

All distant galaxies also have red-shifted spectra, caused by their recession in the expanding Universe; the red shift, and thus the velocity, increases with distance (HUBBLE'S LAW). Most astronomers accept that the large red shifts of quasars are caused by their fast rates of recession, and therefore that they are very remote—hundreds of millions or billions of light-years from our Galaxy. They appear comparatively bright, however, and for them to do so at such an enormous distance, many of them must be a hundred times more luminous than even the most luminous galaxies we know. Moreover, many quasars are variable in light output, halving or doubling their brightness in a few months. This indicates that the light-emitting region can be only about a light-year across, 1/100,000 the size of a galaxy. A quasar is probably, in fact, the very active central region (nucleus) of a galaxy too distant to be seen.

Quasar spectra are very similar to those of the bright nuclei of the much nearer N-GALAXIES, and some quasars show a surrounding "fuzz" on long exposure photographs which appears to be an enveloping galaxy. How such a large amount of energy is produced in a volume only a light-year in diameter remains a mystery; it could be due to a large number of exploding stars (SUPERNOVAE) and their

remnants (PULSARS and BLACK HOLES), or else to super-massive stars.

Some astronomers have speculated that quasars are much nearer, and therefore much less luminous; but the explanation of their red shifts then becomes a problem. One suggestion is that quasars have been ejected from galaxies relatively nearby, and that their red shifts are due to some undiscovered physical law.

Many quasar spectra show absorption lines produced by gas clouds lying between the quasar and the Earth. The number of clouds observed can be as many as 17, always with a lower red shift than the quasar emission lines (which indicate its distance). The clouds may be in intergalactic space along the line of sight to the quasar, or else they may surround the quasar, having been ejected from it some time ago.

A new class of objects, probably related to quasars, was discovered about 1970. They are named *lacertids,* after their prototype BL Lacertae, which was once thought to be a variable star. The lacertids are galactic nuclei which vary even more rapidly than quasars and have no lines in their spectra. They are probably about as far away as the nearest quasars.

R

radar astronomy

The investigation of bodies in the solar system by reflecting radio waves off them. Usually the same radio telescope is used both to transmit a powerful pulse of radio waves and to detect the very faint echo returning moments later. Because the relative amount of energy that returns drops off sharply as increasingly distant objects are investigated, radar astronomy is limited to distances within the solar system.

Radar astronomy began in the 1950s, when "contact" was made with the Moon. Because the speed of the radio waves (the velocity of light) is very accurately known, the time taken for the echo to return allowed a precise determination of the Earth-Moon distance to be made.

Contact was made with Venus in the 1960s. Since the relative sizes of the planets' orbits are known from KEPLER'S LAWS, an accurate distance for Venus led to an extremely precise absolute scale of size for the whole solar system. As a result, the distance from the Earth to the Sun (the ASTRONOMICAL UNIT) is now established with an accuracy of well within a mile.

The way in which radio waves are reflected depends both on the nature of the surface structure and on the presence of any large irregularities like hills or craters. Mercury and Mars, for example, must have loose surface layers, because their radar properties are similar to the Moon's. Radar echoes have also revealed craters on Venus, whose surface is permanently hidden from optical astronomers by clouds.

The rotation of a planet slightly alters the wavelengths reflected from its edges as a result of the DOPPLER EFFECT. In 1962 the rotation period of Venus was first determined by radar techniques; it was found to be 243 days in a retrograde direction (opposite to that of the other planets). Another surprise followed in 1965, when radar astronomers determined that Mercury rotates in 58.6 days, not the 88 days which had long been accepted on the basis of visual observations.

radial velocity

The velocity of a star along the line of sight, expressed as a negative figure if the star is moving toward us, and positive if moving away. It is measured by observing the shift in wavelength of a moving star's spectral lines arising from the DOPPLER EFFECT. Corrections for the Earth's motion are applied to give a radial velocity relative to the Sun. The average radial velocity for stars in the solar neighborhood is about ± 12 miles (20 km) per second.

radiant

The direction from which a METEOR appears to radiate as it enters the atmosphere. The radiant of a sporadic meteor can be determined by triangulation of its path from two sites about 100 miles (160 km) apart. If the meteor's radiant and velocity are known, its orbit around the Sun can be calculated. Meteor streams contain many meteoroids moving in parallel paths with almost identical radiants. Thus stream meteors, which can occur anywhere in the sky, have paths which all appear to radiate from one point. The radiant is actually caused by a perspective effect, similar to that which makes parallel railroad tracks appear to radiate from a distant point.

radiation

See ELECTROMAGNETIC RADIATION.

radiation belt

See VAN ALLEN BELT.

radiation pressure

The tiny force exerted on bodies by a beam of light. The pressure of sunlight on a giant sail could conceivably be used to propel craft through the solar system (this idea is termed "solar sailing"). Radiation pressure helps push the dust in comet tails away from the Sun (the main effect actually comes from the atomic particles of the SOLAR WIND). Larger bodies are not pushed away by radiation pressure, but instead lose momentum and spiral into the Sun (POYNTING-ROBERTSON EFFECT). Radiation pressure from the hot insides of a star is thought to play a part in supporting the star's structure.

radio astronomy

The observation of the Universe at radio wavelengths. In 1931 Karl JANSKY, a physicist working at the Bell Telephone Laboratories, first discovered that some of the "static" that interferes with radio communications comes from space. Jansky and Grote REBER showed that this noise comes from the whole of the MILKY WAY. But radio astronomy developed rapidly only after World War II had pushed forward the development of radio techniques and equipment.

Nature of radio noise from space. Radio waves are a part of the electromagnetic spectrum, with wavelengths roughly a million times longer than those of visible light. The Earth's atmosphere absorbs most wavelengths of electromagnetic radiation; only visible light, some infrared radiation, and some radio waves actually reach the ground from space. Radio waves shorter than about 1 centimeter are absorbed by atmospheric water vapor, and extra-terrestrial radio waves longer than 30 meters are reflected back into space by the Earth's ionosphere. But the wavelengths between 1 centimeter and 30 meters constitute the radio "window" through which a vast amount of astronomical information has been gathered since the 1950s.

All hot objects, such as the Sun, emit radiation over a wide range of wavelengths; such *thermal* radiation accounts for part of the radio radiation received from space. But most of the powerful radio sources are due to the interaction of very fast electrons with magnetic fields. The electrons spiral along the magnetic field, emitting SYNCHROTRON RADIATION. Some gases emit radio waves at certain specific wavelengths ("lines"). The most important is hydrogen, which emits at 21 centimeters wavelength; but fairly complex interstellar molecules have recently been identified in space from their characteristic radio wavelengths (see INTERSTELLAR MOLECULES).

Solar system radio astronomy. An early application of radio techniques in astronomy was to track the trail of hot, electrically conducting gas (IONS) which meteors leave in the atmosphere. This revealed many previously unknown daytime meteor showers.

The Sun and its features such as sunspots and flares can be studied at radio wavelengths. The Moon and most of the planets, however, are quiet at radio wavelengths, except for their weak thermal radiation. Much more can be learned about them by "bouncing" radio waves off their surfaces (RADAR ASTRONOMY). The exception is JUPITER, which has a strong magnetic field and trapped radiation belts of charged particles (like the Earth's VAN ALLEN BELTS), which emit synchrotron radiation, and occasional radio bursts.

The radio sky. The brightest object in the radio sky is the Milky Way. Its radio emission comes not from stars, but from synchrotron emission by the COSMIC RAY electrons moving in the interstellar magnetic fields. Very few stars emit enough radiation to be detected by even the most sensitive radio telescopes. However, the sky is full of individual sources, ranging in apparent size from mere points to disks the size of the full Moon and larger. Many of these sources are within our own Galaxy, and appear close to the line of the Milky Way in the sky. Extragalactic sources, on the other hand, are evenly scattered over the sky.

Radio sources were originally named for the constellations in which they occur, with a letter indicating their order of brightness: for example, the CRAB NEBULA, the brightest radio source in Taurus, was designated Taurus A. Surveys of the radio sky have now revealed so many sources that this system has been dropped in favor of more logical schemes. For the northern hemisphere of the sky, the most widely used catalog is that from the third Cambridge survey (3C), produced at the MULLARD RADIO ASTRONOMY OBSERVATORY; the most extensive surveys of the southern sky have been made at PARKES OBSERVATORY in Australia.

The Parkes survey introduced the now widespread convention of identifying each source by a six-digit number which gives its coordinates in the sky. Thus the source 1954–55 is at right ascension 19 hours 54 minutes and declination −55°. This number is prefixed by letters identifying the observatory which discovered the source—PKS for Parkes, B for Bologna, and so on. (The prefix PSR is commonly used for pulsars, so that the Crab nebula pulsar, for example, is PSR 0531 + 21.) Many early surveys, including the 3C catalog, do not use this logical approach, but simply allocate numbers to the sources in order of right ascension. The 3C catalog runs from 3C 1 to 3C 470.

Galactic radio astronomy. Most strong radio sources within our Galaxy are either the expanding shells of hot gas from SUPERNOVAE (exploded stars), such as the Crab nebula; or else they are the hot gaseous nebulae (H II REGIONS) in which stars are forming. Such sources can also be seen optically, but one class of galactic objects that can be studied virtually only by radio observation are PULSARS, the dense, rapidly rotating remnants of supernova explosions. Since their accidental discovery in 1967 over 100 have been detected, but only one or two are visible optically.

The structure of the Milky Way as a whole can be readily studied by radio astronomy, because radio waves are not absorbed by the interstellar dust that limits the view of optical astronomers. The large-scale distribution of neutral hydrogen gas (H I REGIONS) is revealed by 21-centimeter line observations, which show the hydrogen concentrated into spiral arms. The magnetic field threading the Galaxy can be investigated both by the general synchrotron emission of the Milky Way, and by its effect on the polarization of radio waves from sources outside the Milky Way. The arrival times of pulsar pulses also give a measure of the number of free electrons in interstellar space.

Extragalactic radio sources. Nearby spiral galaxies, like the Andromeda galaxy, appear similar at radio wavelengths to the Milky Way. They show both the general synchrotron radiation and the 21-centimeter line from hydrogen clouds in the spiral arms. Study of the latter are improving our understanding of how stars form from hydrogen clouds, and of the mechanism which produces the spiral arms in galaxies.

The exact wavelength received from a hydrogen cloud depends on its velocity toward or away from us (the DOPPLER EFFECT). Careful measurements reveal the rotation of the gas in a spiral galaxy, enabling its mass to be calculated. The Doppler shift of wavelength also gives the speed of the galaxy along the line of sight (*radial velocity*). This technique can only be applied to galaxies less than about 50 million light-years away because of the limited sensitivity of present-day radio receivers. The Doppler shift of more distant galaxies is found from the optical spectral lines; the distance of the galaxies can then be derived using HUBBLE'S LAW.

One early surprise in radio astronomy was the discovery that one of the brightest sources in the sky (Cygnus A) is associated with a very distant galaxy, 1 trillion light-years away. Nine years later, in 1963, Maarten SCHMIDT found even more remarkably that some radio sources which appear as star-like objects in optical telescopes have very large Doppler shifts; according to Hubble's law they must be as far away as the RADIO GALAXIES like Cygnus A. These QUASARS (or quasi-stellar objects) are now thought to be very powerful explosions at the centers of distant galaxies.

Radio galaxies and quasars emit up to a million times more radio energy than a normal galaxy like the Milky Way, often from vast clouds on either side.

However, in many quasars, the radio emission comes from a central region only a few light-years across, where the explosion is taking place. Some radio galaxies and quasars have both a small central component and extended clouds. The radio emission from the central source usually varies slowly over time, and the rate of change at different wavelengths should help explain how vast amounts of energy are converted into radiation in these sources.

Radio astronomy and cosmology. Radio astronomy has contributed to COSMOLOGY, the study of the whole Universe, in a number of important ways. Radio sources act as probes of the expanding Universe. Even traveling at the velocity of light, the radio waves from the most distant quasars have taken billions of years to reach us, and these distant sources thus give a view of conditions early in the development of the Universe. By counting the numbers of sources of different apparent intensities, cosmologists can determine how the Universe is evolving, and it is found that the number of radio sources in the Universe was greater in the past. This evidence was the first to indicate that the STEADY-STATE THEORY (of an unchanging Universe) is incorrect.

One of radio astronomy's main contributions to cosmology was the 1965 discovery of the BACKGROUND RADIATION, a faint, uniform "hiss" of radio waves from all directions in space, apparently a result of the BIG BANG from which the Universe began.

radio galaxies

Distant galaxies that are very powerful sources of radio waves. Their radio output can be up to a million times that of our own Galaxy. This radiation originates not in the radio galaxy itself, but in two huge clouds, one on either side of the galaxy. In the largest radio source yet discovered, known as 3C 236, these clouds lie along a line 20 million light-years in total length.

The first radio galaxy to be identified was Cygnus A, the brightest radio source in the constellation Cygnus and the second-brightest in the whole sky. In 1954 it was found to coincide with a faint object resembling two galaxies close together. At first, astronomers thought that two galaxies were colliding, and that the radiation originated in the collisions of gas clouds. As more radio galaxies were identified, it became clear that most are single galaxies, not involved in any collision. They are in fact giant elliptical galaxies, containing 10 trillion stars in a roughly spherical volume 300,000 light-years across. The Cygnus A galaxy seems to have a dark dust lane across its center, giving it the divided appearance originally suggesting two close galaxies.

At the center of some radio galaxies (the N-GALAXIES) is a small, highly luminous nucleus, which often produces as much light as all the stars in the rest of the galaxy. When such a small nucleus far outshines the rest of the galaxy, only it will be seen on photographic plates. This apparently star-like object is known as a "quasi-stellar object," or QUASAR.

All types of radio galaxies generally have the two radio-emitting clouds outside the optical galaxy, although the galactic nucleus itself also emits shorter radio wavelengths. The energy supply to the radio clouds must come from the nucleus, where a vast amount of energy is being produced by processes not yet understood. The clouds themselves emit radio waves through a process known as SYNCHROTRON RADIATION, produced when high-speed electrons move in a magnetic field. How the energy reaches the clouds is still a subject of dispute. One theory envisages two continuous streams of electrons and magnetic fields leaving the nucleus, and spreading into clouds where they strike the thin gas between the galaxies. Alternatively, the "slingshot" theory suggests that gravitational interactions in the nucleus can fling out BLACK HOLES in opposite directions, each accompanied by magnetic fields and electrons.

The radio emission from the largest radio galaxy known, 3C 236, is shown to extend over 20 million light-years on this "photograph" of the radio brightness. The galaxy powering the source lies at the center of the cross, and its optical size is only 1/10 the width of the cross.

The second largest radio galaxy known, DA 240, comprises two radio-emitting clouds, totaling 7 million light-years in length. On this "photograph" showing the radio brightness, the bright spot in the center is radiation from the galaxy's nucleus. Other bright spots are distant radio sources, while the arcs are artifacts produced by the telescope.

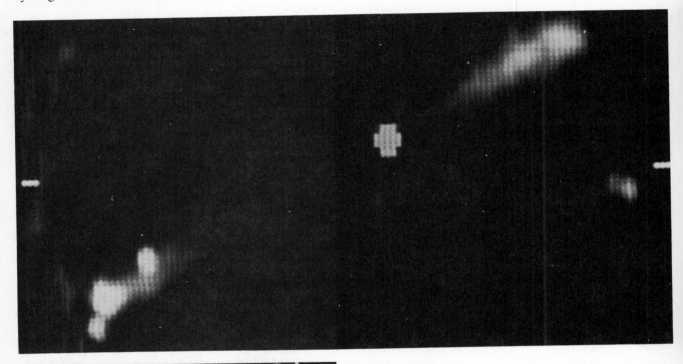

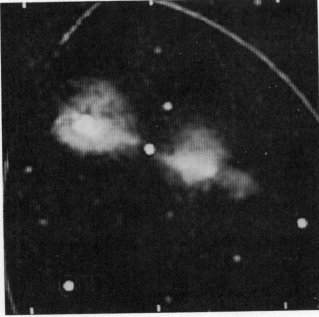

radio telescopes

Instruments for receiving radio waves from space. The telescope includes an antenna, an amplifying system, and some form of recorder to preserve the observation.

An antenna (or dipole) is basically a short metal rod, in which electric current is produced by incoming radio waves. In radio astronomy, radio waves must be collected over a large area, or aperture, because the signals are so weak, and because a large aperture gives a better ability to resolve small details (the telescope is said to have a smaller *beamwidth*). A large number of dipoles spread over a field and connected together forms a very sensitive aerial, and in the MILLS CROSS arrangement the aperture can be as large as 1 mile (1.6 km). Rapid-intensity changes (scintillation) in distant radio sources were first investigated with a 4-acre (1.6-ha) field of dipoles at the Cambridge radio observatory. The high sensitivity of this telescope led to the accidental discovery of the faint, regularly "ticking" PULSARS.

Dish antennae. Most radio telescopes have in fact only one dipole, and the radio waves are collected and reflected onto it by a large metal dish, analogous to the mirror of an optical reflecting telescope.

Because radio waves are about a million times longer than light waves, a radio dish must be much larger than an optical telescope to have satisfactory resolving power. But for the same reason, the mirror surface need not be as perfectly smooth as that of an optical instrument, and by using a steel frame covered with wire mesh relatively large dishes can be built without great difficulty.

Parabolic dishes are the most versatile antennae for radio-telescope systems, useful over a wide range of wavelengths. Radio astronomers often prefer to observe at short wavelengths, however, at which the resolution obtainable with a given aperture is better. The smoothness of the dish sets a limit on the shortest wavelength that can be used. Most dish instruments have the added advantage of being fully steerable (they can be pointed to any part of the sky).

As radio astronomy has developed, technological advances have progressively improved the angular resolution of dish antennae, and some of the largest can now achieve a beamwidth of no more than 1 arc minute (1/30 the apparent diameter of the Moon). At the same time, the shortest wavelengths that can be amplified electronically have been reduced from about 10 centimeters in the early 1960s to a few millimeters today.

As a result, parabolic reflectors can now be divided into two groups. Smaller instruments with very smooth surfaces are used for millimeter wavelength studies, such as the detection of INTERSTELLAR MOLECULES. They were pioneered by a 36-foot (11-m) instrument at the KITT PEAK OBSERVATORY. On the other hand, the giant antennae familiar from the 1950s and 1960s are now used chiefly at centimeter wavelengths. The largest instruments are the 330-foot (100-m) dish at the EFFELSBERG RADIO OBSERVATORY in West Germany, which has a resolving power of 0.8 arc minute at 2 centimeters wavelength; the 250-foot (75-m) JODRELL BANK dish (5 arc minutes at 10 centimeters); and the 210-foot (65-m) dish at PARKES OBSERVATORY in Australia (3 arc minutes at 5 centimeters).

In a class of its own, the giant 1,000-foot (328-m) dish of the ARECIBO RADIO OBSERVATORY was built simply by bulldozing smooth a natural depression in the mountains of northern Puerto Rico and covering it with fine wire mesh. This sacrifices steerability for size, giving a resolution of 0.3 arc minute at 3 centimeters over the strip of sky scanned by the telescope as the Earth rotates. Similar fixed antennae on a smaller scale have been used for Moon studies and for observing the Crab nebula. As an inexpensive means of undertaking particular limited tasks, such instruments are very useful.

Interferometers. The outputs from two dish antennae can be combined electronically to form an INTERFEROMETER system. By moving the antennae relative to one another a radio "picture" of a source can be built up, equivalent to that which would be obtained from a much larger dish. This technique of APERTURE SYNTHESIS was developed by Sir Martin RYLE at the MULLARD RADIO ASTRONOMY OBSERVATORY at Cambridge, England, where the world's largest aperture-synthesis telescope is located. It is equivalent to a single dish 3 miles (5 km) in diameter, and produces radio maps with an angular resolution of 0.01 arc minute, equal to photographs

Each of the identical 90-foot dishes at the Owens Valley radio observatory, California, reflects radio waves on to a receiver in the box supported in front of the dish. Their outputs are combined electronically to produce the resolution of a dish up to 1,600 feet across, the maximum distance between the movable telescopes.

obtained with the best optical telescopes. (The resolving power is comparable to that needed to read newsprint at a distance of over 300 yards.) There are "one-mile" aperture-synthesis telescopes at Cambridge and at the WESTERBORK RADIO OBSERVATORY in the Netherlands. A much larger synthesis instrument, the VERY LARGE ARRAY, is now under construction.

Amplifiers and recorders. The electronic receiving and recording components of a radio telescope must be highly sensitive and capable of amplifying the small voltage fluctuations from the antenna. In practice, all amplifiers introduce a certain amount of receiver noise and prevent very faint radio sources from being detected. Transistors have now completely replaced triode tubes in radio telescope receivers, and new techniques involving parametric amplifiers and MASERS are leading to receivers 10 times more sensitive than those now in use.

The final component of the receiver system is a computer to record the output from the amplifier, to control the movements of the antenna, and to produce a contour map of the radio source as the telescope beam scans across it. The computer can also be set any number of repetitive tasks. In the search for pulsars made at Jodrell Bank using the 250-foot dish, for example, most of the northern hemisphere sky was surveyed in successive 0.5° strips. In each 10 minutes of the search, the computer recorded 16,000 samples of the signal, and while doing so analyzed the previous 10 minutes' data, testing the samples for any of 10,000 periodic variations and printing out any signals suspected of coming from a pulsar.

Ranger probes

A series of American Moon probes. The first Ranger probes were intended to send television pictures as they approached the Moon, and to eject a balsa-insulated instrument package which, after slowing by retro-rockets, would hard-land on the Moon and send back seismometer readings of Moonquakes. All these attempts failed, although the Soviet Union successfully used the same technique in 1966 to land their Luna 9 probe. The Ranger program was reorganized into purely photographic missions to return increasingly detailed television pictures as the probe sped toward the Moon, destroying itself on impact. The final members of the Ranger series successfully produced the first close-ups of the Moon's surface, showing detail far too small to be seen from Earth. The Ranger pictures revealed that even the apparently flattest parts of the Moon (the *mare* areas) are actually pockmarked with tiny craters down to only a few feet in diameter. The detailed views of the Moon returned by the Ranger cameras aided designers of future soft-landing spacecraft.

Probe	Launch date	Remarks
Ranger 1	August 23, 1961	Test launch into Earth orbit
Ranger 2	November 18, 1961	Test launch into Earth orbit
Ranger 3	January 26, 1962	Missed Moon on January 28 by 22,862 miles (36,793 km)
Ranger 4	April 23, 1962	Impacted Moon's far side on April 26; on-board command system failed
Ranger 5	October 18, 1962	Missed Moon on October 21 by 450 miles (724 km)
Ranger 6	January 30, 1964	Impacted Moon February 2; television system failed
Ranger 7	July 28, 1964	Impacted Moon July 31; returned 4,308 photographs
Ranger 8	February 17, 1965	Impacted Moon February 20; returned 7,137 photographs
Ranger 9	March 21, 1965	Impacted Moon March 24; returned 5,814 photographs

Reber, Grote (b. 1911)

American pioneer radio astronomer, the first individual to follow up Karl JANSKY's initial discovery of radio waves from space. Reber built a 31-foot-(9.4-m)-diameter dish aerial in his back garden in 1937, and used it to map radiation from the sky at around a 1-meter wavelength. His first radio maps were published in 1940 and 1942, revealing a radio sky very different from the sky at optical wavelengths. He found a major source of radio noise in Sagittarius, the direction of our Galaxy's center, and he also located several areas of strong radio emission unrelated to visible objects; these have since been identified as sources such as the Crab nebula, a supernova remnant in Cassiopeia, and a radio galaxy in Cygnus. During the war years, Reber was the world's only radio astronomer; his dish aerial was also the world's first specially built radio telescope. His published results attracted the interest of other astronomers, including Jan OORT, who after the war also began to investigate the sky at radio wavelengths. Reber set up a new instrument in Hawaii to investigate longer-wave emissions in 1951. Three years later he moved to the Commonwealth Scientific and Industrial Research Organization in Tasmania, where he continues his radio-astronomy studies.

red dwarf

A star of low surface temperature (2,000–3,000°C) and a diameter about half that of the Sun. Such stars are very faint, with less than 1 percent of the Sun's luminosity, but they are the longest-lived stars in our Galaxy, because of their small energy output. All red dwarfs are still in their MAIN SEQUENCE phase of stellar evolution, shining by converting hydrogen to helium in their interiors; their expected life spans are in excess of 10 billion years. Some 80 percent of the stars in the vicinity of the Sun are known to be red dwarfs, and although they are so faint they cannot be seen at distances much greater than 100 light-years, it is believed that red dwarfs are by far the most numerous stars in the Galaxy. Some (for example, PROXIMA CENTAURI, the Sun's nearest neighbor) are FLARE STARS.

red giant

A star with a low surface temperature (2,000–3,000°C) and a diameter between 10 and 100 times that of the Sun. All stars spend a part of their lives as red giants, once they have converted the hydrogen in their cores to helium, although the period is considerably less than that spent on the MAIN SEQUENCE. Red giants shine by converting helium to carbon (and heavier elements) in a very small, dense core, which is surrounded by an extended, tenuous envelope. Since the gravity at the surface of this envelope is small, matter can easily escape into space in the form of stellar winds and prominences. This process is important in returning matter to the interstellar medium. Red giants are typically 100 times brighter than the Sun, and many of them are long-period VARIABLE STARS of the MIRA type.

red shift

The amount by which the wavelengths of light and other forms of ELECTROMAGNETIC RADIATION from distant galaxies and quasars are increased because of the expansion of the Universe. It is often given the symbol z. This cosmological red shift is one example of the DOPPLER EFFECT, the wavelength change caused by any motion of a light source along the line of sight. The farther the light must travel to reach us, the more the waves are stretched. The wavelengths from distant galaxies are therefore consistently longer than those from nearby galaxies (HUBBLE'S LAW). The shift of SPECTRAL LINES of known wavelength toward the red (longer wavelength) end of the spectrum can be converted to a distance in light-years by using HUBBLE'S CONSTANT.

The fractional change in wavelength is the same for all the wavelengths in an object's spectrum. Thus a galaxy with a red shift of 0.1 has its hydrogen spectral line at 4861 Ångstrom units (Å) increased by 486 Å to 5347 Å, while that at 6563 Å is increased by 656 Å to 7219 Å. The most distant galaxies have red shifts of about 0.7 (corresponding to a distance of 10 billion light-years), but the very bright QUASARS can be seen at much greater distances. The farthest of all quasars has a red shift of 3.53. We see this object, OQ 172, by its very short ultraviolet radiation stretched to visible wavelengths by the expanding Universe.

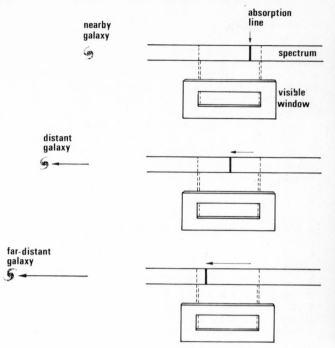

The red shift, or lengthening of the wavelength of light from a receding body, is shown here with reference to the visible "window" on the spectrum to which the eye is sensitive. As galaxies recede at ever increasing speeds, light from them is red-shifted by a corresponding amount, as revealed by the differing position of a specific wavelength marked by a dark absorption line. The red shift causes the entire spectrum from the object to move across the visible window, taking some wavelengths out of sight beyond the red end, and bringing previously invisible wavelengths into view at the blue end. When an object approaches, a similar effect occurs, but in the opposite direction (blue shift).

red spot
The only apparently permanent feature among the swirling cloud bands of Jupiter, first observed by Giovanni CASSINI in 1666. The red spot is elliptical in shape, measuring 20,000 by 8,000 miles (32,000 × 13,000 km) and nestles in a bay on the southern edge of Jupiter's south equatorial belt. Its intensity varies from a striking orange-red to a pallid almost invisible cream, but its shape is constant. The spot was particularly prominent in the late 1870s and again in the mid-1970s. The material in the spot circulates in an ascending spiral, and it is now believed to be a colossal rotating convection storm, similar to tropical thunderstorms on Earth, only driven from below by Jupiter's internal heat rather than by heat from the Sun. The visible spot is an anvil of methane and ammonia cirrus clouds towering some 5 miles (8 km) above the surrounding cloud bands. The spot's color is probably due to material carried up from deep inside the atmosphere, the strength of the updraft

The darkness of Jupiter's Great Red Spot is emphasized in this blue light photograph, taken from 338,000 miles by Pioneer 11. There are indications of a counterclockwise spiral within the spot, supporting the hypothesis that it is a vast hurricane-like storm, with a length $2\frac{1}{2}$ times the Earth's diameter.

regulating the intensity of color. Similar reddish spots have been observed at higher latitudes on Jupiter, but these are less than half the size and are transient in nature.

Redstone rocket
An American rocket for space launches, developed from a short-range ballistic missile. The Redstone rocket was built by Wernher von Braun's team at the U.S. Army Redstone Arsenal in Huntsville, Alabama (later the MARSHALL SPACE FLIGHT CENTER). The Redstone launched America's first two spacemen, Alan Shepard and Virgil Grissom, on short suborbital flights at the start of the MERCURY PROJECT. In a modified form, known as JUNO 1, the Redstone launched America's first satellite, Explorer 1. The basic Redstone was a single-stage vehicle, 70 inches (178 cm) in diameter and 69 feet (21 m) long, with a single engine of 78,000 lb. (35,380 kg) thrust. Because of its low payload capacity, the Redstone was soon superseded by more powerful launchers, such as the ATLAS ROCKET.

reflecting telescope
A telescope in which a concave mirror collects and focuses light from a celestial object. Most modern telescopes are reflectors because they are less

expensive, more compact, and can be made in larger apertures than the refracting telescope, which uses lenses. The heart of any reflecting telescope is the primary mirror, which is concave and reflects light rays to form an image at the primary focus. The distance between the mirror and its prime focus is called the FOCAL LENGTH, and it is affected by the mirror's amount of concavity. For example, the depth of the curved surface of a 12-inch(30-cm)-diameter mirror of focal length 48 inches (120 cm) will be 0.18 inch (4.7 mm). Such a mirror is said to have a FOCAL RATIO (focal length/aperture) of $f/4$. The primary mirror is normally made of glass, coated on its curved front surface with aluminum (see MIRROR). Smaller secondary mirrors are used to divert the light to various observing positions.

In the telescope system known as the *Newtonian,* after its inventor Isaac Newton, the primary image is diverted by a small flat mirror out through the side of the tube. This is the system used in many amateur-owned reflecting telescopes. However, most modern professional reflectors are of the CASSEGRAIN design, in which a convex secondary mirror about a quarter the diameter of the primary mirror reflects the light back through a hole in the primary's center. This convex secondary mirror multiplies the focal length several times, giving the benefit of a long effective focal length in a relatively short tube. The longer the focal length the larger the image. Many telescopes also have a COUDÉ focus, particularly useful in spectroscopy. By swinging the secondary mirror out of the way, the PRIME FOCUS can be used for photography; a photographic plate takes the place of the secondary mirror. The prime focus

The World's Largest Reflecting Telescopes

	Aperture in.	m	Mounting	Observatory	Date
1	236	6.0	Altazimuth	Zelenchukskaya, Caucasus, U.S.S.R.	1976
2	200	5.0	Horseshoe	Mount Palomar, California	1948
3	158	4.0	Horseshoe	Kitt Peak, Arizona	1973
4	158	4.0	Horseshoe	Cerro Tololo, Chile	1975
5	153	3.9	Horseshoe	Siding Spring, Australia	1974
6	144	3.7	Horseshoe	Mauna Kea, Hawaii	*
7	142	3.6	Horseshoe	European Southern Observatory, La Silla, Chile	1976
8	120	3.0	Fork	Lick Observatory, California	1959
9	107	2.7	Cross-axis	McDonald Observatory, Texas	1968
10	102	2.6	Fork	Crimean Astrophysical Observatory, U.S.S.R.	1960
11	102	2.6	Fork	Byurakan, Armenia, U.S.S.R.	1975
12	101	2.6	Fork	Las Campanas, Chile	1975
13	100	2.5	English	Mount Wilson, California	1917

*under construction

provides wide-angle views; correcting lenses are often used at the prime focus to improve the definition over a wide field. An obsolete reflector design, which uses a concave secondary instead of a convex one, is called the GREGORIAN TELESCOPE.

reflection nebula
See NEBULA.

refracting telescope
A telescope in which a large lens or OBJECT GLASS collects and focuses the light from the object being observed. Refractors have lost much of their long-established popularity with the introduction of photography in celestial observation. For photographic work the reflector, in its many varied forms, is on the whole superior because it produces color-free (*achromatic*) images. The image produced by a refracting telescope can never be made absolutely colorless, because no combination of lenses in the object glass can bring light of all colors to precisely the same focus. In principle, the flint glass lens in an achromatic object glass should correct the red-to-violet band of images formed by the crown glass lens, but in practice some of the violet and red light is not correctly focused. A refractor's image therefore exhibits a purplish halo when corrected for normal visual use. The so-called *photographic* correction, in which blue and green light is brought to a sharp focus, is no longer relevant now that photographic emulsions are sensitive to all colors of light, not merely the blue, as was the case in the early days of photography (see ASTROPHOTOGRAPHY).

Refracting telescopes are very much more expensive than reflectors of the same aperture. First, optical (transparent, flaw-free) glass must be used, which is about 10 times as costly as ordinary glass. Furthermore, since the tubes of refracting telescopes are much longer than those of reflectors, the observatory will be large and costly. However, the relatively long FOCAL LENGTH (objective-image distance), and the fact that the tube is completely enclosed, with the light passing along it only once, produce steady images and good definition, while the absence of reflecting surface reduces scattered light and so increases the image contrast.

The main use of the surviving large refracting telescopes today is in ASTROMETRY, a field covering the measurement of close DOUBLE STARS and of stellar positions for determining PROPER MOTIONS and PARALLAX; many visual observers also prefer its use in planetary observation. The current desire for large-aperture telescopes to probe deep into space means that no large refractors have been built for many years, and most of the instruments currently used were built in the 19th century, including the world's largest, the 40-inch (102-cm) at Yerkes Observatory, and the 36-inch (91-cm) at Lick Observatory. The third-largest refracting telescope, the 32.6-inch (83-cm) at the Meudon branch of Paris Observatory, is used regularly for the visual measurement of double stars, as was the 28-inch (71-cm) at the Royal Greenwich Observatory until very recently. Two famous refractors used for both visual and photographic observation, both 24 inches (61 cm) in aperture, were set up at Lowell Observatory, Arizona, and at the Pic du Midi

Observatory in the Pyrenees. The Flagstaff instrument, originally used by Percival LOWELL in his Martian "canal" observations, is now employed photographically in the international planetary patrol project.

regression of nodes
The westward precession of an orbit caused by the gravitational influence of other bodies. The orbit's movement is noted by reference to the *nodes,* the points where the orbit crosses a reference plane such as the ecliptic or celestial equator.

Regulus
The brightest star in the constellation Leo; also known as Alpha Leonis. Regulus, of magnitude 1.36, is a blue-white star 3.8 times the Sun's diameter, 84 light-years away. It has companions of magnitude 7.64 and 13, making it a triple star.

relativity
A theory of the structure of space and time, and its relation to motion and gravity, which was proposed by Albert Einstein.

The special theory (1905) is founded on the principle that the velocity of light is the same for all observers, no matter how they move relative to each other. This apparent paradox was required both theoretically and to explain the unexpected result of the MICHELSON-MORLEY EXPERIMENT, and Einstein showed that our familiar notions of space and time must be abandoned as a result. Among other things, the special theory predicts that time runs at different rates for observers who move at different velocities, and that two events that appear simultaneous to one observer may not do so to another.

In addition, the theory predicts new phenomena in mechanics: among them, the possibility of changing matter into energy, the inability of a material body to reach the speed of light, and the famous "twins paradox" in which a fast-moving space traveler returns to Earth younger than his twin who remains behind. In spite of these apparently bizarre possibilities, the special theory is now an established cornerstone of modern physics. Its predictions have been confirmed in many ways by experiment.

The general theory (1915) is an extension of relativity to situations in which gravity is an important factor. It treats gravity not as a force, but as a consequence of the CURVATURE OF SPACE. In special relativity, Einstein had treated time as an extra dimension, very like the familiar three dimensions of space. Now he showed that the four-dimensional Universe (extending over both space and time) becomes "curved" in the presence of matter. Any body "falling" under gravity is actually following a straight line, at constant speed, relative to the curved background of "space-time."

If space-time is, in fact, curved, then even light should be affected by gravity. Observations have confirmed this, showing that starlight passing close to the Sun is slightly deflected. Moreover, light that leaves the surface of a massive body is shifted to a longer wavelength, an indication that time passes more quickly in an intense gravitational field.

When the force of gravity is weak, Einstein's theory differs little from the conventional Newtonian view of the Universe. The motion of the planets, for example, is hardly affected by relativity. However, a slight disturbance in the predicted motion of Mercury, caused by the effects of relativity, is actually detectable, and the explanation of Mercury's behavior is one of the greatest successes of the theory.

When the force of gravity becomes very strong, strange new effects are predicted. The immense force of gravity exerted by the entire Universe is capable of bending the whole of space. In BLACK HOLES, the space-time around a collapsing star shrinks catastrophically, until even light is trapped within it.

Because gravity is generally such a relatively weak force, general relativity has been experimentally checked in only a very few cases. Although other theories of space-time have been proposed (for example the BRANS-DICKE COSMOLOGY), general relativity is still the most widely accepted.

Relay satellites
Two NASA experimental communications satellites, launched into medium-altitude orbits. The satellites were 172-lb. (78-kg) tapered octagons, studded with solar cells. Receiving on 1725 MHz and retransmitting on 4170 MHz, they could carry 300 one-way voice transmissions, or one black-and-white television channel. The Relay satellites were used to send messages and pictures across the Atlantic and Pacific Oceans, and together with the privately-built TELSTAR satellites proved the value of active-repeater communications satellites.

Relay 1 (launched December 13, 1962) orbited every 186 minutes; Relay 2 (launched January 21, 1964) orbited every 195 minutes.

resolution
The ability of a telescope to distinguish fine detail. The larger the telescope's aperture, the better its resolving power. A telescope's resolution measured in arc seconds is related to its aperture by the famous Dawes' limit (see DAWES, WILLIAM).

If, for example, the components of a double star lie closer together than this limit, they cannot be separated no matter how high a magnifying power is applied. The Dawes' limit does not apply to the resolution of fine extended detail, such as that on planetary disks, where contrast rather than angular width sets the limit.

Reticulum (the net)
A small and faint constellation of the southern hemisphere of the sky, introduced on the star chart of Nicolas Louis de LACAILLE. Zeta Reticuli is a widely separated pair of stars similar to the Sun, of magnitudes 5.2 and 5.5, about 30 light-years from Earth.

retrograde motion
Motion from east to west, opposite to the normal west-to-east direction of motion in the solar system. The four outer satellites of Jupiter, the outermost moon of Saturn, and Neptune's moon Triton, all have retrograde orbits; the planet Venus has a retrograde axial rotation. Planets farther from the Sun than Earth seem to move retrograde (westward against the star background) for a short time as the Earth catches up and overtakes them in its orbit around the Sun.

Rhea

The second-largest satellite of Saturn, sixth in order of distance from the planet. Rhea, discovered in 1672 by Giovanni CASSINI, orbits Saturn every 4 days 12 hours 25 minutes at a distance of 327,590 miles (527,200 km). Its diameter is about 980 miles (1,600 km).

Rigel

The brightest star in the constellation Orion, and the sixth-brightest in the sky; it is also known as Beta Orionis. Rigel is a blue-white supergiant of magnitude 0.08, 78 times the Sun's diameter, and about 850 light-years away. Rigel is a multiple star; it has twin companions of magnitude 7.7, one of which is a SPECTROSCOPIC BINARY of period 9.9 days.

right ascension

The celestial equivalent of longitude on Earth. Right ascension is measured in hours, minutes, and seconds, from 0h to 24h, eastward along the celestial equator from the VERNAL EQUINOX. The symbol for right ascension is α.

Rigil Kent

See ALPHA CENTAURI.

Ritchey, George Willis (1864–1945)

American astronomer and optical instrument maker, coinventor of a modified type of CASSEGRAIN TELESCOPE known as the *Ritchey-Chrétien* design. In 1896 Ritchey joined George Ellery HALE at the Yerkes Observatory, where he made a correcting plate for the 40-inch (101-cm) refractor, thereby turning it into a camera with which he took a famous series of lunar photographs. Ritchey built a 24-inch (61-cm) reflector at Yerkes, and used it to photograph nebulae with a fellow optician and astronomer, Francis Gladheim Pease (1881–1938). Ritchey also ground the mirror for the Mount Wilson 60-inch (152-cm) reflector, with which he photographed faint nebulae; his discovery in 1917 of a nova in one such object, NGC 6946, helped confirm that these were actually distant galaxies, separate from our own Milky Way system. Ritchey later supervised the grinding of the 100-inch (254-cm) Mount Wilson mirror, being joined in 1911 on the telescope's construction by Pease. At Mount Wilson Ritchey began to develop his new optical designs, but his ideas were not adopted. In 1919 he left Mount Wilson and later moved to the optical institute at Paris Observatory. In 1931 he returned to the U.S. Naval Observatory, where he designed and built a 40-inch reflector.

Roche's limit

The distance from a planet inside which a satellite will be torn apart by tidal forces. The disrupting forces occur because the portion of the satellite nearest the planet is tugged more strongly by gravity than the portion farthest away. Small meteorite-sized lumps of rock and man-made satellites can hold together well inside Roche's limit, but where the planet and its satellite are made of similar material, the limit is $2\frac{1}{2}$ times the planet's radius. For satellites of lesser density, Roche's limit is correspondingly farther away. The tiny particles of Saturn's rings all lie within Roche's limit; if the Moon came too close to our Earth, it too would end up as a set of rings. Roche's limit is named for the French astronomer Edouard Roche (1820–1883), who calculated its existence in 1848.

rocket

An engine used for propulsion which works on the reaction principle, forward propulsion being achieved by reaction to a jet of gases streaming backward out of a nozzle. It differs from the jet engine, which also works by reaction, in that it carries an oxygen supply (usually liquid oxygen) to burn its fuel, unlike the jet engine, which obtains oxygen from the air. Rockets can therefore function in airless space better than in the atmosphere, where air pressure and drag impede progress. The fuel and the oxidizer are carried in separate tanks, and they are pumped into the combusion chamber and ignited by a spark. (Some small rocket motors use other propellants, such as HYPERGOLIC FUELS, which ignite spontaneously on contact and need no ignition system.)

Origin and development. The rocket was probably invented about the start of the 13th century by the Chinese; it is known that rockets were used at the Battle of K'ai-fung-fu in 1232. These were made with gunpowder, and were similar to today's firework rockets. News of the invention spread to Europe, but rockets were superseded for military purposes by the development of guns. However, Indian soldiers used rockets so effectively against the British at the Battles of Seringapatam in 1792 and 1799 that a British artillery officer, William CONGREVE, decided to investigate their possibilities. Congreve greatly improved the range, weight, and accuracy of rockets, which became a standard part of British artillery in the early 19th century. Soon, rocket battalions were also used by other European nations.

A Congreve rocket was stabilized by a long stick, like a firework. In 1843 the British inventor William Hale (1797–1870) drilled angled vents around the base of his rockets, so that the escaping exhaust gases caused the projectile to spin; this spin stabilization did away with the need for sticks. By 1865, Hale had perfected a design in which the rocket was spun by curved vanes inserted into its exhaust stream. Hale rockets superseded those of Congreve by 1867. However, by the 20th century the increasing range and accuracy of guns again overtook rocket development, and it was not until World War II that rockets returned to the field of battle.

Pioneers of astronautics. Toward the end of the 19th century, visionaries began to glimpse the value of rockets for space travel. In 1881 the Russian Nikolai Ivanovich Kibalchich (1853–1881), in prison awaiting execution for the assassination of Tsar Alexander II, conceived a man-carrying rocket, driven by the explosion of a series of gunpowder cartridges. In about 1890 the German Hermann Ganswindt (1856–1934) proposed a manned spaceship powered by the thrust of artillery-type shells. However, the true founder of astronautics was the Russian Konstantin TSIOLKOVSKY, who by 1898 had worked out the mathematical formulae governing a rocket's operation. In subsequent papers he established the need for liquid fuels, which are more controllable and have greater power than solid fuels, and analyzed the theory of the multistage, or step, rocket, which allows a much higher final speed. But

Tsiolkovsky was only a theoretician. The first man to experiment with liquid-fueled rockets was the American Robert H. GODDARD, who flew the world's first liquid-fueled rocket in 1926. At the same time in Germany, the writings of Hermann OBERTH inspired rocketeers. Europe's first liquid-fueled rocket was flown in 1931 by the German Johannes Winkler (1897–1947). The German Society for Space Travel, founded in 1927, also flew liquid-propellant rockets in 1931, one of the experimenters being the young Wernher VON BRAUN. The following year this work came under German army sponsorship, which led in 1937 to the setting up of the Peenemünde experimental station, where the V-2 was developed. This advanced liquid-fueled rocket, a 1-ton warhead, was a forerunner of today's ballistic missiles. It both changed the pattern of warfare for the future, and brought the world to the threshold of the Space Age.

Rockets in the space age. At the end of World War II, Wernher von Braun and his team surrendered to the U.S. Army, with whom they continued their rocket experiments on a small scale, while the U.S. Air Force developed the more powerful ATLAS, TITAN, and THOR missiles. During this time the Soviet Union began to develop a missile that could cross the Atlantic. Following Tsiolkovsky's inspiration before the war, Soviet engineers such as Friedrich TSANDER had made considerable progress in rocket development. Tsander's coworkers Sergei KOROLEV, Valentin Glushko, and Mikhail TIKHONRAVOV, headed Soviet missile development after the war. Their first intercontinental missile flew in August 1957; in October that year it was used to launch the first Earth satellite, Sputnik 1. With upper stages added, the same booster launched early Soviet Moon probes, and the Vostok manned spacecraft. Not until 1967 was this booster publicly unveiled, in the version that launched the first Soviet cosmonauts; consequently, it is usually referred to as the VOSTOK LAUNCHER (an alternative name is Standard launcher). In improved form, with new upper stages, it launches manned Soyuz spacecraft.

After the development of the REDSTONE and JUNO missiles for early American space launches, von Braun's team began to design the SATURN rocket family specifically for manned space applications, principally the Apollo program. This led to the giant Saturn V, the world's most powerful rocket, first successfully fired in 1967. Two years previously the Soviet Union had introduced its PROTON rocket, of under half Saturn V's power, which was intended to send a manned Soyuz spacecraft looping around the Moon. Trial shots of unmanned ZOND craft were launched, but the Soviet lunar goal was deferred after Apollo 8 made the first manned lunar orbit in 1968. Because of technical difficulties the Proton launcher has never been considered safe for manned launchings, and its uses have been confined to launching unmanned space stations and lunar and planetary probes.

The Soviet Union has also attempted to build a giant rocket, known in the West as the G type, more powerful than the Saturn V. This would be used to launch large space stations into Earth orbit, or to send substantial payloads to the Moon for the first Russian lunar landing. The G-type booster is believed to be a three-stage vehicle with a total height of about 340 feet (105 m), first-stage diameter 52 feet (16 m), and with first-stage thrust about 11,000,000 lb. (5,000,000 kg). It could put over 100 tons into Earth orbit, send 60 tons to the Moon, or 25 tons to the planets. Reports suggest that one rocket exploded on the launchpad in mid-1969; another broke up during flight in mid-1971; and a third vehicle was destroyed by ground command after two minutes in late 1972. Further development of the booster, initially for a manned lunar landing, is expected.

Rockets in the future. In 1980 the standard disposable type of rocket will be largely displaced by the reusable SPACE SHUTTLE, a vehicle boosted away from the Earth by conventional rockets, but which will

The design of a multistage rocket is shown well by the basic launcher used by the USSR for Soyuz and other manned spacecraft. The first stage, at the bottom, has four strap-on motors and an additional central motor. Above this, the four motors of the second stage can be seen through a separating framework. Above the dark colored section is the third stage, with the spaceship above that. At the top is the escape tower, which can pull the manned spaceship clear of danger in case of a launch disaster.

glide back to Earth like an aircraft. The Shuttle's power is provided by rockets similar to those in current use, because only chemical fuels produce the sudden surge of thrust necessary to break away from the Earth's gravity. But rockets of the future, designed to perform in space alone, will use forms of propulsion more efficient than bulky and heavy chemical fuels. One idea is the so-called ion rocket, or electric rocket, in which atoms of an element such as the metal cesium are heated to the point that their atoms are stripped of electrons; these electrically charged particles (*ions*) can then be accelerated out of the rocket by electric fields to produce a high-speed exhaust. Although the thrust of such an engine is very low, it can operate for long periods of time, thus building up a very high final speed. Small electric rockets have been tested in space.

Another suggestion is for a fission rocket, in which atomic power from a reactor like that in a submarine is used to heat a liquid such as liquid hydrogen or even water, turning it into a gas which is expelled from the rocket. Eventually, rockets may be able to tap nuclear fusion, duplicating the atomic processes that power stars. A dream for the distant future is the nuclear ramjet, which would scoop up hydrogen atoms from gas clouds between the stars to feed an on-board fusion reactor. The most exotic propulsion system of all is the photon rocket, which would use a beam of light particles—photons—to produce thrust. The exhaust speed of such a rocket would be the fastest speed possible—the speed of light. But such a propulsion system is completely beyond the capabilities of our current or foreseeable technology.

Roemer, Ole or Olaus (1644–1710)

Danish astronomer who in 1676 attempted the first measurement of the speed of light. In 1672 Roemer was appointed to the new Paris Observatory. While there, he noted that the times between successive eclipses of Jupiter's moons varied; as the Earth was approaching Jupiter the time intervals diminished, and as the Earth and Jupiter drew apart the intervals increased. He surmised that the difference was due to the time taken for light to cross the space between Earth and Jupiter, and from his observations announced that the speed of light was 140,000 miles (225,000 km) per second. Although about 25 percent too small, this was the first demonstration that light in fact had a finite speed. In 1681 Roemer returned to Denmark, becoming director of Copenhagen Observatory, where he is credited with inventing the transit instrument.

Rosse, Lord (1800–1867)

Irish astronomer (full name, William Parsons, third earl of Rosse), who built a famous 72-inch (183-cm) reflecting telescope, the world's largest until the opening of the Mount Wilson 100-inch (254-cm) in 1917. Rosse at his family seat of Birr Castle, Parsonstown, began in 1827 to improve the design and construction of large reflecting telescopes, which were still at an experimental stage. He developed a new alloy of copper and tin that would take the maximum polish (telescope mirrors were then made exclusively of metal), and in 1828 invented a steam-driven engine for automatic mirror polishing. In 1839 he cast a trial 36-inch (91-cm) mirror and used it for observing, before embarking on his

intended giant telescope. In 1842 and 1843 he successfully made two 72-inch (182-cm) mirrors of 54-foot (16.5-m) focal length, weighing 4 tons each; they were used in rotation, one being repolished while the other was mounted in the telescope. The giant telescope, called the Leviathan of Parsonstown, had a tube 7 feet (2.1 m) wide and 58 feet (17.7 m) long. It began operation in 1845, mounted between two walls 56 feet (17 m) high which acted as windbreaks, but which restricted its field of movement to 10° either side of the meridian. With the telescope, Rosse discovered the spiral shape of certain nebulae, beginning in 1845 with the so-called Whirlpool nebula, M51; 75 years later these spiral nebulae were recognized to be distant galaxies. Rosse noted the ring-like structure of planetary nebulae, and in 1848 gave the CRAB NEBULA its name, because he drew it as a pincer-like shape. Rosse employed distinguished observers, such as Ralph Copeland (1837–1905) who, from 1889 to 1905, was astronomer royal for Scotland, and J. L. E. DREYER. After Rosse's death the telescope was little used; it was dismantled in 1908, although the walls and tube remain.

Royal Greenwich Observatory

Britain's national astronomical observatory, located at Herstmonceux Castle near Hailsham, Sussex. The observatory was founded at Greenwich, London, by King Charles II in 1675. Its original purpose was to make accurate observations of the Moon and stars in order to aid navigators in establishing longitude. The first director was John FLAMSTEED; he and his successors bore the title ASTRONOMER ROYAL until 1972, when the post of astronomer royal was separated from the observatory's directorship. The first observatory building, designed by Sir Christopher Wren (1632–1723), himself an amateur astronomer, was opened in 1676 and is now called Flamsteed House. In addition to basic positional astronomy, the work of the observatory progressed to include the accurate measurement of time; in 1880 GREENWICH MEAN TIME was adopted as legal time in Britain, and in 1884 the meridian passing through Greenwich was chosen as the world's prime meridian (0° longitude). Both the *Nautical Almanac* Office, producing catalogs for astronomers, navigators, and surveyors, and the Time Department, are important sections of the observatory. Because of deteriorating observing conditions in London, the observatory was moved from Greenwich to Herstmonceux after World War II, although it retains the name Royal Greenwich Observatory (RGO); the move was completed in 1958. The observatory, however, no longer lies exactly on the Greenwich meridian. Its main telescope is the 98-inch (250-cm) Isaac Newton reflector, installed in 1967. The observatory also contains a 38-inch (96.5-cm) and a 36-inch (91-cm) reflector, two 30-inch (76-cm) reflectors, and a 26-inch (66-cm) refractor. The Royal Greenwich Observatory plans to establish a new observing station, the Northern Hemisphere Observatory, on La Palma in the Canary Islands, with a 177-inch (450-cm) reflector as its main telescope. Also at the observatory will be a 100-inch (254-cm) reflector (the Isaac Newton telescope with a new mirror), and a 39-inch (100-cm) reflector. The Northern Hemisphere Observatory should be in operation by 1980.

RR Lyrae variables

Pulsating stars (see VARIABLE STARS) whose average luminosity is exactly the same (absolute magnitude $+0.5$) and which can thus be used to calibrate distances throughout the Galaxy. Belonging to stellar population II (older stars), they are found in the galactic nucleus and halo and also in GLOBULAR CLUSTERS (they are often called "cluster variables"). RR Lyrae stars belong to SPECTRAL TYPES A to F and have periods of between 0.3 and 0.9 days varying by about a magnitude. They are named for their seventh-magnitude prototype, RR Lyrae. In 1920, Harlow SHAPLEY used the apparent magnitudes of RR Lyrae stars in globular clusters to determine the clusters' distances, showing that the Galaxy was much larger than had previously been thought. RR Lyrae stars in the galactic nucleus have also been used to determine the distance to the center of the Galaxy. It is believed that RR Lyrae stars represent a short phase (lasting about 80 million years) in the final evolution of population II RED GIANT stars.

Russell, Henry Norris (1877–1957)

American astronomer, codiscoverer with Ejnar HERTZSPRUNG of the relation between a star's brightness and color, now graphically displayed on the HERTZSPRUNG-RUSSELL DIAGRAM, which Russell first produced in 1913. From this diagram, Russell proposed a scheme of stellar evolution in which stars began as red giants, shrinking and increasing in temperature, ending their lives as cool red dwarfs. However, by the 1930s it was realized that stars derive their energy from nuclear reactions in their interiors, not from simple contraction. Russell was a leading pioneer in the study of eclipsing variable stars. In 1912 he published a major paper on the analysis of eclipsing-binary light curves, in which he showed how to calculate the sizes of the stars and their orbits. From this, Russell was able to compute their distance from Earth. His work on star sizes and distances led in 1913 to the discovery, independent of Hertzsprung, that stars fall into two brightness classes, which Hertzsprung had termed giants and dwarfs. Russell's plot of brightness against spectrum type showed the band of dwarf stars now known as the MAIN SEQUENCE. In 1929 Russell published an analysis of the Sun's spectrum, noting the presence of 56 elements; he measured the relative abundances of the elements, showing that hydrogen was the major constituent. Termed the Russell mixture, this was the first detailed breakdown of a typical star's true composition.

Ryle, Sir Martin (b. 1918)

English radio astronomer who developed the technique of APERTURE SYNTHESIS, by which signals from several small radio dishes are combined to synthesize the performance of a much larger dish. This was pioneered by Ryle at the MULLARD RADIO ASTRONOMY OBSERVATORY, Cambridge, which he founded and became director of in 1957. With a succession of increasingly powerful radio telescopes, Ryle has been able to catalog progressively fainter radio sources, showing that the number of sources increases with distance before falling away abruptly at the observable edge of the Universe. This suggests that the Universe has been evolving from a definite origin, which supports the BIG-BANG theory of cosmology. Ryle became Britain's twelfth astronomer royal in 1972, in recognition of the importance of his work in radio astronomy. In 1974 he shared the Nobel Prize in physics with his colleague Antony HEWISH.

S

Sagan, Carl Edward (b. 1934)

American astronomer and biologist, who has done pioneering work on the possibility of life elsewhere in the Universe. He demonstrated processes leading to the origin of life by bombarding mixtures of gas with ultraviolet light in order to simulate the effect of the Sun's radiation on the atmosphere of the early Earth. The experiments produced amino acids, the building blocks of protein. Sagan suggested in 1960 that water vapor in the clouds of Venus reinforced the GREENHOUSE EFFECT of its carbon-dioxide atmosphere, preventing heat from escaping and building up the planet's high observed temperature. He proposed a scheme for making Venus habitable by using algae to break down the carbon dioxide and release oxygen, and has also suggested a similar scheme for Mars. In 1966 Sagan discovered the existence of high elevations on Mars by analysis of radar echoes from the planet. He also suggested that wind-blown dust causes the seasonal changes in the dark areas on Mars, rather than growing vegetation as was previously supposed; photographs from the Mariner 9 probe confirmed this view. With the radio astronomer Frank DRAKE, Sagan has investigated the problem of interstellar communication, devising the message plaque fixed to the Pioneer 10 and 11 spacecraft, and the radio message briefly transmitted from the Arecibo radio telescope. Drake and Sagan have also listened with the Arecibo dish for radio messages from nearby galaxies (see LIFE IN THE UNIVERSE).

Sagitta (the arrow)

A small constellation lying in the Milky Way near Aquila, visible during the northern hemisphere summer. Despite its faintness, it is one of the 48 constellations that were listed by PTOLEMY almost 2,000 years ago. The constellation's most interesting feature is the recurrent nova WZ Sagittae, seen to flare up in 1913 and again in 1946.

Sagittarius (the archer)

A constellation of the zodiac, lying in the south celestial hemisphere and best seen from the northern hemisphere during summer. The center of our Galaxy lies in the direction of Sagittarius, giving rise to the dense Milky Way star fields that are a feature of the constellation. The actual nucleus of the Galaxy is believed to be marked by the radio source Sagittarius A. Famous bright nebulae in Sagittarius include the Lagoon nebula (M8, NGC 6523), the Omega nebula (M17, NGC 6618), and the Trifid nebula (M20, NGC 6514). The constellation also includes numerous open and globular star clusters. The Sun passes through Sagittarius from mid-December to mid-January, reaching its farthest point south (the winter SOLSTICE) on about December 22.

Salyut

Soviet space station, measuring 39 feet (12 m) in length in the shape of three connected cylinders $6\frac{1}{2}$ feet (2 m), 10 feet (3 m), and $13\frac{1}{2}$ feet (4 m) in diameter. Salyut has a volume of about 3,500 cubic feet (100 m³) roughly a quarter that of the American SKYLAB, and weighs $18\frac{1}{2}$ tons. It is launched by the Soviet PROTON rocket. The first Salyut, launched in April 1971, drew power from four wing-like solar cells, two at each end of the station. Subsequent Salyuts were of modified design, using three rotable solar panels mounted amidship. The Salyuts contain equipment for astronomical and Earth observation, and for biological studies. Salyuts 1 and 4 were believed to be mainly scientific in purpose; and Salyuts 2 and 3 were principally military surveillance stations. Salyut 3 operated semi-automatically, telemetering data and ejecting a film package. Crews are ferried up to the space station by Soyuz craft; a simplified Soyuz design, omitting solar panels, has been introduced for Salyut missions. The Salyut program has been dogged by failures both with Soyuz craft and the station itself. A Salyut launch on July 30, 1972, suffered a rocket failure; Salyut 2 broke up in orbit; and another failed Salyut, launched to replace it on May 11, 1973, was disguised under the name Cosmos 557.

Mission	Launch date	Results
Salyut 1	April 19, 1971	Soyuz 10 crew docked on April 24 but did not enter, probably because of hatch failure. Soyuz 11 crew spent 23 days aboard in June, but died due to spacecraft pressure failure during reentry. Salyut 1 reentered October 11
Salyut 2	April 3, 1973	Disintegrated in orbit; reentered May 28
Salyut 3	June 24, 1974	Soyuz 14 docked and transferred crew for 14-day mission. Rendezvous attempt by Soyuz 15 failed. Salyut 3 reentered January 24, 1975
Salyut 4	December 26, 1974	Soyuz 17 docked and transferred crew for 29-day mission. Subsequent Soyuz 18 crew completed 63-day mission in space station. Unmanned Soyuz 20 docked automatically.

Sänger, Eugen (1905–1964)

Austrian pioneer of astronautics, who developed the concept of a reusable space transporter, now known as the SPACE SHUTTLE. Sänger's proposals for a space transporter were born in the late 1920s; he published his first description of a rocket plane in 1933. In the 1930s at the University of Vienna he developed rocket motors using diesel fuel with burning times of up to 30 minutes. From 1936 to 1945 he directed a rocket research institute for the Luftwaffe, and while there Sänger produced detailed plans for a long-range winged bomber, boosted into the air by a rocket-powered sled and accelerated by its own engines to the edge of the atmosphere; it would skip in and out of the atmosphere to extend its range before finally dropping toward its target. This work provided a basis for many postwar studies on reusable rocket systems, and shortly before his death he drew up designs for such a two-staged winged space transporter. Sänger foresaw the use of such vehicles to ferry men and materials to orbiting space stations, the role now outlined for the space shuttle.

San Marco

An Italian rocket launching site based on two converted oil rigs in Ngwana Bay off the Kenyan coast, operated by Rome University's Aerospace Research Center. Being 2° south of the equator it is excellently placed for launches into equatorial orbit. The San Marco launch platform is supported by

Satellite	Launch date	Remarks
San Marco 1	December 15, 1964	First Italian satellite; launched by NASA from Wallops Island. Atmospheric studies
San Marco 2	April 26, 1967	First orbital launch from San Marco platform. Atmospheric studies
San Marco 3	April 24, 1971	Launched from San Marco platform. Atmospheric studies
San Marco 4	February 18, 1974	Launched from San Marco platform. Atmospheric studies

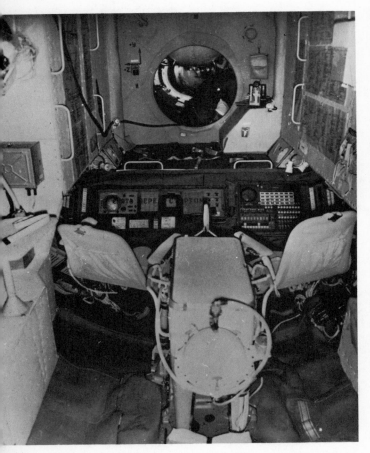

The interior of the Soviet Salyut 1 space station, the first space laboratory. Numerous hand holds make maneuvering in free fall easier. In the background is the hatchway for access to the Soyuz spaceship.

20 legs resting on the seabed; it measures 295 by 88½ feet (90 × 27 m). Nearby is the triangular Santa Rica platform, which houses the launch control center and launch personnel. The San Marco platform is used for launching American Scout rockets and sounding rockets. It became operational in 1966, and made its first orbital launch with the second of Italy's San Marco series of satellites for atmospheric studies. The platform has also been used to launch America's series of Small Astronomy Satellites (the first U.S. satellites to be launched by another country), and Britain's Ariel 5 X-ray satellite.

saros

The period of 6,585.3 days (18 years 11⅓ days) after which the Earth, Moon, and Sun return to their same positions relative to each other and eclipses repeat themselves. During the saros there are 223 lunations.

satellite

Any small object orbiting a larger body. The world's first artificial satellite was SPUTNIK 1, launched on October 4, 1957. The Sputniks were superseded in 1962 by the COSMOS SATELLITES. America's first Earth satellite was EXPLORER 1, launched on January 31, 1958. By the end of 1974, 1,606 Earth satellites had been launched, 674 of which were still in orbit. See also COMMUNICATIONS SATELLITE; NAVIGATION SATELLITE; WEATHER SATELLITE; and the names of individual satellite programs.

satellites of the planets

See under separate planets, and under the names of the individual satellites.

Saturn

The sixth planet from the Sun, renowned for its unique ring system. Of all the planets in the solar system Saturn is surpassed in size only by Jupiter. Like its huge neighbor it is composed mostly of gas and wreathed by dense clouds. Saturn's shape is flattened by its rapid rotation, which varies from as little as 10 hours 14 minutes at the equator to 10 hours 40 minutes at latitude 60°; the axis is inclined at 26° 44'. Saturn moves at 6 miles (9.6 km) per second in an elliptical path between 9.01 and 10.07 astronomical units from the Sun (average distance 886,703,000 miles; 1,427,010,000 km), taking 29½ years to complete one orbit.

Structure. Viewed through a medium-sized telescope, Saturn's disk shows bands and belts like those on Jupiter. On Saturn, however, they are much more diffuse, fewer in number, and rarely showing much detail. Colors are also much less pronounced, with a golden yellow predominating. The equatorial band is yellow-white, with an orange-yellow equatorial belt and yellow tropical and temperate zones where orange belts are sometimes faintly glimpsed. The polar regions which appear greenish, are the darkest part of the surface. The cloud tops are a bitter −170°C, some 50° colder than on Jupiter, because of the planet's greater distance from the Sun. As a result of the lower temperature, there are fewer outbursts of activity in the clouds, and there are no distinct markings like the great red spot or white ovals on Jupiter. The most celebrated outburst was the great white equatorial spot which appeared in August 1933,

Saturn and its rings photographed with the 100-inch Mount Wilson reflector. The bright outer rings, separated by Cassini's division, are easily visible; but the two inner rings are too faint to be seen on this photograph. Light and dark bands, similar to Jupiter's, can be seen on Saturn's disk.

and which slowly spread until the whole equatorial zone brightened. High-latitude white and dark spots are less conspicuous but more common, appearing every few months.

Saturn's atmosphere is so cold that most of the ammonia freezes into clouds at lower levels. The outer atmosphere therefore consists mainly of hydrogen, helium, and methane, the latter forming particularly strong absorption bands in the planet's spectrum. Saturn's overall composition resembles that of Jupiter, but with a somewhat smaller proportion of hydrogen. Saturn is believed to have a rocky core about 12,000 miles (20,000 km) in diameter, about half the size of Jupiter's core, covered with a 3,000-mile (5,000-km) coating of ice and surrounded by a 5,000-mile- (8,000-km)-thick layer of metallic hydrogen. The rest is hydrogen gas, which gives the planet its low average density. The rocky core accounts for only about 15 percent of Saturn's mass. Because Saturn is less than one-third the mass of Jupiter, its center is not compressed as greatly. But pressure at the center is still some 50 million atmospheres, and the temperature is more than 15,000°C. Like Jupiter, Saturn is radiating slightly more heat than it receives from the Sun. Either the planet is still slowly contracting, or the heat is left over from its formation.

The rings. Saturn's rings are perhaps the most beautiful in the solar system. They form a complete unbroken band around the planet's equator and have a maximum diameter of about 170,000 miles (275,000 km) from rim to rim. There are three main rings (A, B, and C), together with two fainter ones. Ring B is the wide and brilliant ring, flanked on the inside by the elusive crepe ring C, and bordered on the outside by ring A. Rings A and B are separated by the dark CASSINI DIVISION. In 1969 a much fainter ring D was discovered inside ring C, almost touching the planet's surface, while a ring of very scattered debris seems to exist outside ring A.

The rings were recorded almost as soon as the

telescope was invented, but early observers were unable to interpret the poor images produced by their primitive telescopes. Christiaan HUYGENS first realized the nature of the rings in 1655 and also explained their changes in appearance. During one 29½-year orbit of Saturn around the Sun, there are two periods when the rings are tilted toward us edge-on. When they are presented edge-on, the rings disappear from view, which implies that they must be very remarkably thin—measuring no more than about a kilometer in thickness.

The true nature of the rings was shown in 1895 by the American astronomer James Keeler, who found by the Doppler shift in light from the rings that they rotate like a swarm of particles on independent orbits, rather than as a solid disk. Careful observation of the rings has revealed several bands in which the density of ring particles is lower than elsewhere. These are believed to be due to the gravitational effects of Saturn's nearest moons, particularly Mimas, which perturb ring particles into different orbits. A ring particle moving in an orbit with a period which is an exact fraction of a satellite's period will suffer repeated gravitational tugs by the satellite at the same point in its orbit. These perturbations will produce a thinning of particle density. The gaps in the particle distribution are direct analogies of the KIRKWOOD GAPS in the asteroid belt, which are caused by perturbations from Jupiter.

The planet's globe is clearly visible through the crepe ring C, which confirms that the ring is made of particles. In addition, when the rings pass in front of a bright star, the star is visible in the Cassini division, visible slightly dimmed through ring A, and even seen faintly through ring B. Also, moons which pass into the shadow of the rings are still faintly visible even when shaded by the densest part of ring B. The ring particles reflect about 70 percent of the light hitting them, and the rings can actually be brighter than the rest of Saturn.

The true nature of the ring particles is still something of a puzzle. They are probably blocks about the size of bricks, coated with frozen water. The rings lie inside ROCHE'S LIMIT for Saturn, within which a satellite would be pulled apart by Saturn's gravitational force. The rings may be shattered remains of a moon that strayed too close, or they may be the building blocks of a satellite that never formed.

Satellites. Saturn has ten moons, eight of which move in near-circular orbits close to the planet in the plane of the rings. Beyond these eight, Iapetus moves in an orbit inclined at 15°, and Phoebe, much more distant, moves in a retrograde elliptical orbit.

Saturn rockets

A family of large space launchers developed for manned applications by a team under Wernher VON BRAUN at the MARSHALL SPACE FLIGHT CENTER, Huntsville, Alabama. Planning began in 1958 on a booster originally called Juno V, which had eight propellant tanks from REDSTONE rockets clustered around a central large JUPITER tank, and used eight engines developed from Thor and Jupiter rockets. In 1959 the name Saturn was adopted, and the project was officially taken up by NASA; by the following year its role as a launcher for Apollo had been defined, an S-IV second-stage design was chosen, and there were plans for an S-V third stage which never materialized. On October 27, 1961, the first Saturn 1 was test-flown from Cape Canaveral. By then, the Apollo Moon-landing goal had been announced and plans were laid for more powerful successors to Saturn 1. Largest of these was the so-called Nova class of launchers, with a lift-off thrust of 12,000,000 lb. (5,500,000 kg), eight times that of the Saturn 1. Eventually, however, the designers settled on the three-stage Saturn V. In 1964 the Saturn 1 test flights put dummy Apollo capsules into orbit, concluding in 1965 with the launch of three PEGASUS micrometeorite-detection satellites.

Saturn 1B. For actual manned Apollo flights an uprated Saturn 1, called the Saturn 1B, had begun development in 1962, with a modified upper stage called the S-IVB. This improved booster was first flown on February 26, 1966, and on October 11, 1968, launched Apollo 7, the first manned Apollo. Saturn 1B was subsequently used for launching Skylab crews and the American half of the Apollo-Soyuz mission. For these later launches, the Saturn 1B was erected on a pedestal to bring its total height to that of the larger Saturn V, so that it could use the same launch facilities. Saturn 1B's first stage was 80 feet 4 inches (24.5 m) long and 21 feet 5 inches (6.5 m) in diameter, producing 1,640,000 lb. (744,000 kg) thrust from eight H-1 engines. The S-IVB second stage, 58 feet 5 inches (17.8 m) long and 21 feet 8 inches (6.6 m) in diameter, produced up to 225,000 lb. (102,000 kg) thrust from a single engine called the J-2. It was topped by a ring-shaped instrument unit containing the electronic guidance equipment to control the rocket's operation. Saturn 1B's overall height with payload was 224 feet (68 m), and it weighed 650 tons loaded.

Satellites of Saturn

	Discoverer	Diameter (km)	Orbit radius (10^3 km)	Inclination	Eccentricity
Janus	Dollfus (1966)	300	159	0°	0
Mimas	Herschel (1789)	500	186	1.5°	0.02
Enceladus	Herschel (1789)	600	238	0°	0.0045
Tethys	Cassini (1684)	1,000	295	1.1°	0
Dione	Cassini (1684)	800	378	0°	0.002
Rhea	Cassini (1672)	1,600	527	0.3°	0.001
Titan	Huygens (1655)	5,800	1,222	0.3°	0.029
Hyperion	Bond (1848)	500	1,483	0.6°	0.104
Iapetus	Cassini (1671)	1,600	3,560	14.7°	0.028
Phoebe	Pickering (1898)	200	12,951	150° (retrograde)	0.163

A Saturn V rocket lifts the Apollo 11 spacecraft off the pad at the Kennedy Space Center. This is the full Saturn configuration weighing a total of 2,850 tons (with fuel) and capable of lifting a 150-ton payload into orbit or sending a 50-ton payload to the Moon.

Saturn V. Saturn 1B allowed the Apollo program to progress until the larger Saturn V Moon rocket was available. The design for this three-stage vehicle was outlined in 1961, and construction began the next year. Its first stage was based on the F-1 engine developed by the Rocketdyne Corporation, with a thrust of 1,500,000 lb. (680,000 kg); this is equal to the entire first-stage thrust of the original Saturn. The Saturn V's first stage (called S-1C) had five such engines. Its S-II second stage contained five J-2 engines, and the S-IVB third stage had one J-2 engine. The first Saturn V was launched on a test flight on November 9, 1967, sending a dummy Apollo command module into orbit; it was the most powerful rocket ever launched. This flight also initiated the new launch complex 39 at Kennedy Space Center, from which the Moon missions were to leave. After one more test flight the Saturn V was used for a manned launch, the Apollo 8 mission in December 1968. Subsequent Moon flights, and the Skylab space station, were also carried by Saturn V.

The rocket's overall height, including spacecraft, was 363 feet (111 m). Its first stage was 138 feet (42 m) long and 33 feet (10 m) wide; its total thrust was about 7,600,000 lb. (3,450,000 kg). The S-II second stage was 81 feet 7 inches (24.9 m) high, 33 feet (10 m) in diameter, with a thrust of up to 1,160,000 lb. (526,000 kg). The S-IVB top stage was 58 feet 7 inches (17.9 m) long, 21 feet 8 inches (6.6 m) in diameter, and produced up to 230,000 lb. (104,000 kg) thrust. Total loaded weight of the Saturn V at lift-off was about 2,850 tons. Because of improving technology and different mission requirements, engine performance and total weight varied slightly at each launch. Saturn V's capacity was about 150 tons in Earth orbit, or 50 tons to the Moon.

scattering

The deflection of light and other forms of radiation. The most familiar form of scattering is that produced by molecules and dust particles, as in the Earth's atmosphere. This is named *Rayleigh scattering,* for the English physicist Lord Rayleigh (1842–1919), who in 1871 first explained its operation. Shorter-wavelength (blue) light is scattered more than red, which is the reason the sky is blue. Another form of scattering is the *Compton effect,* named for the American physicist Arthur Holly Compton (1892–1962), who described it in 1923. In this process, particles of light (photons) collide with electrons, thereby transferring some of their energy to the electrons. The photons then have less energy, and therefore a longer wavelength, and they are scattered in random directions. The *inverse Compton effect,* which is important in astronomy, occurs when the electrons themselves have very high energies and are moving close to the speed of light—around quasars or supernovae, for example. In this case, energy is transferred from the electrons to the photons, which gain energy. This can account for the production of X rays from quasars and supernovae.

Schiaparelli, Giovanni Virginio (1835–1910)
Italian astronomer who first reported "canals" on Mars. Schiaparelli noted these straight markings at the near approach of Mars in 1877 (the same occasion on which Asaph HALL discovered the moons of Mars). He named them *canali,* meaning channels; but the word was mistranslated as canals, implying they were artificial. This aroused the interest of Percival LOWELL, who founded the once-popular view of life on Mars. Schiaparelli did not subscribe to Lowell's theories. He continued mapping Mars, and originated the current nomenclature of Martian surface features. Schiaparelli also made extensive observations of Mercury and Venus, concluding that they both kept the same face turned to the Sun, which has since been disproved.

Schiaparelli is perhaps best known for his discovery that meteor showers follow the same orbits as comets, and he proposed that meteors were produced by the disintegration of comets. He showed that the Perseid meteors were associated with a comet seen in 1862 (1862 III), and the Leonids with comet 1866 I; his ideas were confirmed in 1872 when Biela's comet, which had disintegrated on a previous approach to the Sun, was replaced by the Andromedid meteor shower.

Schirra, Walter Marty (b. 1923)

The only astronaut to fly in all three types of American spacecraft: Mercury, Gemini, and Apollo. Schirra was one of the original seven astronauts selected in 1959. He flew in *Sigma 7*, the fifth Mercury mission, on October 3, 1962, completing six orbits of the Earth in 9.2 hours. In December 1965 he commanded the Gemini 6 mission, which, with Gemini 7, made the world's first space rendezvous. In October 1968 he commanded Apollo 7, the Earth-orbital maiden flight of the Apollo capsule. Schirra left the astronaut corps in 1969 to enter private business.

Schmidt, Maarten (b. 1929)

Dutch-born American astronomer, the first to interpret the spectrum of a QUASAR, establishing the enormous speed at which these objects are receding. In the early 1960s, Jesse Greenstein, Allan Sandage, and Schmidt used the Hale 200-inch (508-cm) telescope to study the optical counterparts of some compact radio sources. These faint, star-like objects with strange spectra were called "quasi-stellar sources" (later shortened to *quasars*). In 1963, Schmidt showed that the enormous RED SHIFT in the spectrum of the brightest quasar, 3C 273, meant it was moving away from us at over 25,000 miles (40,000 km) per second. Other quasars were subsequently found to have even greater red shifts. Schmidt has spent the last decade making detailed studies of these exciting but poorly-understood objects.

Schmidt telescope

A wide-angle photographic telescope first constructed in 1930 by the Estonian optician Bernhard Voldemar Schmidt (1879–1935). Normal reflecting telescopes have a field of view of not more than about $\frac{1}{2}°$ (equivalent to the apparent diameter of the Moon), while wide-field camera lenses suffer from CHROMATIC ABERRATION and distortion. Schmidt's wide-field reflector eradicated the optical defects of ordinary reflectors by using a spherical mirror with a correcting lens in front.

Although a mirror with a spherical curve is afflicted with SPHERICAL ABERRATION, which destroys the sharpness of its images, by placing in front of it a thin lens or *plate*, polished with a double curve, Schmidt was able to correct the aberration and obtain star images of superb quality across a previously unheard-of field of view. His first camera, using a 10.2-inch (36-cm) correcting plate in conjunction with a 17.3-inch (44-cm) mirror, covered a field of view of 16° with excellent definition.

The difficulty of making the correcting plate delayed the development of the Schmidt camera until after its inventor's death, but its eventual acceptance was rapid. One of the largest Schmidt cameras in the world is the 48 × 72 inch (122 × 183 cm) at Mount Palomar, installed as a scouting instrument for the 200-inch (508-cm) reflector; in 1973, a similar instrument was installed at the Anglo-Australian Observatory in New South Wales. The largest Schmidt, with a 79-inch (200-cm) mirror, was installed at the Karl Schwarzschild Observatory, Tautenberg, East Germany, in 1960. These instruments have between them now mapped the entire sky, covering stars down to the 20th magnitude and lower. Another large Schmidt, the 39-inch (100-cm) at the European Southern Observatory, La Silla, Chile, also took part in the work.

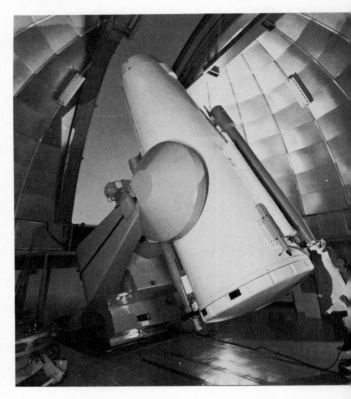

The 48-inch (1.22-m) Schmidt telescope at Siding Spring, Australia, is carrying out a survey of the southern skies to match that of the northern heavens made with the famous Mt Palomar Schmidt of the same size in the early 1950s. Like the Palomar instrument, it photographs areas of the sky 6° square on plates 14 inches (36 cm) square.

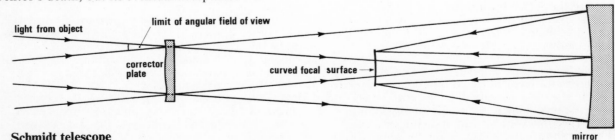

Schmidt telescope

light from object

limit of angular field of view

corrector plate

curved focal surface →

mirror

Schwarzschild, Karl (1873–1916)

German astronomer, noted for his pioneering work on the theory of stellar structure, particularly the transport of heat by radiation in the outer layers of stars. Arthur EDDINGTON extended this work to the interior of stars, producing the first modern view of a star's structure. Schwarzschild developed the technique of measuring star brightnesses by photography, comparing their visual and photographic magnitudes to obtain their COLOR INDEX. He worked on the development of relativity theory, predicting the existence of the SCHWARZSCHILD RADIUS around high-density objects. His son, the German-born American astronomer Martin Schwarzschild (b. 1912), helped create the modern understanding of a star's life-history, as outlined in his classic 1958 book *The Structure and Evolution of the Stars*. Martin Schwarzschild originated and directed Project Stratoscope, a balloon-borne camera-telescope for celestial observations, which paved the way for current astronomical satellites.

Schwarzschild radius

The distance from an extremely dense or massive body at which the ESCAPE VELOCITY equals the velocity of light. The theory of RELATIVITY forbids any greater speeds than the velocity of light, and Karl SCHWARZSCHILD first showed that neither matter nor radiation from within such a radius can escape to the rest of the Universe. The surface of the imaginary sphere around a body at the Schwarzschild radius is called an *event horizon* or BLACK HOLE.

The Schwarzschild radius for a mass equal to the Sun's is only 2 miles (3 km); a star's core must be compressed to about this size by a SUPERNOVA explosion before a black hole is formed. By comparison, a body of the Earth's mass would have a Schwarzschild radius of only 0.4 inch (1 cm); no known process can reduce a planet to this size and density.

scintillation

The "twinkling" of a celestial object. Rapid variations in brightness of a star or other point source are caused by the Earth's atmosphere—generally the lowest 5 miles (8 km), which is rarely completely steady. Changes in air temperature produce turbulence, which varies the amount by which the atmosphere bends light passing through it. The resultant scintillation is visible through a telescope as the bad "seeing," which mars the steadiness of an image. To overcome scintillation, observatories must be sited high on mountains in areas of particularly stable air. Radio sources also scintillate, because the radiation from them undergoes similar effects when passing through clouds of charged particles in space.

Scorpius (the scorpion)

A constellation of the zodiac lying in the southern hemisphere of the sky, visible in the northern hemisphere during summer. Its brightest star is ANTARES. Beta Scorpii is a multiple star of magnitudes 2.9, 5.06, and 9.7; the brightest of the three is also a spectroscopic binary of period 6.8 days. Mu^1 Mu^2 Scorpii form an optical (unrelated) pair, as do $Omega^1$ $Omega^2$ Scorpii and several other apparent doubles. Other objects of note are the bright globular cluster M4

NASA's smallest complete launcher, Scout, used for orbiting small payloads (up to 300 lb.) into Earth orbit. Its four stages are all solid fueled, unlike other launchers which use liquid fuel This adds to its versatility, since it can be fired from small launch sites.

(NGC 6121) and the open cluster M6 (NGC 6405), called the jewel-box cluster. The brightest X-ray source in the sky, Sco X-1, is identified with a faint star in the constellation. The Sun passes through Scorpius briefly during the last week of November.

Scott, David Randolph (b. 1932)

Commander of the Apollo 15 mission, which landed on the Moon on July 30, 1971. With fellow astronaut James Benson Irwin (b. 1930) he explored the lunar surface for over 18 hours during a $2\frac{3}{4}$-day stay, traveling 17 miles (27 km) in the Lunar Roving Vehicle. Scott, a qualified aeronautical engineer, flew with Neil ARMSTRONG on the Gemini 8 mission in March 1966, which accomplished the world's first space docking. In 1969 he was command module pilot on the Apollo 9 mission, which tested the lunar module in Earth orbit. Scott was selected as an astronaut in 1963. In 1975 he left the astronaut corps to become director of NASA's Flight Research Center at Edwards, California.

Scout rocket

An American solid-fuel space launcher for orbiting small satellites. It is NASA's smallest launch rocket, and the only one employing solid fuels on all stages. Scout is a four-stage rocket, 75 feet (22.9 m) tall with payload attached, weighing 47,300 lb. (21,450 kg). It was introduced in 1960, and has since launched numerous scientific satellites, including members of the Explorer series, from Wallops Island, the Western Test Range, and the San Marco platform. Modern Scout rockets can place payloads up to 390 lb. (175 kg) in Earth orbit. Scout's Algol II first stage is 30 feet (9 m) long and 40 inches (1 m) in diameter, with a thrust of 105,000 lb. (47,600 kg). An alternative first stage, Algol III, has an increased width of 45 inches (1.14 m), giving a thrust of around 140,000 lb. (63,500 kg) for heavier payloads. Its Castor II second stage, 20 feet (6 m) long and 30 inches (76 cm) wide, has a thrust of 61,000 lb. (27,500 kg). Antares II, the third stage, is 10 feet (3 m) long and 30 inches (76 cm) wide, with a thrust of 21,000 lb. (9,500 kg). The fourth stage, Altair III, is 6 feet (1.8 m) long and 20 inches (50 cm) wide, with a thrust of 6,000 lb. (2,700 kg). A fifth stage is now also available for high-orbit missions.

Sculptor (the sculptor)
A faint constellation in the southern hemisphere of the sky, below Cetus, introduced by Nicolas Louis de LACAILLE. The south galactic pole lies in Sculptor. The constellation also contains a nearby dwarf galaxy in our local group.

Scutum (the shield)
A faint constellation of the southern hemisphere of the sky introduced by the Polish astronomer Johannes HEVELIUS, who originally called it Scutum Sobieskii, or Sobieski's shield, in honor of a patron. It is visible between Aquila and Serpens during the northern hemisphere summer. Delta Scuti varies from magnitudes 4.9 to 5.2 every 0.194 days, and is the prototype of a rare class of pulsating variable stars. The cluster M11 (NGC 6705) contains about 200 stars.

seasons
Divisions of the year caused by the varying presentation of the Earth's northern and southern hemispheres to the Sun. If the Earth's axis were upright with respect to the plane of its orbit, the Sun would always seem to pass directly overhead to an observer on the equator. Since the axis is in fact inclined from the upright at an angle of about $23\frac{1}{2}°$, the north and south poles are alternately tilted toward the Sun. The Sun's altitude therefore increases and decreases during the year, producing the seasons. Spring in the northern hemisphere starts at the spring EQUINOX, about March 21. Summer begins at the summer SOLSTICE (around June 21), autumn at the autumnal equinox (roughly September 23), and winter at the winter solstice (about December 22).

selenocentric
Term meaning centered on the Moon. A *selenocentric orbit* is an orbit around the Moon.

selenology
Study of the lunar surface, equivalent to geology on Earth. A *selenologist* is a lunar geologist.

Serpens (the serpent)
A large constellation straddling the celestial equator, representing a serpent wound around OPHIUCHUS. The constellation is split into two parts: Serpens

Caput, the head, and Serpens Cauda, the tail; it is best seen during the northern hemisphere spring and summer. Delta Serpentis is a double star, of magnitudes 5.16 and 4.23. M5 (NGC 5904) is a bright globular cluster, 27,000 light-years away. M16 (NGC 6611) is a famous star cluster and nebula of gas which contains dark GLOBULES.

service module
Unmanned section of a spacecraft containing engines for course corrections and reentry, air and water supplies, and electrical power for the spacecraft's instruments. America's first manned spacecraft, Mercury, was self-contained, but an embryonic service module, called the equipment module, was introduced on Gemini flights. In the Soviet Vostok and Voskhod craft the last stage of the launch rocket went into orbit attached to the capsule and acted as a service module. The true Service Module (SM), with a large propulsion engine, was introduced on the Apollo spacecraft. A similar rear section, though with a less powerful engine, is attached to the Soviet Soyuz craft.

Sextans (the sextant)
A faint and insignificant constellation of the equatorial region of the sky, lying between Leo and Hydra, visible in the northern hemisphere spring. It was introduced on the 1690 star map of Johannes HEVELIUS.

Seyfert galaxy
A galaxy with a very small, bright nucleus, containing high-speed clouds of gas indicating recent explosions. Such galaxies were first studied by Carl K. Seyfert (1911–1960) in 1943, and about 100 are now known. One percent of SPIRAL GALAXIES seem to have these bright nuclei, but there are also a few Seyfert galaxies which are not spirals.

A Seyfert galaxy nucleus contains a very small core, less than a light-year across, which emits predominantly blue light as well as ultraviolet, infrared, and sometimes radio waves. This core seems to be a "mini-quasar," over 100 times fainter than normal QUASARS. Seyfert galaxies may therefore provide a link between normal galaxies and quasars. Around the core are clouds of hot gas ionized by the ultraviolet light from the core, and moving at hundreds of miles per second. Filling the rest of the nucleus (100 to 1,000 light-years across) is much more tenuous gas that emits optical FORBIDDEN LINES.

The light from the core of Seyfert galaxies varies in brightness over a period of months, and is probably produced by electrons moving in strong magnetic fields (SYNCHROTRON RADIATION). The infrared radiation from some Seyferts also varies, and may be produced by the same synchrotron process. In other Seyferts the infrared radiation comes from dust grains in the nucleus that have been heated by the light from the core. The amount of radiation emitted at infrared wavelengths from the nucleus is often considerably greater than the luminosity of the entire galaxy at optical wavelengths.

Shapley, Harlow (1885–1972)
American astronomer, who discovered that our Galaxy is much larger than was previously supposed, and that

This short exposure photograph shows only the bright central regions of the Seyfert galaxy NGC 4151; the faint outer parts would cover the entire photograph. Even so, the galaxy's center is overexposed, and the very small bright nucleus appears larger than its true size. The barred spiral, **top left**, is a very distant background galaxy.

the Sun is not centrally placed within it. Shapley made his discovery from a study of CEPHEID VARIABLE stars in globular clusters around our Galaxy. In 1912, Henrietta LEAVITT had discovered the relationship between period of variation and average brightness of Cepheid variables. In 1916–1917 at Mount Wilson Observatory, Shapley photographed similar variable stars in globular clusters, and carefully calibrated the period-luminosity law to determine the clusters' distances. Shapley also noted that most globular clusters are concentrated in one part of the sky, toward the direction of Sagittarius. If globular clusters are in fact scattered symmetrically around our Milky Way system, then this apparent direction of concentration must mean that the Sun is not centrally placed in the Milky Way. Shapley's bold conclusion overthrew all previous thinking about the Sun's place in the Universe—a revelation in many ways comparable to the theory of Copernicus, which ejected the Earth from the center of the heavens. From his calculations of the clusters' distances, Shapley deduced that the Galaxy was about 10 times larger than previously estimated, and that the Sun lay about 50,000 light-years from the center. However, the effect of interstellar gas and dust, dimming stars and making them appear too faint, made his estimates too high. By 1930 he had reduced his figures to a diameter of 100,000 light-years, with the Sun about 30,000 light-years from the center. This is in accord with modern values. These results were independently confirmed by the Dutch astronomer Jan OORT. Shapley's conclusions led in 1920 to the so-called Great Debate, a public discussion with the American astronomer Heber Doust Curtis (1872–1942) about the nature of faint nebulae. Although Shapley believed they were either part of our own enlarged Milky Way system or near

neighbors of it, Curtis suggested they were separate star systems, or galaxies, far distant from the Milky Way. On this point, Shapley was defeated.

From 1921 until his retirement in 1952, Shapley was director of the Harvard College Observatory. As a Harvard student in 1914, Shapley had shown that Cepheid variables were giant stars whose light changes were caused by actual pulsations, a theory later developed by Arthur EDDINGTON. Shapley continued his studies of variable stars; and in the 1930s he discovered the first two dwarf galaxies, both members of our LOCAL GROUP, which lie in the constellations of Sculptor and Fornax; he also showed that galaxies occur in clusters, which he called *metagalaxies*.

Shatalov, Vladimir Alexandrovich (b. 1927)
Soviet cosmonaut, pilot of the Soyuz 4 spacecraft, which docked with Soyuz 5, in January 1969. While the craft were docked, two space-suited cosmonauts walked in space from Soyuz 5 to Soyuz 4, returning to Earth with Shatalov in the Soyuz 4 command module. In October 1969 Shatalov was overall commander of the first joint flight of three manned spacecraft; he flew in Soyuz 8, which maneuvered in space with Soyuz 6 and Soyuz 7. Shatalov's third flight came in April 1971 as commander of Soyuz 10, which docked with the Salyut 1 space station. Despite expectations, the cosmonauts did not enter the space station but undocked after $5\frac{1}{2}$ hours and returned to Earth; some reports spoke of a hatch failure that could have proved fatal. Shatalov, one of the Soviet Union's most experienced spacemen, subsequently became director of cosmonaut training.

Shepard, Alan Bartlett (b. 1923)
The first American to be launched into space, a naval pilot selected as an astronaut in 1959.

Shepard was boosted in the *Freedom 7* capsule on a suborbital flight lasting 15 minutes 22 seconds on May 5, 1961. He reached a maximum altitude of 116.5 miles (187.5 km), covering a distance of 303.8 miles (488.9 km). Shepard was temporarily grounded because of an ear disorder, and became chief of the astronaut office. After an operation in

1969 he was fit to fly again, and commanded the Apollo 14 mission, which landed on the Moon on February 5, 1971. He and Edgar Dean Mitchell (b. 1930) made two Moon walks, lasting a total of $9\frac{1}{4}$ hours, during which they ascended to the rim of Cone crater to take samples. Shepard left the astronaut corps in 1974.

shooting star
The popular term for a METEOR.

sidereal day
The time the Earth takes to rotate on its axis with respect to a fixed point in space, such as is indicated by a star. Unlike the apparent SOLAR DAY, which is subject to considerable fluctuations, the length of the sidereal day was long thought to be constant. The rotation of the Earth, however, is now known to be slowing down by about 0.003 second a year, which means that the apparent sidereal day is gradually lengthening. The length of the *mean* sidereal day is 23 hours 56 minutes 4 seconds (see DAY).

sidereal period
The time taken by a planet or satellite to go once around its orbit as measured against the star background. The sidereal period of a planet can be said to be its "year." It differs from the period observed from the Earth, because the Earth itself is moving. For example, the Moon's sidereal period is 27.32 days, but its cycle of phases appears to last 29.53 days. The latter, the *apparent* period, is known as the SYNODIC PERIOD.

sidereal time
Time as determined by the stars, used for finding the position of a celestial object in the sky. A sidereal day is divided into 24 hours of sidereal time, which therefore represents the apparent rotation of the CELESTIAL SPHERE (the imaginary stellar vault) around the Earth.

sidereal year
The time taken for the Earth to revolve once around the Sun with respect to the star background. This period is equivalent to 365.26 SOLAR DAYS, or 366.26 SIDEREAL DAYS.

siderite
An iron meteorite; its name is derived from the Greek *sideros*, meaning "iron." Siderites are composed of an iron-nickel mixture with small amounts of cobalt, copper, phosphorus, carbon, iron sulfide, and traces of other elements and minerals. Siderites are solid masses with a density about 7.8 times that of water. Their variable nickel content divides them into three distinct types. Siderites with 5 to 6 percent nickel are named *hexahedrites*; those with 6 to 20 percent nickel are called *octahedrites*; and siderites with more than 20 percent nickel form *ataxites*. All iron meteorites were formed deep inside a parent body where an iron-nickel core slowly cooled (at a rate of less than 250° per million years) and crystallized. Siderites must have originated in about 6 to 10 asteroid-like parent bodies 130 to 320 miles (210–515 km) across, and were released in collisions between 1,000,000 and 100,000 years ago.

siderolite
A stony-iron meteorite, containing about equal amounts of nickel-iron metal and stony minerals. There are two main types of siderolites, although the name is often also applied to meteorites that are difficult to classify. The *pallasites* consist of either complete or shattered crystals of the mineral olivine, averaging about 5 millimeters in diameter, embedded in nickel-iron. The olivine solidified while the metal was still molten, and it must have formed near the edge of the parent body's molten core. The *mesosiderites* contain a near-equal mixture of nickel-iron and the minerals pyroxene and plagioclase. The metal may be interlaced between the stone fragments or may appear as separate grains. Siderolites seem to have cooled extremely slowly, at less than 1° per million years, and must have originated deep in their asteroid-like parent planets.

Siding Spring Observatory
Astronomical observatory located at an elevation of 3,822 feet (1,165 m), 18 miles (29 km) from Coonabarabran, New South Wales. It was founded in 1965 as a field station of MOUNT STROMLO OBSERVATORY, and has reflectors of 40-inch (104-cm), 26-inch (66-cm), and 24-inch (61-cm) aperture. In 1973 the United Kingdom's Science Research Council opened a 48/72-inch (122/183-cm) Schmidt telescope at Siding Spring. At the same site, the British and Australian governments jointly operate the 153-inch (390-cm) Anglo-Australian Telescope (AAT), inaugurated in 1974.

Sirius
The brightest star in the sky, of magnitude −1.47; it is 8.7 light-years away, the fifth-closest star to the Sun. Sirius lies in the constellation Canis Major, and is also known as Alpha Canis Majoris. It is a white star, 1.76 times the Sun's diameter and 23 times as luminous; its mass is 2.35 times that of the Sun. Sirius is actually double, with a magnitude 8.4 white dwarf companion orbiting it every 50 years; the white dwarf is 0.99 times the Sun's mass, 0.022 times its diameter, and 0.002 times its luminosity. Sirius' HELIACAL RISING was a familiar sign to the ancient Egyptians that the Nile floods were about to begin; the Egyptians used Sirius observations to measure the length of the year.

Skylab
First American space station, modified from the third stage of a Saturn V rocket. Skylab, weighing 75 tons (about four times that of the Soviet SALYUT), was launched into a near-circular orbit about 270 miles (435 km) high on May 14, 1973. About a minute into the launch the space station's micrometeoroid shield deployed prematurely and was ripped away, destroying one solar panel and jamming another. The meteoroid shield had been painted white to reflect the Sun; with it gone, the space station began to overheat. The Skylab mission was saved by the first crew, who erected a sunshield, and during a space walk cut free the jammed solar panel. Skylab contained four sections. Largest was the orbital workshop 10,644 cubic feet (301 m³) in volume, 48.1 feet (14.7 m) long and 21.6 feet (6.6 m) in diameter; this was divided into living quarters and a work section.

Skylab. America's space laboratory, seen from an Apollo transit craft in June 1973. One solar paddle is missing, having been torn off accidentally, the makeshift sunshield which was used to prevent the spaceship overheating can be seen. The cross-shaped solar paddles generated power for the Apollo Telescope Mount.

The airlock module, 17.6 feet (5.4 m) long and varying in diameter from 5.5 feet (1.7 m) to 10 feet (3 m), contained equipment for controlling the station's operation, and a hatch for space walks. The 622-cubic-foot (17.6-m³) airlock module allowed astronauts to leave Skylab without depressurizing the entire station. The multiple docking adapter, 17.3 feet (5.3-m) long and 10 feet (3 m) in diameter, contained an Apollo docking port at its forward end and a reserve or rescue port in its side. This 1,140-cubic-foot (32.3-m³) section contained controls for the Earth-resources scanners, a furnace and vacuum chamber for experiments in processing materials, and controls and display console for the Apollo telescope mount. The telescope mount contained six telescopes for solar observation, and was powered by a windmill-shaped array of four solar panels. At launch it was lined up with the rest of Skylab, but swung to one side when in orbit. The three Skylab crews showed that men can successfully work for long periods in weightless conditions, and demonstrated the value of manned space stations for observations of Earth and sky and for developing new industrial techniques. Total returns from the missions were 45 miles (72 km) of magnetic tape, 2,500 square feet (232 m²) of Earth-resources photographs, and over 200,000 images of the Sun. Skylab is expected to remain in orbit until the 1980s.

Mission	Launch date	Results
Skylab 1	May 14, 1973	World's largest payload; orbited by two-stage Saturn V
Skylab 2	May 25, 1973	Crew of Charles Conrad, Joseph P. Kerwin, and Paul J. Weitz. Deployed parasol sunshade to cool space station, and freed solar wing during EVA. Crew returned June 22, setting new space duration record of 28 days, 49 minutes
Skylab 3	July 28, 1973	Crew of Alan L. Bean, Owen K. Garriott, and Jack R. Lousma. Installed new Sun shield during space walk, made extensive Earth surveys and solar observations that provided new data on formation of solar flares. Returned September 25, setting new duration record of 59½ days
Skylab 4	November 16, 1973	Crew of Gerald P. Carr, Edward G. Gibson, and William R. Pogue. 7-hour EVA on December 25 by Pogue and Gibson to change film in the telescope mount and observe comet Kohoutek. Returned February 8, 1974, after record 84-day mission

Slayton, Donald Kent (b. 1924)
One of America's original seven astronauts, selected in 1959. Slayton was named to fly the second Mercury Earth-orbital mission, but was grounded in 1962 because of a heart murmur. Instead, he became director of flight crew operations until 1972 when he was eventually passed as fit to fly. Slayton was docking module pilot on the Apollo-Soyuz mission.

Slipher, Vesto Melvin (1875–1969)

American pioneer of spectroscopy, who discovered that many galaxies are receding at high velocities. This led to Edwin HUBBLE's conclusion that the Universe is expanding. Slipher began to measure the radial velocities (motions in the line of sight) of galaxies in 1913. These motions were revealed by shifts in spectral lines, and Slipher was the first man to detect them. Slipher's observations helped show that galaxies were indeed separate star systems from our own, but their full meaning became clear only in the 1920s, when Hubble found that a galaxy's speed of recession corresponds to its distance. This has provided the foundation for all current cosmological studies.

Slipher spent his working life at the Lowell Observatory, joining it in 1901 and serving as director from 1916 to 1952. He made spectroscopic studies of the rotation of planets, and discovered dark bands in the spectra of the major planets that revealed the components of their atmospheres, notably ammonia and methane. He showed that calcium and sodium are widely distributed between the stars, since they appear superimposed on stellar spectra. In 1913 he found that the nebulosity around the Pleiades shines by reflecting starlight; this was the first discovery of a reflection nebula, as distinct from a bright-line emission nebula like that in Orion (see NEBULA), and it proved that dust as well as gas exists between the stars. Slipher detected the rotation of a spiral galaxy in 1913; his later studies showed that galaxies rotate with their arms trailing. After Percival Lowell's death, Slipher directed the photographic search that led to the discovery of the planet Pluto by Clyde TOMBAUGH in 1930.

Slipher's brother, Earl Carl Slipher (1883–1964), also worked at Lowell Observatory, and was a noted observer of Mars. He produced two famous photographic atlases, one of Mars and one of the planets.

Small Astronomy Satellites (SAS)

Three American satellites in the EXPLORER series, designed to observe the sky at short wavelengths blocked from the ground by the Earth's atmosphere. They were launched from the SAN MARCO platform off the coast of Kenya. The first Small Astronomy Satellite, also called Explorer 42, was launched on Independence Day in Kenya and was named Uhuru, the Swahili word for Freedom. The 315-lb. (143-kg) satellite made the first systematic scans of the sky at X-ray wavelengths, work continued by the OAO-3 and ARIEL-V satellites. The 410-lb. (186-kg) second SAS, otherwise known as Explorer 48, carried a spark chamber to study gamma rays. The 430-lb. (195-kg) SAS-C, also called Explorer 53, was launched for a more detailed investigation of individual X-ray sources in our Galaxy.

Smithsonian Astrophysical Observatory

An astrophysical research institute of the Smithsonian Institution, founded in 1890 by Samuel Pierpont Langley (1834–1906). In 1955 Smithsonian Astrophysical Observatory (SAO) headquarters moved to the grounds of the HARVARD COLLEGE OBSERVATORY, and in 1973 the two institutions established a joint Center for Astrophysics. The SAO has a worldwide network of satellite-tracking cameras, and operates the Mount Hopkins Observatory, 35 miles (56 km) south of Tucson, Arizona. On a ridge 7,600 feet (2,316 m) high at Hopkins, a telescope 39-feet (10-m) wide, made of 248 hexagonal mirrors, was installed in 1968 to observe faint light emitted by gamma-ray impacts with the atmosphere; in 1970 a 60-inch (152-cm) reflector came into operation. A multiple-mirror telescope, consisting of six 72-inch (183-cm) mirrors whose light-collecting area equals that of a single 176-inch (450-cm) mirror, is being installed on the mountain's 8,585-foot (2,617-m) summit by the SAO in conjunction with the University of Arizona.

solar apex

The point on the CELESTIAL SPHERE toward which the Sun appears to be moving; the antapex is the point directly opposite in the sky. Systematic motion of the Sun relative to other stars is revealed by analyzing their PROPER MOTIONS and RADIAL VELOCITIES. In 1783, Sir William HERSCHEL used this method to locate the position of the apex in the constellation Hercules. It was later shown, however, that the position of the apex, and the velocity of the Sun toward it, depend on the group of stars chosen as a reference frame. Modern measurements indicate that the Sun travels toward the constellation Lyra with a velocity of 12 miles (20 km) per second, relative to nearby stars within 60 light-years.

solar constant

The amount of energy received from the Sun on a given area on the edge of the Earth's atmosphere, when the Earth is at its average distance from the Sun. The accepted value of the solar constant is 2 calories per minute per square centimeter, equal to about 1.3 kilowatts per square meter. The solar "constant" varies by one or two percent because of changing solar activity. The solar constant was first measured by the American astrophysicist Charles Greeley Abbot (1872–1973).

solar cycle

The period of about 11 years over which changes in the SUN's surface activity appear to go through a cycle. The periodic behavior of SUNSPOTS was established in 1843 by Heinrich Schwabe, who found a period of 10 years between the successive times at which the number of sunspots reached a maximum. It was later established that the average period was closer to 11 years, but the length of an individual cycle can be anything from 7 to 17 years.

F. G. W. Sporer discovered that at the beginning of a new cycle (after a minimum), the first spots always appear between latitudes 30° and 45° on either side of the solar equator. Later, spots appear progressively closer to the equator, increasing in number for $4\frac{1}{2}$ years, and then declining for the remaining $6\frac{1}{2}$ years of the cycle. Before the last spots of the old cycle have disappeared, the first spots of the new cycle can be seen at higher latitudes. Successive cycles of sunspots have opposite magnetic field directions, so that from a magnetic standpoint the solar cycle repeats itself only after 22 years.

The number of sunspots is related to the number and intensity of solar active regions; thus variations in other types of solar activity go through the same cycle. These include the number of PROMINENCES, FLARES, and PLAGES in the PHOTOSPHERE and CHROMOSPHERE, and the intensity of streamers in the CORONA. The latter are related to the SOLAR WIND of particles blowing into space, which is in turn responsible for magnetic storms and aurorae on the Earth. An 11-year cycle for these geomagnetic effects has in fact been known for a century. More recent research suggests that even the Earth's weather may follow the solar cycle, though the reasons are still unknown.

solar day

The time the Earth takes to rotate on its axis with respect to the Sun. On average, this is equal to 1.002738 mean SIDEREAL DAYS. The difference between the two periods occurs because the Earth is also revolving around the Sun as it rotates on its axis. The Earth must spin approximately an extra 1/365th of a revolution each day to bring the Sun facing the same hemisphere again.

solar system

The group of planets, comets, and asteroids orbiting the Sun, whose gravitational pull dominates space in all directions out to a distance of 2.4 light-years. The Sun makes up more than 99.95 percent of the mass of the solar system and is its only significant source of light and heat.

The planets can be divided into two groups: the terrestrial planets and the giant planets. The first group stretches from 0.3 to 1.7 ASTRONOMICAL UNITS from the Sun (an astronomical unit, or a.u., is the average distance from the Earth to the Sun). Five bodies make up the terrestrial planets: Mercury, Venus, the Earth-Moon system and Mars. All are similar in nature, being made predominantly of rock, some with central metallic cores. The presence or absence of atmosphere is dictated by the planet's surface gravity and temperature.

From 1.7 to 4.9 a.u. there is a gap in the system, beyond which comes the giant planet group stretching from 4.9 to 30.3 a.u. from the Sun. This consists of four essentially similar bodies, Jupiter, Saturn, Uranus, and Neptune, all globes of light-weight gases, probably with rocky cores. Jupiter is the dominant planet, making up 75 percent of the mass of the entire planetary system. Beyond Neptune is Pluto, a small frozen terrestrial-type planet which may be either an escaped satellite of Neptune or perhaps the largest of a group of small planets that occur beyond 35 a.u. It is unlikely that further giant planets exist beyond Neptune.

The gap between Mars and Jupiter contains a vast swarm of small rocky bodies known as the asteroids, with diameters from 600 miles (1,000 km) down to a few inches. These have orbits which scatter from 0.2 to 15 a.u., with the main swarm at 2 to 4 a.u. The inner solar system also contains the periodic comets, comets which have been perturbed by the planets' gravity into short-period elliptical orbits. The vast majority have aphelia close to Jupiter's orbit, but some are associated with the other giant planets. Finally, beyond all the planets, a great swarm of tiny, frozen comet nuclei is thought to exist at distances from 20,000 to 60,000 a.u. A small proportion of these are perturbed into elongated orbits so that they move toward the Sun. The comet cloud is so far from the Sun that it is only loosely bound to the solar system and many comets must be perturbed into interstellar space by the gravitational pull of stars passing within a few light-years of the Sun.

The space between the planets is not empty, but packed with areas of fine dust particles and gas molecules. The asteroids constantly suffer collision and form debris, while the comets steadily shed both gas and dust as they are heated by the Sun. Fine dust and gas are ejected from the system by the SOLAR WIND, a stream of atomic particles flowing from the Sun. Dust too heavy to be eliminated in this way slowly spirals into the Sun over a period of several million years.

solar time

Time used for all ordinary purposes, measured in terms of the Earth's rotation with respect to the Sun rather than to the stars (SIDEREAL DAY). *Apparent* solar time, as indicated by a sundial, does not run at a constant rate because of the Earth's slightly eccentric orbit and the inclination of its axis (see EQUATION OF TIME). *Mean* solar time, in which these variations are smoothed out, is therefore always used.

solar wind

A continuous stream of protons (hydrogen nuclei) and electrons, together with a few helium nuclei, which is constantly emitted by the Sun in all directions. It can be regarded as the uppermost part of the Sun's CORONA, forced away into interplanetary space by energy traveling up from the solar surface (the PHOTOSPHERE). The particles travel away from the Sun at speeds between 220 and 500 miles (350 and 800 km) per second; the solar wind has an average density of 5 protons and 5 electrons per cubic centimeter as it passes the Earth.

Although a solar wind had long been suspected for a number of reasons, among them the link between solar activity and changes in the Earth's magnetic field, direct measurements were not possible until the advent of probes exploring interplanetary space. The first extensive measurements were made by MARINER 2 on its voyage to Venus in 1962, and the latest results extend from only 0.3 a.u. from the Sun, as recorded by the HELIOS PROBE, to the 8 a.u. plumbed by the Jupiter probe PIONEER 10. The wind probably extends out to about 100 a.u., where it becomes so weak it is stopped by interstellar gas.

Many effects of the solar wind can be observed indirectly. Radio waves from distant sources are made to scintillate ("twinkle") as they pass through irregularities in the wind. Solar wind particles striking the nucleus of a COMET ionize molecules, and propel them into a straight tail pointing away from the Sun. "Knots" in the tail are seen to accelerate under the continuous pressure of the wind.

The Earth's magnetic field traps solar wind particles into the MAGNETOSPHERE, especially into the region of the VAN ALLEN BELTS. The particles in turn affect geomagnetic activity, such as magnetic storms, AURORAE, and radio fadeouts. These follow a 27-day cycle, as the more powerful wind from above the Sun's most active regions is swept past the Earth by the 27-day rotation of the Sun. Magnetic effects also vary with the 11-year SOLAR CYCLE, which governs changes in solar wind strength.

solstice

The moment when the Earth's axis is inclined at its maximum ($23\frac{1}{2}°$) toward the Sun. The north pole is tilted sunward at its maximum about June 21 (producing the beginning of the northern summer and the southern winter) while the maximum sunward inclination of the south pole, giving the opposite conditions, is about December 22. At these times the Sun is at its greatest angular distance, $23\frac{1}{2}°$, north and south of the *celestial equator*.

South African Astronomical Observatory

An observatory opened in 1973 at an altitude of 6,004 feet (1,830 m), 9 miles (14 km) from Sutherland in Cape Province, South Africa. The SAAO is an amalgamation of the CAPE OBSERVATORY, where its headquarters lie, and the former Republic Observatory, Johannesburg; it is operated jointly by the British and South African governments. It contains a 39-inch (100-cm) reflector, transferred from the Cape Observatory, together with a 20-inch (51-cm) reflector from the Republic Observatory. In 1975, the 74-inch (188-cm) reflector of the Radcliffe Observatory, Pretoria, was moved to the SAAO.

Southern Cross

See CRUX.

Soyuz

Soviet manned spacecraft for long-duration flights and rendezvous and docking missions. Soyuz is built in three parts. At the front is a near-spherical orbital compartment 8.7 feet long by 7.3 feet wide (2.65 m × 2.25 m) used for working in space. The central section is a bell-shaped command module 7.2 feet long by 7.1 feet wide (2.2 m × 2.15 m), in which cosmonauts sit during take-off and reentry. At the rear is a cylindrical service module 7.5 feet long by 7.2 feet wide (2.3 m × 2.2 m) which contains supplies, maneuvering engines, and retro-rockets. Soyuz has a habitable volume of 360 cubic feet (10.2 m³) and weighs 14,750 lb. (6,690 kg). The Soyuz service module has two wing-like solar panels attached to generate electricity, but these are removed in the simplified Soyuz design that ferries crews to and from the SALYUT space station. The Soyuz docking tunnel is located at the forward end of the orbital compartment, through which cosmonauts must crawl to transfer between docked craft. The orbital compartment can be used as additional space for resting or for scientific equipment. Both the orbital compartment and the service module are jettisoned before reentry, burning up in the atmosphere.

Soyuz was originally intended to carry three men without space suits. But this was changed after the Soyuz 11 mission, when a pressure loss prior to reentry killed the three crewmen. Soyuz crews now wear space suits during launch and reentry; but the weight and space that this takes up means that Soyuz can now hold only two cosmonauts. During the first Soyuz flight, the craft went out of control in orbit and crashed to Earth after reentry when its parachute lines became twisted, killing its test pilot. Soyuz may have been intended for a manned round-the-Moon flight; for such missions it would have had the orbital compartment removed. Unmanned test flights to the Moon of the Soyuz command and service modules were made under the name of Zond, and for lunar missions Soyuz would have been launched by the PROTON booster. However, for all flights into Earth-orbit Soyuz has been launched by a rocket using the same lower stages as the VOSTOK LAUNCHER, but with a more powerful upper stage of about 270,000 lb. (122,000 kg) thrust.

A Soyuz spaceship, as modified for use in the the Apollo-Soyuz Test Project. The smaller compartment is the orbital module; on its end can be seen part of the docking mechanism. At the other end is the instrument module, with its solar panels; in the middle is the descent module, the only part which returns to Earth.

Mission	Launch date	Results
Soyuz 1	April 23, 1967	Vladimir Komarov killed during reentry on April 24 after 18 orbits
Soyuz 2	October 25, 1968	Unmanned target for Soyuz 3
Soyuz 3	October 26, 1968	Georgi Beregovoi maneuvered close to Soyuz 2 but did not dock with it
Soyuz 4	January 14, 1969	Vladimir Shatalov docked with Soyuz 5; Yevgeny Khrunov and
Soyuz 5	January 15, 1969	Alexei Yeliseyev transferred into Soyuz 4 by a space walk, leaving Boris Volnyov to return to Earth alone in Soyuz 5
Soyuz 6	October 11, 1969	Georgi Shonin and Valeri Kubasov made joint maneuvers with Soyuz 7 and 8, and conducted welding experiments
Soyuz 7	October 12, 1969	Anatoli Filipchenko, Vladislav Volkov, and Viktor Gorbatko carried out joint maneuvers with Soyuz 6 and 8

Mission	Launch date	Results
Soyuz 8	October 13, 1969	Vladimir Shatalov and Alexei Yeliseyev commanded group flight of Soyuz 6, 7, and 8; first flight involving three craft, seven cosmonauts
Soyuz 9	June 2, 1970	Andrian Nikolayev and Vitaly Sevastyanov made record 17½-day flight
Soyuz 10	April 23, 1971	Vladimir Shatalov, Alexei Yeliseyev, and Nikolai Rukavishnikov docked with space station Salyut 1 but did not enter, possibly due to hatch problem
Soyuz 11	June 6, 1971	Georgi Dobrovolsky, Viktor Patsayev, and Vladislav Volkov docked with Salyut 1 and transferred for record 23 days; crew members were killed during reentry because of capsule pressure loss
Soyuz 12	September 27, 1973	Vasily Lazarev and Oleg Makarov made two-day test flight of simplified Soyuz for space-station ferry missions
Soyuz 13	December 18, 1973	Pyotr Klimuk and Valentin Lebedev made week-long scientific flight
Soyuz 14	July 3, 1974	Pavel Popovich and Yuri Artyukhin docked with Salyut 3 space station for 16-day mission

Mission	Launch date	Results
Soyuz 15	August 26, 1974	Gennady Sarafanov and Lev Demin failed in attempts to rendezvous automatically and dock with Salyut 3
Soyuz 16	December 2, 1974	Anatoly Filipchenko and Nikolai Rukavishnikov made 6-day rehearsal for Apollo-Soyuz mission
Soyuz 17	January 11, 1975	Alexei Gubarev and Georgi Grechko docked with Salyut 4 space station for 29-day mission
Soyuz	April 5, 1975	Vasily Lazarev and Oleg Makarov failed to reach orbit because launch rocket upper stage failed to separate. Soyuz capsule returned to Earth safely, but was not numbered
Soyuz 18	May 24, 1975	Pyotr Klimuk and Vitaly Sevastyanov completed 64-day mission aboard Salyut 4 space station
Soyuz 19	July 15, 1975	Alexei Leonov and Valeri Kubasov performed joint docking with American Apollo in Apollo-Soyuz Test Project
Soyuz 20	November 17, 1975	Unmanned test of automatic shuttle craft; docked with Salyut 4 on November 19 and transferred fuel into space station. Landed automatically February 16, 1976

space probe

Any instrument-carrying device sent from the Earth to examine another celestial body, or to take measurements of conditions in space. The first space probe was LUNA 1 in 1959, which flew past the Moon. Starting in February 1961 the Soviet Union began its series of VENUS PROBES, and in November 1962 launched the first of its MARS PROBES. However, the first probes successfully to send results from other planets were those in the American MARINER series. For other American planetary exploration series see PIONEER and VIKING. For American Moon probes see RANGER, SURVEYOR, and LUNAR ORBITER.

Space Shuttle

Reusable winged space transporter, which is launched like a conventional rocket but which glides back to Earth like an aircraft. The concept originated in the 1930s with the rocket-plane design of the Austrian spaceflight pioneer Eugen SANGER, and was developed during World War II at Peenemünde by Wernher VON BRAUN and his group, who drew up plans for a multistage winged rocket. Following the war, several designs for winged, recoverable launch systems were outlined, and trial flights began during the 1960s with simple aerodynamic shapes known as lifting bodies. However, non-reusable spacecraft were developed for the first manned flights because of their relative simplicity and lower initial cost.

Modern Space Shuttle. NASA officially began investigating Space Shuttle designs in 1968. Although initially envisaged with a winged reusable booster as well as a winged reusable orbiter, development costs meant that by 1972 the winged, fly-back booster

was discarded in favor of conventional rockets. As finally defined, the Orbiter is 122 feet (37.2 m) long and 57 feet (17.4 m) high, with a wingspan of 78 feet (23.8 m). It has three main engines of 470,000 lb. (213,200 kg) thrust each, fed by propellants from an external tank 154 feet (46.9 m) long and 27.5 feet (8.4 m) in diameter. This tank is jettisoned just before reaching orbit and reenters the atmosphere; it is the only part of the system not planned for reuse. Two solid-fuel strap-on boosters are also ignited at lift-off; these are 149 feet (45.4 m) long and 12 feet 2 inches (3.7 m) in diameter, and each produces a thrust of 2,650,000 lb. (1,202,000 kg). They drop away at a height of 27 miles (43 km) and parachute back for recovery and reuse. Total weight of the Shuttle at launch is approximately 4,400,000 lb. (2,000,000 kg). The Orbiter, which is the size of a modern jetliner, is designed to carry up to four crew members and three passengers into orbit. Its cargo bay measures 60 feet (18.3 m) long by 15 feet (4.6 m) wide, capable of carrying as much as 65,000 lb. (29,500 kg). It will be able to bring back up to 32,000 lb. (14,500 kg) from orbit; one of the advantages of the Shuttle is that it can be used to retrieve satellites for repair.

Shuttle missions. The Shuttle's large capacity will be used to take many satellites into space simultaneously, to launch major scientific payloads such as the LARGE SPACE TELESCOPE, and to ferry men and material into orbit for purposes such as space station construction. A small space station known as Spacelab is being developed by the EUROPEAN SPACE AGENCY, to be carried in the Shuttle's payload bay for international scientific and engineering experiments in orbit.

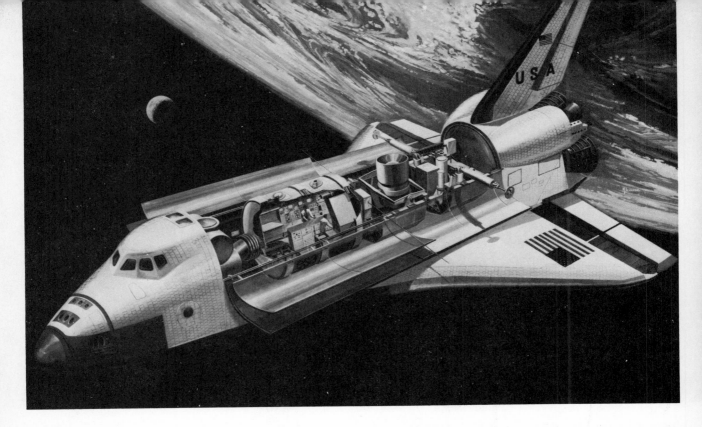

The Space Shuttle, shown here in a cutaway drawing, will enable scientists to travel into space to use the Spacelab, carried inside the Shuttle's payload bay. Behind the manned compartment is a telescope which could be used for extensive studies of the sky in infrared and ultraviolet—impossible to observe from the Earth's surface because the atmosphere blocks them. The scientific staff travel in the forward cabin, but transfer into the Spacelab compartment in orbit, where they work in a "shirtsleeves environment" without space suits.

The Shuttle can spend from one to four weeks in orbit before returning to Earth for a runway landing; construction of a 15,000-foot (4.6-km) runway began at Kennedy Space Center in 1974. After about two weeks the Orbiter will be ready to fly again with a new payload. The Shuttle will be assembled and launched from Cape Canaveral, using facilities modified from the Saturn-Apollo program. Each Orbiter is expected to make 100 or more missions. Reuse will slash the cost of space travel by about 90 percent, and the introduction of the Shuttle is expected largely to replace conventional rockets. The first Orbiter is scheduled to make a test landing in mid-1977, being released from a Boeing 747 Jumbo jet. The first manned launch of the Shuttle should take place in mid-1979.

space station
A large orbiting spacecraft for long-term human habitation. Space stations allow detailed astronomical observations to be made above the blurring effect of the Earth's atmosphere; they also act as platforms for Earth surveys, and permit experiments to take place in a total vacuum and at zero gravity, conditions unattainable on Earth. The idea was current as early as 1869, when the American

author Edward Everett Hale (1822–1909) wrote a story entitled "The Brick Moon," in which he proposed the launching of an artificial satellite (made of brick) into a 4,000-mile (6,400-km) polar orbit to aid navigation. The moon was to be hurled aloft by rolling against a rapidly-spinning flywheel. Unfortunately, the brick moon slips too early, and is propelled into orbit along with 37 construction workers and their families, who continue to live there.

Scientific proposals for a space station were made in 1923 by the German astronautical pioneer Hermann OBERTH, who envisaged space stations as refueling posts for space rockets. Grandiose schemes since have envisioned large space stations being constructed from parts ferried up from Earth, perhaps built in a wheel shape and spinning to provide artificial gravity.

In the mid-1960s the U.S. Air Force began development of the Manned Orbiting Laboratory, a two-man space station for military reconnaissance, intended to be launched atop a Titan 3 rocket with the astronauts in a modified Gemini capsule. The project was canceled in 1969. About that time, NASA was studying designs for a 12-man scientific space station to be launched during the late 1970s as a follow-on to the Apollo program. Plans envisaged eventual extension of the station to hold 50 men. The Soviet Union is also believed to have similar advanced designs under consideration. However, NASA's plans were curtailed because of budgetary cuts.

The first space station to be launched was the Soviet Union's SALYUT, which went into orbit on April 19, 1971, and was boarded by its first crew in June of that year. Several subsequent Salyuts were launched. America's larger SKYLAB station was launched on May 14, 1973; its third and final crew returned to Earth on February 8, 1974. Both Salyut and Skylab were somewhat rudimentary space stations, being

190

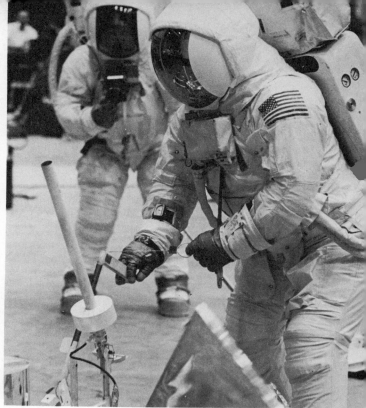

Giant space stations of the future may be similar to the conjectural 1,000-foot (305-m) "Space Station Five" devised for the film *2001—A Space Odyssey*. The station spins slowly, producing the effect of normal Earth gravity directed toward the rim of the wheel, which would appear to be the floor. At the hub, bodies would be in zero gravity.

Apollo 11 astronauts practice their lunar surface activities in a Houston laboratory while wearing full space suits and back-packs.

A space suit without its outer "beta cloth" layer, being tested by Apollo 15 astronaut Alfred M Worden.

converted from the upper stages of rockets. No further American space-station missions are planned until the European-built Spacelab becomes available in the early 1980s. This module will travel in the cargo bay of the SPACE SHUTTLE. Eventually, the Shuttle may be used to ferry into orbit components for a larger permanent space station.

space suit

A device worn by astronauts to protect them from the airless conditions of space. A space suit is a kind of personal spacecraft for the astronaut. It supplies air for him to breathe, insulates him from the extremes of hot and cold, and provides a vital pressurized container. The basic components of a space suit are an inner pressure garment made of coated nylon, covered with insulating layers of aluminum-coated plastic film and glass-fiber cloth. A space suit becomes very rigid when inflated, and careful design is needed to allow a space-suited astronaut some mobility.

The first Mercury space suits were modified from high-altitude aviation suits, and were only intended to be inflated in an emergency, if the spacecraft lost pressure. Soviet Vostok space suits were concealed by a loose-fitting cloth garment, but were probably similar in design. For Gemini missions a more advanced suit was designed, pressurized by an umbilical cord from the spacecraft. Wearing such space suits, American astronauts made their first space walks (EVAs). For the first Soviet EVA, in the Voskhod 2 craft, Alexei Leonov wore a suit with a back-pack similar to those later used in the Soyuz 4 and 5 missions for crew transfers. A back-pack allows the astronaut to move independently of the spacecraft, as was required for the Apollo Moon-landing missions. For Moon walks a special Moon suit was devised with a back pack called the Portable Life-Support System (PLSS).

Apollo suits were made more flexible by introducing bellows-like joints. They also had additional layers to withstand the more extreme conditions of walking and working on the Moon. Next to the astronaut's skin was a liquid cooling garment that prevented him from overheating. Around this was the pressure garment, a restraint layer to prevent the suit from ballooning, layers of insulation, and a micrometeorite-protection layer, topped with an abrasion-proof outer covering. Helmet and gloves were attached by air-tight rings. The plastic helmet, with a movable visor to shield against solar glare, was fixed in position, and the astronaut was free to move his head inside; he could take sips from a water bag mounted in the neck of the suit. The gloves had special insulation against heat and abrasion. For Moon walks rubber-soled overshoes were slipped on and the suit contained a urine-collection bag. The Portable Life-Support System provided air-conditioning for the space suit and supplied water for the cooling garment; it also embodied a radio transmitter. The PLSS allowed astronauts to spend up to eight hours on EVAs. The Apollo Moon suit's total weight was 180 lb.; similar suits were used on Skylab spacewalks.

spallation
The erosion of surfaces by the impact of small particles and high-energy radiation.

spec.
Abbreviation for the Latin SPECULUM, meaning mirror.

specific impulse
A measure of a rocket engine's performance. Specific impulse is calculated by dividing the rocket's thrust by the weight of fuel used per second. The result gives the time in seconds for which a unit of fuel produces a unit of thrust. The higher a rocket's specific impulse, the more efficient it is.

speckle interferometry
A technique for reconstructing star images that have been distorted by atmospheric turbulence, or poor seeing. Light from a star is distorted by constantly moving "cells" in the lower atmosphere, usually 4 to 12 inches (10–30 cm) in size. The image of a star seen through a large telescope thus consists at any instant of a group of "speckles," each a poor-quality image of the star, like the multiple images produced by a fly's eye. To reconstruct the star image the speckle pattern is first photographed in a limited wavelength band, using an IMAGE INTENSIFIER to allow a short exposure time. In one technique, the speckle pattern is then illuminated by a LASER, so that the combined interference of the individual poor images produces one single image. Several photographs of such single images are combined to reconstruct the final high-quality image. Alternatively, individual speckles

are combined using a computer to produce a final good image. By such methods, resolutions better than 0.01 arc second can be achieved, thus revealing the disks of large stars, or separating close doubles.

spectral lines
The narrow lines observed when an object's light is dispersed into a spread of wavelengths (a SPECTRUM). Each line represents light of one particular wavelength. Spectral lines may either be bright (emission lines), or occur as dark ABSORPTION LINES against the bright background of a continuous spectrum of all wavelengths. Gaseous nebulae show emission-line spectra, while ordinary stars have absorption-line spectra. The lines in the Sun's spectrum were discovered in 1814 by Joseph Fraunhofer, and are known as FRAUNHOFER LINES. The spectral lines of other astronomical bodies were first detected by William HUGGINS and Pietro Secchi in the 1860s.

Each type of atom has its own unique set of spectral lines. These reveal the elements present in the star's surface or in the nebula, like a spectral "fingerprint." The positions of the lines in the spectrum are the same whether the lines are absorption or emission.

The spectrum of hydrogen contains only a few lines, while a metal like iron has thousands. For any one element the number of lines, their wavelengths, and their relative strengths all depend on properties of the atom that can be calculated or measured in the laboratory. The relative strengths of a hydrogen line and an iron line, for example, depend on three factors: the atomic parameters, the proportions of each element in the star, and the star's surface temperature. Since the majority of stars have nearly the same relative abundance of elements, the effects of temperature allow stars to be readily classified by their spectral appearance into a sequence of SPECTRAL TYPES, governed by the star's surface temperature.

A comparison of the strengths of the various lines of any particular element ("curve of growth analysis") reveals the exact abundance of that element, when the atomic parameters and the surface temperature are known.

A detailed study of the way a particular spectral line decreases in strength from its center out to the continuous spectrum on either side (the *profile*) produces more information about the physical conditions at the star's surface. Spectral lines are made wider by higher temperatures or higher pressures, and both these quantities can be measured from a good profile. In particular, a giant star can be distinguished by its very narrow spectral lines, due to the low pressure at the surface of such a large object, where the force of gravity is relatively small. The nature of a star's magnetic field can also be determined, since it produces a splitting of the spectral lines through the ZEEMAN EFFECT.

spectral type
The category into which a star can be classified on the basis of the SPECTRAL LINES that occur in its spectrum.

From the 1860s, when William HUGGINS and Pietro SECCHI first observed the stars spectroscopically, it had been noticed that stellar spectra differ considerably, both in the number of lines present and in the darkness of the ABSORPTION LINES relative to

Right Satellite data can be presented in a variety of ways. This view of Arabia and the Red Sea was transmitted by TV and reconstructed in the laboratory. Instead of reflectance or surface brightness being displayed in various shades of gray, as in a conventional picture, colors are used. This makes it easier to pick out regions of similar reflectance.

Below This full frame from a LANDSAT satellite shows the region of Florida from West Palm Beach in the north to Key Largo in the south. Miami is halfway down the coast-line. The original consisted of four separate images, taken in green, red and two regions of the infrared. Each image was assigned a color shifted by the same amount in the visible spectrum, so that the normally invisible infrared data could be seen—the "false color" technique. Clouds reflect all colors equally, so they show up as white. Inland from Miami, the Everglades appear green.

Below right A "false color" view of SE England, almost cloud-free. The urban area of London shows up clearly in the center of the frame; to its left, two small parallel light features are the runways of Heathrow Airport. Near the top, a straight line strip shows the area around drainage channels. Though the area shown is 120 miles (192 km) on a side, features only 250 yards (230 m) across can be seen clearly.

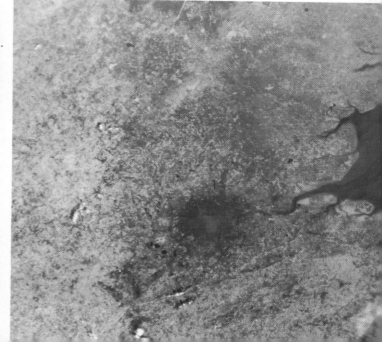

Left A true color view of the Earth, taken by Apollo astronauts between the Earth and Moon. The outlines of Africa and Arabia are easily recognized. Fine cirrus cloud, too faint to show up, cuts down the visibility of the Sahara region. Clouds rim many coastlines, while hurricanes rage in the southern oceans. The south polar cap is also cloud-covered.

Below An artist's conception of the Viking Mars lander heading for touch down on the Martian surface. The view is to the West, with the Earth about 20 degrees below the Sun. The parachute carries the aeroshell from which the lander detaches when about 20,000 ft. from the surface.

the bright background (the *continuum*). Spectra were first classified systematically by E. C. PICKERING at the Harvard College Observatory at the beginning of this century, and this work culminated in the monumental *Henry Draper Catalogue* of 225,300 stars published in 1924. Originally the letters of the alphabet had been used to classify spectra in order of increasing complexity. But it was soon shown that the most meaningful order is that representing the stars' surface temperatures, which can be deduced from the stars' colors. The consequent rearrangement of the letters produced the sequence as used today, which runs (from hottest to coolest) O B A F G K M. The peculiar, extremely hot WOLF-RAYET stars are often included at the beginning of the list as class W. There are also the classes R, N, and S, which are about the same temperature as the K and M stars, but differ from the standard sequence in having an unusually high concentration of certain HEAVY ELEMENTS. R and N stars have an excess of carbon, and are often called "carbon stars" (class C), while S stars contain zirconium and yttrium.

Stars at the beginning of the sequence, up to class G, are referred to as "early-type," while those after G are "late-type." These names are purely historical in origin, and do not imply that stars evolve along the sequence.

The spectra change from one class to the next. For example, between classes F and G the spectral lines of hydrogen become less pronounced, while those of calcium become stronger. This enables each class to be subdivided into ten *types*. The Sun is thus type G2, one-fifth the way between a G0 and a K0 spectrum.

Classification of spectra is done by eye alone, and involves the study of certain easily recognizable lines. The most important are those of hydrogen, the H and K FRAUNHOFER LINES of calcium, and the numerous lines due to such heavy elements as carbon and titanium oxide.

When the spectra of giant and dwarf stars of the same type were compared in detail, slight differences were found. This prompted W. MORGAN to add a LUMINOSITY classification to the Harvard star classification. Luminosity class V corresponds to a normal dwarf (MAIN SEQUENCE) star like the Sun, III to a giant, and I to a supergiant. (A few stars fall into the intermediate categories IV and II.) As an example, the Sun is G2 V, Aldebaran is K5 III, and Rigel is B8 I. This MORGAN-KEENAN CLASSIFICATION (MK system), published in 1943, is now used universally to categorize stars by their spectral lines.

spectroscope

A device for observing the SPECTRUM. Strictly speaking, a spectroscope is an instrument used by the naked eye, but in practice the term can be loosely employed for all devices which record the spectrum.

A simple prism or diffraction grating used to observe a white light-source will split the image into the rainbow colors of the spectrum. This is in effect a continuous series of overlapping images of the source, spread out by wavelength. If a small band of wavelengths are deficient, as in an ABSORPTION LINE, they will not be seen unless they cover a greater width of the spectrum than a single image of the source—in other words, they will not be resolved. To improve the resolving power, the light to be analyzed

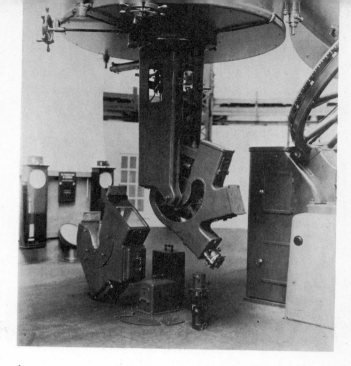

A spectrograph attached to the 73-inch (1.88-m) reflector at Victoria, British Columbia. By attaching the plate holder shown on the lower branch to either of the other two, a higher dispersion spectrum may be obtained.

is passed through a slit which is imaged by a lens system. The width of the slit compared to the dispersion of the spectrum determines the narrowest feature that can be seen. Thus the resolving power depends on the spectroscope, rather than on the size of the telescope. Visual spectroscopes are of this basic type. Substituting a camera for the eye gives a permanent record of the spectrum, and the instrument becomes a *spectrograph,* the resulting photograph being called a *spectrogram*.

If the spectroscope has a scale for reading off the wavelengths, it is strictly speaking a *spectrometer*. This name is becoming widely used for devices in which a PHOTOELECTRIC CELL either scans the spectrum or views it through filters with limited transmission bandwidths, thus giving an electrical read-out corresponding to the light intensity. Such an instrument should correctly be termed a *spectrophotometer*.

When observing the Sun, it can be inconvenient to see only the spectrum of a single strip, as represented by the slit. Consequently, the slit is arranged to scan the solar disk, and another slit, moving in conjunction isolates the spectral line of interest. An image of the Sun in one wavelength only can be built up. Where a photograph is taken, the device is called a *spectroheliograph*. If an observer wishes to view the disk in the light of one line only, the slits must scan rapidly so that persistence of vision enables the whole disk to be seen at once; the device is known as a *spectrohelioscope*. An alternative to this system now widely used is to view the Sun through a narrow band-pass interference filter in the light beam of a telescope; no other apparatus is necessary. With bandwidths of the order of an angstrom, the disk is seen in one wavelength only.

spectroscopic binary

A DOUBLE STAR in which the motion of one star around the other is detected by a spectroscope. Although the two stars in a spectroscopic binary are too close to be seen individually, the movement of one star around the other produces a DOPPLER EFFECT in its spectral lines (provided the orbit is not exactly at right angles to the line of sight). As the star alternately approaches the Earth and then recedes, its spectral lines shift first one way and then the other, revealing its orbital period. The first spectroscopic binary discovered was MIZAR, by E. C. Pickering in 1889. If the orbit is edge-on to us, one star will periodically obscure the other, causing an ECLIPSING BINARY, such as ALGOL.

If the two stars in a spectroscopic binary are of nearly equal brightness, two spectra can be seen (the pair are known as a *two-line spectroscopic binary*). In this case, astronomers can deduce the mass of both stars. (If the plane of the star's orbit is at right angles to the line of sight so that no Doppler effect is visible, the two spectra may still be individually detectable. This is termed a *spectrum binary*.) Usually, however, the light from the brighter star swamps the light from the fainter, and only one spectrum can be seen (the pair are a *single-line spectroscopic binary*). In this case the astronomer can only deduce the minimum possible mass for the unseen secondary star. In the case of the single-line spectroscopic binary Cygnus X-1, the minimum mass of the unseen companion is six solar masses. It is therefore too massive to be a white dwarf or neutron star (which could be no larger than about 1.5 solar masses, the Chandrasekhar limit). Consequently, many astronomers think that the unseen companion of Cygnus X-1 is a BLACK HOLE.

spectroscopy

The study of the spectra of heavenly bodies. By using a SPECTROSCOPE white light can be analyzed into its component colors, producing the visible SPECTRUM of the rainbow, from dark red through violet. There are three general types of spectra: *continuum, emission,* and *absorption*. A continuum is the complete band of color, with no other features; an absorption spectrum is a continuum crossed by certain dark lines; and an emission spectrum consists of bright colored lines only.

The German scientists Gustave Robert Kirchhoff (1824–1887) and Robert Wilhelm Bunsen (1811–1899), put forward the basic principles governing spectra in 1859. What has become known as Kirchhoff's law of radiation states that, for a given temperature and wavelength, the ratio between energy radiated and energy absorbed by a body is fixed. Thus, if a body emits all wavelengths, it must also absorb all wavelengths. This leads to the somewhat surprising result that an object which radiates perfectly also absorbs perfectly, and will therefore be perfectly black—a so-called *black body*. Furthermore, a body which emits light at a certain wavelength only, also absorbs light at the same wavelength only.

The other basic principle of spectroscopy is that an incandescent body—either solid, liquid or gaseous—will emit at all wavelengths; that is, a continuum is emitted. A gas at comparatively low temperature, however, will emit only a line spectrum.

Putting these two principles together shows that the hot surface of a star will emit a continuum, just as will the glowing filament of a light bulb or molten metal. The atmosphere of the star, however, is cooler, and it will absorb certain wavelengths characteristic of the gases present. If these gases could be viewed by themselves, they would be seen to be emitting light at the same wavelengths.

Only certain lines are permitted because the electrons of atoms can only exist in specified energy levels. If a certain quantity of energy is added to the atom, causing it to be in an "excited" state, the electrons can increase their energy only by fixed amounts, between energy levels. If the atom loses the energy again, it can only do so by radiating at the same energy levels—that is, at particular wavelengths of radiation.

Each atom or molecule has its own distinct set of energy levels. Taking the hydrogen atom, the simplest example because it has only one electron, the lowest energy level, in which the atom is completely neutral, is called the ground state. An electron jumping between this and the next energy level—in either direction— will produce a line of wavelength 1,216 angstroms, in the ultraviolet.

Jumps (or *transitions*) up to higher energy levels produce lines of greater energy, further into the ultraviolet. The difference between levels soon becomes small, however, and the lines become closer together until they reach a series limit. Transitions between the next energy level up and higher ones produce a similar set of lines in the visible part of the spectrum called the Balmer series, which are easily recognized in stellar spectra (see SPECTRAL TYPE). If the jumps are upward, then energy has been added to, or absorbed by, the atom, and so absorption lines result. If the jumps are downward, then the atom is releasing its energy and produces emission lines.

spectrum

The entire range of ELECTROMAGNETIC RADIATION, from gamma rays to radio waves. More usually, the visible wavelengths only—the colors of the rainbow from violet to red—are known as the spectrum. Astronomers often also use the word to refer to a photograph of a star's visible-light spectrum, more properly called a *spectrogram*.

speculum

Latin word for mirror. The first mirrors for reflecting telescopes were made from a special alloy, two-thirds copper and one-third tin, called speculum metal because it was designed to take a high polish. However, metal mirrors do not keep their figure accurately, and soon tarnish so that they must frequently be repolished. About the turn of the century metal mirrors were superseded by glass mirrors with a thin reflective coating of silver or aluminum (see MIRROR).

spherical aberration

Loss of sharpness in an image when different parts of a lens or mirror do not bring light rays to the same focus. A spherical concave mirror produces a blurred image of a distant object, such as a star, because the central part of the mirror has a longer focal length than the outer zone. The same effect is produced by a simple lens, although the focal point is harder to define because of the added effect of CHROMATIC ABERRATION. For astronomical use, the spherical aberration of a mirror is eliminated by deepening the center slightly, and thus shortening the focal length; the curve of a mirror so deepened is a PARABOLA. The effect can also be counteracted in special optical systems such as the SCHMIDT TELESCOPE. The usual method of curing the spherical aberration of a lens is to combine it with another lens, which is also necessary to remove chromatic aberration (see OBJECT GLASS).

spheroid

A regular, rounded object, shaped like a slightly squashed sphere. A spheroid can be thought of as the solid surface formed by rotating an ellipse about either its longer or its shorter axis. If about its shorter axis, the result is a *prolate spheroid*; if about its longer axis, the result is an *oblate spheroid*. The Earth is an example of an oblate spheroid—it is slightly flattened at the poles.

Spica

The brightest star in the constellation Virgo, and one of the brightest in the entire sky; it is also called Alpha Virginis. Spica, a blue-white star eight times the Sun's diameter, has a magnitude of 0.96, and is 260 light-years away. It is a spectroscopic eclipsing binary with a period of four days; the masses of the two stars are 10.9 and 6.8 times that of the Sun.

spicules

Vertical, jetlike features of the solar CHROMOSPHERE, giving it a fine, hairlike appearance where it merges with the CORONA. These jets of hot gas, at temperatures of 10,000°, shoot 6,000 miles (10,000 km) up into the corona and subside again in less than five minutes. There are perhaps 500,000 spicules on the Sun at any given time, aligned with the magnetic fields around the edges of supergranulation cells in the chromosphere. These cells, each 20,000 miles (35,000 km) across, are part of a network of convective motions in the chromosphere, and it has been suggested that spicules are involved in carrying energy from the lower to the upper chromosphere.

Although spicules appear to be governed by magnetic fields, they are absent over active regions where the field is particularly strong. Recently, observations at ultraviolet wavelengths from Skylab have revealed the presence of "macrospicules," giant spicules 20,000 miles (35,000 km) long, which last for up to 40 minutes.

spiral galaxy

A type of galaxy in which many of the stars and nebulae lie along spiral arms that appear to wind out from the center. Lord ROSSE discovered the spiral pattern in some fuzzy "nebulae" with his 72-inch (180-cm) reflecting telescope in 1845, but not until the 1920s were these proved to be external galaxies comparable to our MILKY WAY. Photographs taken with large telescopes show that about three-fourths of the galaxies in the Universe are spirals.

A spiral galaxy has a central, rounded nucleus (noticeably elongated in a *barred spiral*) composed of old stars, surrounded by a spherical halo also made up of old stars, many of which are concentrated in GLOBULAR CLUSTERS. Young stars, and gas which has not yet formed stars, lie in a disk, about 100,000 light-years in diameter but only 2,000 light-years thick. This material orbits the galactic center just as the planets move around the Sun, with the more distant stars traveling more slowly (*differential rotation*). The rate of the disk's rotation reveals the galaxy's total mass, which is typically a hundred billion times that of the Sun.

The spiral arms prominent on photographs are regions of the disk where stars, gaseous nebulae, and dust clouds are particularly concentrated. The spiral pattern is thought to rotate around the galactic nucleus as a whole, so that the outer parts must travel

A closeup of the chromosphere in the red spectral line of hydrogen showing spicules aligned with the magnetic field. These short-lived jets of hot gas, each 6,000 miles (10,000 km) long, can be seen to form a pattern, following the borders of the supergranulation cells.

faster than the inner. Since the opposite is true of individual stars, gas and dust, whose rotation speed decreases with distance from the center, the material comprising a spiral arm is constantly changing. It seems that the disk material becomes bunched together as it orbits through that part of the disk where the spiral arm is. The bunching is probably caused by the gravitational attraction of the material already comprising the spiral arm; although this material soon orbits out of the arm, the newly bunched material continues the process. The result is a self-perpetuating arm, of a type known as a "density wave."

Spiral galaxies are classified by the prominence of their spiral structure. A galaxy with a nearly "smooth" disk and a bright nucleus is termed Sa (or SBa if it is a barred spiral); a galaxy nucleus is Sc, or SBc. Intermediate types are Sb or SBb. The Milky Way is a typical spiral galaxy, in type between Sb and Sc.

NGC 2811 Sa

NGC 3031 Sb
NGC 1300 SBb

NGC 628 Sc

Four spiral galaxies, showing the variety of structure among some of the types. NGC 2811 (Type Sa) has a large nucleus and tightly-wound arms; NGC 3031 (Type Sb) has more open arms; while NGC 628 (Type Sc) has loose, clumpy arms surrounding a small nucleus. NGC 1300 (Type SBb) is a moderately open-armed barred spiral, in which the arms spring from a central bar.

sporadic
A meteor which does not belong to a recognized meteor stream, but orbits the Sun in its individual path. Sporadic meteors can be seen on any clear dark night, their numbers rising from 6 per hour at 6 P.M. to 14 per hour at 6 A.M. The numbers increase because the Earth's leading edge (its morning hemisphere) sweeps up more particles from space than does its evening side. Sporadic meteors tend to be caused by meteors; they are remains of ancient meteor streams or debris from collisions between larger meteoroids.

spring tide
See TIDES.

Sputnik

A series of Soviet satellites; Sputniks 1 and 2 were the Earth's first two artificial satellites. Sputnik 1 was a 184.3-lb. (83.6-kg) aluminum sphere 23 inches (58 cm) in diameter, filled with nitrogen gas to maintain an even temperature, and with four aerials of length 8 feet to 9½ feet (2.4–2.9 m). The Sputnik transmitted "beep-beep" signals as it orbited Earth every 96 minutes between 142 and 588 miles (228 and 947 km); the frequency of the signals indicated the on-board temperature. Contrary to many reports, Sputnik 1 carried no other on-board instrumentation. Propagation of the radio signals from the Sputnik revealed characteristics of the Earth's ionosphere, and tracking of the satellite and the rocket stage that carried it into orbit gave information about the density of the upper atmosphere. Sputnik 2, weighing an impressive 1,121 lb. (508 kg), carried the dog Laika into space, the first living creature to orbit the Earth. Laika lived in a cylindrical compartment for 10 days, dying in orbit before the craft burned up in the atmosphere. The dog's flight showed that living organisms could survive in space. Sputnik 3, weighing 2,926 lb. (1,327 kg), was the first extensively instrumented Soviet satellite. It confirmed the existence of the Van Allen radiation belts discovered by the American Explorer 1. Sputnik 4 was the first of a series of unmanned tests of the Vostok spacecraft (the so-called Korabl Sputniks), and Sputniks 7 and 8 were associated with the first Soviet Venus probes. After Sputnik 10, Soviet Earth satellites were renamed COSMOS SATELLITES. The Sputnik series were launched by the so-called VOSTOK LAUNCHER, without an upper stage for the first three satellites but with upper stages for the remaining Sputniks.

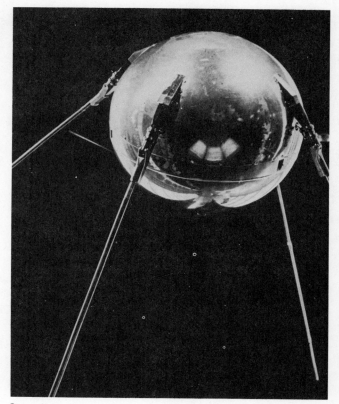

Sputnik 1, the world's first artificial satellite, launched on October 4, 1957. The 184-lb. (83.6-kg) satellite orbited for just three months transmitting data on its internal temperatures and pressures.

Mission	Launch date	Results
Sputnik 1	October 4, 1957	First artificial satellite. Reentered January 4, 1958
Sputnik 2	November 3, 1957	Carried dog Laika. Reentered April 14, 1958
Sputnik 3	May 15, 1958	Scientific satellite. Reentered April 6, 1960
Sputnik 4	May 15, 1960	Vostok test flight. Intended recovery failed when retro-rockets sent capsule into higher orbit
Sputnik 5	August 19, 1960	Vostok test; dogs Belka and Strelka recovered in capsule on August 20 after 18 orbits
Sputnik 6	December 1, 1960	Vostok test flight. Dogs Ptsyolka and Mushka perished when spacecraft burned up after 17 orbits due to incorrect angle of reentry
Sputnik 7	February 4, 1961	Failed Venus probe
Sputnik 8	February 12, 1961	Launched Venus 1 from Earth-orbit
Sputnik 9	March 9, 1961	Vostok test flight. Dog Chernushka recovered after one orbit
Sputnik 10	March 25, 1961	Vostok test flight. Dog Zvezdochka recovered after one orbit

Stafford, Thomas Patten (b. 1930)

One of the few spacemen to fly on four missions, commander of the Apollo spacecraft in the Apollo-Soyuz joint docking mission in July 1975. Stafford was selected as an astronaut in 1962, and in December 1965 flew in the Gemini 6 craft that rendezvoused with Gemini 7. In June 1966 he commanded the Gemini 9 mission, with Eugene CERNAN. In May 1969 Stafford commanded the Apollo 10 mission with Cernan and John YOUNG; this was a rehearsal for the first lunar landing attempt. After the Apollo-Soyuz flight, Stafford left the astronaut corps to command the Air Force's Flight Test Center at Edwards Air Force Base.

star

A self-luminous ball of gas; the Sun is a typical star. The night-sky stars appear as points of light because they are much farther away than the Sun; light reaches us in 8.3 minutes from the Sun, but takes 4.3 years from the nearest star. Under clear skies, about 3,000 stars are visible to the naked eye at any given time; normally, however, the effects of atmospheric pollution and street lighting mean that far fewer stars can be seen. Astronomers have cataloged over one million stars, but little is known about most of them except their approximate brightness and position. In the Palomar Sky Survey, the 48-inch (122-cm) Schmidt telescope at Mount Palomar photographed the sky from the north

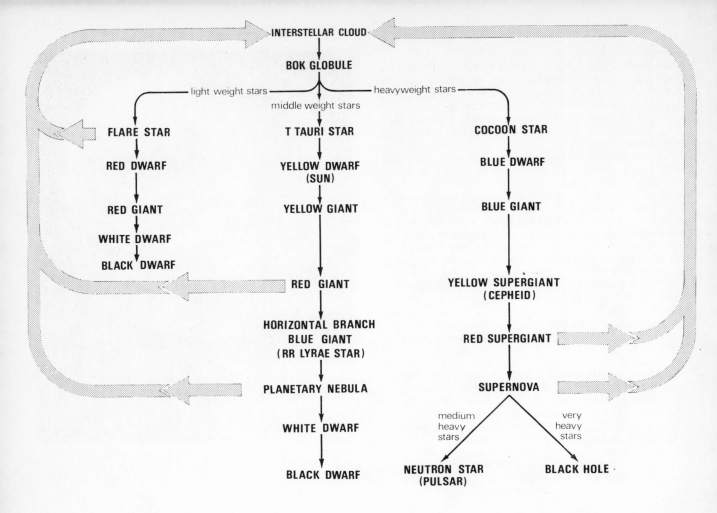

INTERSTELLAR CLOUD

BOK GLOBULE

light weight stars — heavyweight stars

middle weight stars

FLARE STAR — T TAURI STAR — COCOON STAR

RED DWARF — YELLOW DWARF (SUN) — BLUE DWARF

RED GIANT — YELLOW GIANT — BLUE GIANT

WHITE DWARF

BLACK DWARF — RED GIANT — YELLOW SUPERGIANT (CEPHEID)

HORIZONTAL BRANCH BLUE GIANT (RR LYRAE STAR) — RED SUPERGIANT

PLANETARY NEBULA — SUPERNOVA

WHITE DWARF — medium heavy stars — very heavy stars

BLACK DWARF — NEUTRON STAR (PULSAR) — BLACK HOLE

pole to 33° south of the equator; about 800 million stars are estimated to be shown on these photographs. However, our Galaxy (the Milky Way) is estimated to contain at least 100 billion stars.

Nature of stars. Stars vary widely in size and temperature. The largest are the red SUPERGIANTS, up to 1,000 times the size of the Sun. The smallest known stars are NEUTRON STARS like the one in the Crab nebula, which can be as small as 10 miles (16 km) in diameter. The surface temperatures of stars range from up to 100,000°K down to less than 2,000°K. Variations in temperature produce corresponding differences in colors, from the hot blue of VEGA, through the warm yellow of CAPELLA, to the cool red of BETELGEUSE. Astronomers learn about the nature and composition of stars by studying their light (SPECTROSCOPY). Atoms in the star's atmosphere absorb some light from the star at specific wavelengths; the relative strengths of these absorption lines in a spectrum reveal the density, temperature, and chemical composition of a star's outer layers. Most stars consist largely of hydrogen, with some helium and a sprinkling of certain other elements.

Formation of stars. Stars are formed by the collapse of the great clouds of gas and dust like the ORION NEBULA, in our Galaxy. Often the collapse is triggered by compression, as clouds bump into each other, or by changes in the heating of the cloud by nearby stars. The cloud then collapses under the effect of its own gravity, breaking up into smaller

Simplified diagram showing the main stages in the evolution of stars from interstellar clouds. Mass loss back to the interstellar medium is indicated with grey arrows. Protostars are shown following the Bok Globule, and the long-lived main sequence stages after the protostars. The planetary nebula and the supernova represent the principal explosive stages, while the stellar deaths appear at the very last level.

subcondensations, which continue to collapse until their internal temperature rises above about 10 million degrees. The pressure of the gas then increases rapidly and halts the contraction; a star has been born. A fragmenting cloud will give rise to a star cluster like the PLEIADES. The smaller subcondensations are believed to look like the small, dark GLOBULES seen in various parts of the Galaxy. Often, the rotation of a globule will make it split into two or more parts, perhaps forming a DOUBLE STAR. Other globules may give rise to a star and PLANETS.

Inside a star. The energy of the stars comes from nuclear reactions in their very hot central regions. Atoms are stripped down to their nuclei and crushed together, releasing energy. In stars like the Sun, hydrogen combines to form helium directly by the collision of hydrogen nuclei in the PROTON-PROTON CHAIN; in hotter stars, with central temperatures above about 15 million degrees, the main process is the CARBON-NITROGEN-OXYGEN CYCLE, in which hydrogen is added to carbon nuclei to form nitrogen, which then breaks up into carbon and helium. At temperatures around 100 million degrees helium is turned into carbon and oxygen, and then at still higher temperatures the matter inside the star is successively converted into neon, magnesium, silicon, and iron, each reaction releasing energy to heat the star.

Evolution of stars like the Sun. When a star like our Sun first forms out of an interstellar gas cloud, it is about 50 times the size of the present Sun and 500 times as luminous. The central regions are too cool for nuclear reactions, so that the star contracts, drawing on its gravitational energy to make good the loss by radiation into space. As the star contracts, its luminosity decreases, and the central regions heat up. After 30 million years the center is hot enough for nuclear reactions to take place, and the contraction begins.

The star remains much the same for about 10 billion years, the major part of its lifetime, with a luminosity and size similar to that of our Sun. This phase of hydrogen burning is the period when the star is on the MAIN SEQUENCE; it lasts for so long because there is an enormous energy reserve in the hydrogen of which the star is primarily made.

Eventually the star will use up its hydrogen fuel in the hot central regions, producing an inert helium core. Since there is plenty of fuel left in the rest of the star, the nuclear burning advances outward until the hydrogen is burning in a shell surrounding the core. When this happens, the star begins to increase in size and luminosity, slowly at first but then at an increasing pace until the star becomes a vast RED GIANT producing 1,000 times the light output of the Sun. Not only does the light output increase, but the star expands to become as much as 100 times the size of the Sun. When our Sun reaches this size, it will have engulfed the Earth.

As the star evolves, the helium in the center increases in temperature until it becomes hot enough to burn. The onset of this is very sudden, and the whole star must readjust to the new energy source. In a matter of days it contracts, decreasing in light output to about 100 times that of the Sun. But as the helium burns, the star again begins to grow in size and brightness, burning up its helium to form a carbon core and then burning helium and hydrogen in shells around the core. The star expands to even greater size than in its previous giant phase, up to 400 times the size of the Sun with 10,000 times its light output. It becomes so large its own gravity can hardly hold it together, and eventually the surface layers are blown off into space. This ejected matter forms a shell expanding away from the hot remnant central core, as we see in a PLANETARY NEBULA. The central core remains as a hot WHITE DWARF, which slowly cools to a cinder-like black dwarf star as it radiates the rest of its heat into space.

Evolution of massive stars. Stars more massive than the Sun undergo the same early evolution. The centers of these stars are so hot that most energy comes from the carbon-nitrogen-oxygen cycle. The star changes little while it is burning the hydrogen in the core; it is also a main sequence star and spends most of its life in this stage. But it is considerably brighter than the Sun: a star 10 times the Sun's mass has a luminosity about 10,000 times

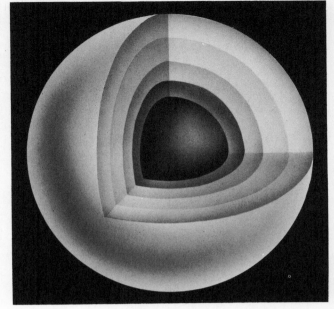

The interior of a large mass star before it becomes a supernova. The star has shells of different material. In the center neon is being converted into iron. When the star has an iron core it collapses, the outer shells heat up and the star explodes.

that of the Sun. Such large stars evolve much more rapidly since they burn up fuel at a greater rate; their main-sequence stage lasts only 10 million years.

When the central store of hydrogen is exhausted, hydrogen burning moves out into a shell surrounding the burnt-out helium core. However, this core contracts, heating up to a temperature of 100 million degrees, at which it starts to be converted into carbon and oxygen. The star swells up to become a giant or supergiant, remaining like this for the rest of its life. When the helium in the center is all converted into carbon and oxygen, the core again contracts until it is hot enough to burn carbon into magnesium. When this is burnt out, the magnesium burns into neon which itself burns into iron. The evolution of the center of the star consists of burning up one fuel, contracting and heating up to ignite the next fuel, while the lower-temperature reactions are taking place in successive shells away from the center.

When an iron core has been formed, the star is near the end of its life. Iron is the most stable element; to convert iron into heavier elements consumes energy, instead of releasing it. The required energy supply can only come from gravitational contraction; therefore, the core collapses. As the central core does so, the outside falls in on top of it, rapidly increasing its temperature and accelerating the nuclear burning in the regions where there is still some fuel. The sudden release of energy makes the star explode violently, producing a SUPERNOVA. In the explosion numerous elements are produced and scattered into space. In some cases the star may blow itself entirely to bits; in others, the star's compressed core may be left as a neutron star. In still other cases the remnant core may be so massive that it continues to contract, disappearing from sight as a BLACK HOLE.

Kappa Crucis, a young galactic (or "open") star cluster in the Southern Cross. It comprises about a hundred stars which show a great variety of colors, and have earned it the name of the "Jewel Box." Fifteen light-years across, this cluster is believed to be only sixteen million years old.

star cluster

An aggregation of stars, held together by their own gravity. Two distinct types of star clusters are observed, the huge, round GLOBULAR CLUSTERS of about 1 million stars, and the much smaller and sparser OPEN CLUSTERS.

Globular clusters formed early in the history of our Galaxy (10 billion years ago), before it settled down to the flat disk-like arrangement of stars which we now see. As a result, the globular clusters, about 125 in number, occur all around the Milky Way, and not merely in the disk. They orbit around the center of our Galaxy in highly elongated ellipses, with a period of about 100 million years. The number of globular clusters increases toward the galactic center, and most therefore are observed in the direction of the constellation Sagittarius, where the Milky Way's nucleus lies.

The brightest globular clusters, Omega Centauri and 47 Tucanae, are easily visible to the naked eye in the southern hemisphere. Large telescopes can readily resolve the outer parts of most globulars into individual stars. Their distances (10,000 to 150,000 light-years) can be determined from the apparent magnitudes of stars known as RR LYRAE VARIABLES, which always have the same intrinsic luminosity. They are usually about 100 light-years in diameter, and they contain only stars composed almost entirely of hydrogen and helium (POPULATION II stars).

Open clusters occur in the flat disk of our Galaxy, and are far more numerous than the globulars. About 1,000 are known, and it is estimated that the Milky Way may contain a total of 18,000. Open clusters are difficult to detect beyond 5,000 light-years, both because they contain far fewer stars than the globulars, and because the dust in the Milky Way's disk dims the light from them (INTERSTELLAR ABSORPTION). The nearest is the HYADES (in the constellation Taurus) at 130 light-years, while the PLEIADES (also in Taurus) and Praesepe (in Cancer) are 400 light-years distant. Open clusters average 10 light-years in diameter, and they contain only a few

hundred stars, similar in chemical composition to the Sun (POPULATION I stars).

Open clusters have formed continuously since the galactic disk appeared. The oldest are of almost the same age as the globular clusters, while the youngest are only a few million years old. Open clusters are being created in places where stars condense out of gas clouds, for example in the ORION NEBULA. Most of the stars in these newly formed STELLAR ASSOCIATIONS will disperse among the other stars of the disk, but those near the center of the nebula may remain bound by their gravity to form clusters that will remain together for billions of years.

star streaming

The tendency of stars that occur over a comparatively large area of the sky to be traveling in the same direction in space. The term was first used by J. C. KAPTEYN in 1904, when an investigation into the PROPER MOTIONS of nearby stars showed that their directions of motion are not random, and that two particular directions in space are favored by a large number of them. If a correction is made for the Sun's motion, these two directions point toward and away from the center of our own Galaxy. Kapteyn's two star streams are thus reducible to one, with stars moving in both directions along it. When astronomers found that the Sun is orbiting the center of the Milky Way, Karl SCHWARZSCHILD and Bertil LINDBLAD showed that this star stream is merely a reflection of the fact that the nearby stars travel in elliptical rather than circular orbits around the galaxy.

Today the term "star stream" is used in a different sense. It refers to small, sparsely scattered groups of stars moving in the same direction. Five of the stars in the Plow, together with Sirius, are part of one stream, while others surround OPEN CLUSTERS of stars like the Hyades.

steady-state theory

A theory of COSMOLOGY, now credited by very few astronomers, which regards the Universe as essentially unchanging over time. In the BIG-BANG cosmological model, now generally accepted, the Universe has a birth, growth, and perhaps death. The steady-state theory, as proposed by Hermann BONDI and Thomas GOLD (1948) took a very different standpoint, beginning with the PERFECT COSMOLOGICAL PRINCIPLE of an unchanging Universe. Although individual stars and galaxies pass through their life-cycles, it was suggested that new galaxies are constantly forming out of freshly created matter. As the Universe expands, this new matter appears to fill in the "gaps" left by the receding galaxies in the EXPANDING UNIVERSE, and thereby maintains a constant overall density.

The CONTINUOUS CREATION rate is very low (about one atom per cubic mile of space per hour), and Fred HOYLE, one of the theory's main protagonists, showed that Einstein's theory of relativity could be modified to explain the creation of matter from a reservoir of negative energy (the C-field) which fills the Universe.

The steady-state Universe has no beginning and no end, and this feature made it a very popular choice among cosmologists. By 1965, however, support had dwindled. The theory was unable to account for either the radio astronomers' "source counts" data (which showed many more radio sources in the past), or the newly discovered microwave BACKGROUND RADIATION. The latter can be simply explained according to the big-bang theory as the "echo" of the big bang itself at the beginning of an evolving Universe. Moreover, the strongest reason for the original introduction of the perfect cosmological principle, the fact that the Universe appeared to be younger than the Earth, had been removed in 1952, when Walter BAADE showed that the size and age of the Universe had been grossly underestimated.

Hoyle experimented with variations on the original steady-state theory, postulating that we live in a temporary expanding "bubble" in a Universe which is constant on a much larger scale. The background radiation he attributed to dust grains in space. Hoyle eventually abandoned the theory altogether, though his coworker J. V. Narlikar still supports it.

stellar association
An aggregation of 10 to 1,000 stars which are close together merely because they have not moved far from their common place of origin. Unlike the stars in STAR CLUSTERS, those in associations are moving too fast for their gravity to hold them together permanently.

Although a few of the stars may in the end stay together as multiple stars or OPEN CLUSTERS, most members of an association disperse into the general background of stars in our Galaxy in about 10 million years, only 1 percent of our Galaxy's age. As a result associations can be recognized by the concentration of young stars, the very hot and highly luminous O and B SPECTRAL TYPES, or by the occurrence of the even younger T TAURI STARS.

The OB associations and T associations were first investigated in 1947 by V. A. AMBARTSUMIAN, who found that they often occur together. For example, many of the stars forming the constellation of Orion are part of an OB association of 1,000 stars that surrounds a T association containing a few hundred members. At the center, more stars are still condensing out of the gas of the ORION NEBULA.

step rocket
See ROCKET.

Steward Observatory
An astronomical observatory operated by the University of Arizona at an elevation of 6,811 feet (2,076 m) on Kitt Peak (see KITT PEAK NATIONAL OBSERVATORY). The Steward Observatory was founded at Tucson in 1922, with a 36-inch (91-cm) reflector, and moved to Kitt Peak in 1963. A 90-inch (229-cm) reflector was opened in 1969.

Stonehenge
A prehistoric circle of giant stones near Salisbury, southern England, once used as an astronomical observatory of remarkable sophistication. Stonehenge, which has been carefully excavated and studied, appears to have been built at four different periods. Stonehenge I, consisting of an outer circle of 56 stones and bank together with an outlying "heel" stone, was constructed about 1800 B.C. In Stonehenge II, built sometime during the 17th century B.C., the avenue containing the heel stone was

continued some 2 miles (3 km) eastward to the River Avon, and large igneous stone pillars—"bluestones"—were brought from over 150 miles (240 km) to form two circles. Stonehenge III, set up soon after 1600 B.C. and Stonehenge IIIb, entailed some changes and the addition of a circle of stone uprights with a continuous ring of stone lintels above them. These stones, some of which weigh 50 tons, came from a site 20 miles (30 km) away. Final refinements—Stonehenge IIIc—were made about 1400 B.C.

The heel stone and central uprights were aligned with the sunrise at the summer solstice, and the outer 56 ring holes—the "Aubrey" holes—could be used as a "computer" for calculating, with the help of observations, calendar dates and, possibly, eclipses. Recent research has shown that Stonehenge was part of a network of megalithic observatories spread over Britain and parts of France.

Strömgren, Bengt Georg Daniel (b. 1908)
Swedish astronomer, who showed that bright nebulae are made to glow by radiation from hot, young stars embedded within them, which ionizes the hydrogen gas of the cloud to form a so-called H II REGION. During the 1930s he suggested the existence of what became known as the *Strömgren sphere,* an area in the gas cloud made luminous by the star's radiation, sharply bounded by a cool, dark area of unionized gas (an H I REGION). The size of the luminous Strömgren sphere could be calculated from the density of the gas and the temperature of the embedded star. This theory accounted for the observed shapes of glowing H II regions in our Galaxy, and allowed astronomers to estimate the true sizes of H II regions in other galaxies, vital knowledge in attempts to determine galactic distances. Strömgren also showed that the scatter of stars about the MAIN SEQUENCE was due to their differing chemical composition. Strömgren succeeded his father Svante Elis Strömgren (1870–1947) as director of Copenhagen Observatory in 1940. In 1951 he became joint director of Yerkes and McDonald observatories. There he pioneered the classification of stellar spectra by carefully measuring the star's brightness at specific selected wavelengths.

Struve family
A succession of four generations of astronomers, founded by Friedrich Georg Wilhelm von Struve (1793–1864), a German who in 1817 became director of Dorpat Observatory (now Tartu in Estonia). He equipped the observatory with a $9\frac{1}{2}$-inch (24-cm) refractor by Joseph FRAUNHOFER, which was then the largest instrument of its kind in the world. Struve then began a series of measurements of binary stars, surveying the heavens from 15° south of the celestial equator to the north celestial pole. He examined 120,000 stars and increased the number of known binary systems from 700 to 3,112, accurately measuring them all. In 1835 Struve was invited by Tsar Nicholas I to supervise the building of a new observatory at PULKOVO near St. Petersburg (Leningrad), and here, using a new refractor 15 inches (38 cm) in aperture, he continued his binary-star work, also measuring the PARALLAX of Vega.

In 1864 F. G. W. Struve was succeeded at Pulkovo

by his son Otto Wilhelm Struve (1819–1905), who himself measured some 500 other binaries and, in the mid-1880s, equipped the observatory with a new 30-inch (76-cm) refractor. O. W. Struve had two sons, Hermann and Ludwig. Hermann Struve (1854–1920) became director of the Berlin Observatory, and made his name with observations of Saturn's rings and binary stars. His son, Georg Struve (1886–1933), studied planetary satellites and completely remodeled the observatory at Neubabelsberg, of which he became director. Ludwig Struve (1858–1920) was appointed to the chair of astronomy at Kharkov, and his son, Otto Struve (1897–1963), also became a noted astronomer. Otto left Russia in 1921 for the United States. At Yerkes Observatory, he discovered the existence of interstellar calcium from the absorption lines it made in stellar spectra. In 1932 he became joint director of Yerkes and McDonald observatories. He discovered the fast rotation rates of many hot, large stars, and was an influential champion of the view that many stars like the Sun may have planetary systems.

Sun

The star that is the central body of the SOLAR SYSTEM. The Sun contains 99.9 percent of the mass of this system, and it is by far the most important member, controlling the motions of all the other bodies.

The Sun is, however, only one of some 100 billion stars which make up our Galaxy, the MILKY WAY, and is located in the galactic disk at a distance of about 30,000 light-years from the center. It is a typical POPULATION I (comparatively young) star, situated on the inner edge of a spiral arm. Even with a velocity of 150 miles (250 km) per second, the Sun takes 225 million years to orbit the Galaxy.

Overall properties. At a distance of only 93,000,000 miles from the Earth, the Sun is some 270,000 times closer than the next nearest star (PROXIMA CENTAURI). It presents a disk measuring just over half a degree of angular diameter; it is therefore the only star whose surface features can be studied. The Sun is a very average star; a yellow dwarf of SPECTRAL TYPE G2, which corresponds to a surface temperature of 6,000°. Although its apparent magnitude in our skies is −26.5, the Sun's ABSOLUTE MAGNITUDE—the brightness it would appear to have if it were situated at a standard distance of 10 PARSECS (33 light-years)—is only +4.8. The absolute magnitudes of other stars range from −8 to +19; the Sun falls in about the middle of this range. It is, however, quite a small star, with a diameter of 865,000 miles (1,392,000 km), only 109 times that of the Earth; although its volume could contain a million Earth-sized bodies. The Sun is 330,000 times heavier than the Earth, with a mass of 1.99×10^{27} tons; its average density is only about a quarter that of the Earth's, indicating a fundamental difference in structure and composition.

The Sun's rotation is not uniform like that of a solid planet, but becomes slower toward the poles. The period of rotation is 24.7 days at the equator, 28.2 days at 45° latitude, and about 34 days near the poles. The adopted mean value, that which occurs at latitude 15° is 25.38 days, called the sidereal period of rotation. The mean period observed from Earth is 27.27 days, because the Earth revolves around the

Sun in the same direction as the rotation. This is known as the synodic period of rotation. The solar equator is inclined to the ECLIPTIC at an angle of 7° 15′.

Efforts have recently been made to detect whether the Sun is an exact sphere or a slightly oblate SPHEROID. The relatively slow rotation rate of the outer layers would not be expected to give rise to an equatorial bulge, but Robert DICKE has suggested that certain phenomena, such as the advance of MERCURY's perihelion (usually explained by RELATIVITY), might arise if the Sun has a small, fast-rotating core. Dicke has reported an oblateness of 5 parts in 100,000, which he claims to be consistent with this theory; but the measurement is difficult, and other observations have contradicted this result. Searches for oblateness have recently shown, however, that the Sun pulsates very slightly with a period of 2 hours 40 minutes. This dramatic discovery is expected to give new information about the Sun's interior, particularly the change in chemical composition toward the center.

Surface layers and atmosphere. The temperature at the center of the Sun is estimated to be 15 to 20 million degrees, but the "surface," from which we receive most of its light, is only about 6,000°. This PHOTOSPHERE is not a true solid surface, but a layer of relatively dense gas a few hundred miles thick, whose temperature falls from 9,000° at the bottom to 4,300° at the top. The upper level marks the transition to the CHROMOSPHERE, a far more rarefied and nonhomogeneous layer 12,000 miles (20,000 km) in thickness, in which temperature rises rapidly with height to 1 million degrees. At this point, the CORONA is reached, a very low-density region of completely ionized gas (PLASMA). Some parts of the corona may reach 4 million degrees. The outermost layers of the corona are in fact streaming away from the Sun into space and constitute the SOLAR WIND, which has been detected even beyond the orbit of Jupiter.

The spectrum of the Sun is crossed by thousands of dark ABSORPTION LINES and bands (FRAUNHOFER LINES), from which the chemical composition of the solar photosphere can be determined. The relative abundances of elements are a good guide to the composition of the gas from which the Sun formed, although they will differ from abundances in the interior, which have been altered by nuclear reactions. The photosphere consists of 90 percent hydrogen and 8 percent helium. It also contains small amounts of all the other elements (the HEAVY ELEMENTS), whose proportions are remarkably close to those observed on Earth.

It was previously possible to obtain the spectrum of the chromosphere and corona only during a total solar eclipse, when the dazzling light of the photosphere is blocked off. Bernard LYOT's development of the coronagraph, which artificially blocks off the photosphere, now allows observations to be made at any time. More recently, results at X-ray and ultraviolet wavelengths from rockets and satellites have yielded new information about the coronal spectrum. It contains many emission lines from elements such as iron and calcium, which have been stripped of nearly all their electrons by the extremely high temperatures. Many of these are so-called FORBIDDEN LINES, indicating a very low density for the corona. The cooler, denser

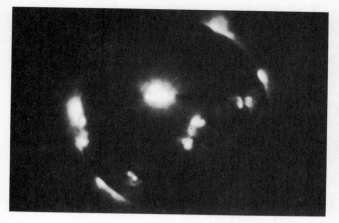

A photograph of the Sun taken at X-ray wavelengths from a rocket above the Earth's atmosphere. The bright regions are hot spots in the corona where the low-density gas is at temperatures of several million degrees. The surface we see at optical wavelengths (photosphere) is too cool to emit X rays, and appears black in this picture.

chromosphere also shows several emission lines, which generally correspond to the absorption lines in the photospheric spectrum. Its pinky-red color (visible during a total eclipse) is due to emission from a strong hydrogen line (Hα) at 6563 Å. The element helium was discovered through its emission lines in the solar chromosphere.

Solar activity.   The Sun is by no means an undisturbed or unchanging body. The photosphere is threaded with a weak magnetic field measuring about 1 gauss. For reasons that are still not certain, this field can be compressed and intensified so that loops of magnetic field are forced up through the surface, giving rise to an "active region." The field strengths may become as high as 4,000 gauss, and are thought to prevent the usual motions of photospheric gas over areas several hundreds or thousands of miles across. These SUNSPOT regions are cooler and therefore darker than the undisturbed photosphere; they may last for many weeks. An active region also reveals its presence in the chromosphere and corona. In the lower levels of the chromosphere, PLAGES and PROMINENCES (both quiescent and eruptive) are associated with active regions, even though sunspots are not always present on the photosphere beneath. When a particularly complex spot group is visible, highly energetic FLARES occur in the upper chromosphere and lower corona, resulting in the emission of radiation and charged particles which interact with the Earth's MAGNETOSPHERE. It is believed that flares are a result of discharges between intense magnetic fields of opposite polarity.

The level of solar activity varies over a period of about 11 years, the SOLAR CYCLE. Similar activity has also been detected on several cooler stars, which are thought to have sunspots and flares on a much larger scale than the Sun.

Energy generation.   The energy output from the Sun is 3.8×10^{33} ergs per second; each square centimeter of the Sun's surface is continuously radiating energy at the rate of a 9-horsepower engine.

Since the Earth is small and far away, it receives only a small proportion of this energy, yet this tiny fraction amounts to 4,690,000 horsepower per square mile (see SOLAR CONSTANT). The source of the Sun's enormous energy output was a mystery for many years. If the energy were derived solely from heat stored internally, the Sun's brightness would last only a few thousand years. Although many theories had been proposed, it became apparent in the 1930s that the only form of energy generation which would persist for billions of years was nuclear energy. Two mechanisms have been suggested—the PROTON-PROTON CHAIN and the CARBON-NITROGEN-OXYGEN-CYCLE. Both are thermonuclear processes occurring at extremely high temperatures, but the proton-proton chain is thought to be by far the more common process occurring in the Sun. In both processes, a mass of hydrogen is transformed into a slightly smaller mass of helium. The residual mass is transformed into energy according to Einstein's formula $E = mc^2$ (where c is the velocity of light). One gram of material therefore produces 22 trillion calories of energy. This means the Sun is losing 4 million tons of mass each second, but this amount will correspond to only 7 percent of its total mass in 1 trillion years. The process of thermonuclear fusion as we understand it is compatible with an age for the Sun of about 5 billion years. The Sun is at present in a stable state; the outflowing of energy tends to make it expand, but this is balanced by gravitation which tends to make it collapse. The Sun's supply of hydrogen will last at least another 5 billion years. At the end of that period it is believed the Sun will expand to become a RED GIANT star. This description of solar energy production predicts that the thermonuclear reactions should produce neutrinos as a byproduct. Being small and electrically neutral, these particles would pass unimpeded through the Sun and eventually reach the Earth. Neutrinos should therefore provide an insight into the interior of the Sun that is not given by electromagnetic radiation. The search for neutrinos has, however, been so far unsuccessful, even though the predicted flux should easily have been detected. Since the basic theory of hydrogen fusion is unlikely to be seriously wrong, astronomers are currently investigating modifications which would yield a lower neutrino flux.

sunspot

A relatively cool, dark area on the solar PHOTOSPHERE. Individual spots range in size from small "pores" about 900 miles (1,500 km) across, to huge, complex spots 100 times as large. Generally, sunspots form in groups, which may cover hundreds of millions of square miles. Each spot has a dark central area, the umbra, surrounded by a lighter penumbra, which has a fine, radial structure; both regions are cooler than the photosphere, the umbra by about 1,600°, the penumbra by 500°. The lower temperatures permit the formation of molecules, which would not otherwise exist in the photosphere. These can be studied with a spectroscope, which also reveals that the sunspot gases are in motion, streaming from the umbra into the penumbra with velocities of about 1.2 miles (2 km) per second (the Evershed effect).

It is believed that spots are regions where the normal

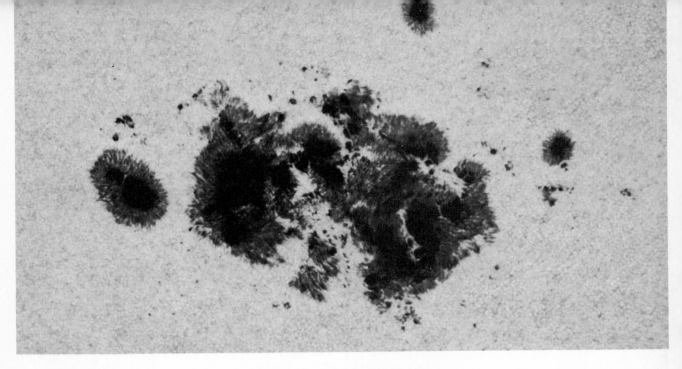

The giant sunspot group of May 17, 1951, 125,000 miles (200,000 km) in length. This photograph, taken with the 60-foot Mt. Wilson tower telescope, shows small-scale features such as pores, and fine radial structure in the penumbra. The larger leading spot, **right**, has a very complex structure with bright bridges crossing its umbra.

convective motions of the photosphere have been inhibited by strong magnetic fields. Splitting of spectral lines (the ZEEMAN EFFECT) reveals fields of up to 4,000 gauss at the centers of sunspots, with the lines of force directed upward in the umbra, but becoming more nearly horizontal toward the penumbra. Like all other magnetic phenomena, the frequency of spots is governed by the SOLAR CYCLE; there are 100 times more spots on the disk at sunspot maximum than at minimum. The earliest spots of a cycle are found at latitudes 40° north and south of the solar equator, but as the cycle proceeds, they form progressively closer to the equator. It is rare, however, to find spots within 10° of the equator, or more than 40° away.

The first indication of the birth of a sunspot is the appearance of a bright PLAGE region in the lower CHROMOSPHERE. Pores develop on the photosphere below, which grow rapidly as the magnetic field increases. After a week, the spot group is fully developed, and the "leading" spot (which "leads" in the direction of the Sun's rotation) is usually the largest. The two main spots in a group always have opposite magnetic polarity, and all leading spots in one hemisphere are of the same polarity. In a large spot group, the lines of force are very complex, and the energy released by the reconnection of field lines to produce a more simple configuration gives rise to FLARES. Most spot groups die away slowly after 10 days, taking about a month to decrease in area and disappear. The magnetic field remains until the spot has almost completely vanished.

sunspot cycle
See SOLAR CYCLE.

supergiant stars
The very brightest, largest stars, formed when stars heavier than the Sun evolve into old age. The red supergiant ANTARES is so large that if it were at the center of the solar system it would engulf the orbit of Mars and reach almost to the asteroid belt. Supergiants are rare, because they burn out very quickly. But they are so bright that most of the patchy features seen in photographs of the spiral arms of external galaxies are in fact associations of supergiants. Sixteen of the hundred brightest stars are supergiants, five in the Orion region (Rigel, Alnilam, Alnitak, Saiph, and Betelgeuse).

superior conjunction
The instant at which Mercury or Venus is directly behind the Sun as seen from Earth.

superior planet
A planet whose orbit is farther from the Sun than the orbit of the Earth. Mars, Jupiter, Saturn, Uranus, Neptune, and Pluto are the superior planets.

supernova
A star that explodes and ejects most of its mass at very high velocities. The energy released is a million times that of a NOVA, an explosion in a star's atmosphere that leaves the star fundamentally unchanged. After a supernova explosion the star is either totally destroyed, or its central core collapses into a very dense NEUTRON star, or possibly even a BLACK HOLE.

In a supernova explosion, a star becomes as bright as a small galaxy for a few days. Over 400 supernovae have been discovered in other galaxies since systematic searches were begun by Fritz ZWICKY in 1937. Most fall into two main categories.

Type I supernovae brighten to an absolute magnitude of −19.4, decrease by 3 magnitudes in a month, and then fade at a constant rate of 1/70 magnitude per day. Since all Type I supernovae are of similar intrinsic brightness, their apparent magnitude

Three views of a type I supernova, **arrowed**, in the galaxy IC 4182, showing its decline in brightness from maximum in 1939. The galaxy and the foreground stars appear more prominently in the later pictures (taken in 1940 and 1942) as progressively longer exposures are required to show the fading supernova.

reveals the distance of their parent galaxy. The spectrum of a Type I supernova shows that about one solar mass is ejected, at a speed of about 7,500 miles (12,000 km) per second, and that this material contains no hydrogen. The pre-supernova is probably the core of an old red giant that has lost its hydrogen-rich envelope.

Type II supernovae have more varied light curves. Their maximum absolute magnitude is about −17.6, and most of them drop to an almost constant magnitude for several weeks before fading further. Their spectra show that up to 10 times the Sun's mass is ejected, at a velocity of about 5,000 miles (8,000 km) per second. The pre-supernova must thus have at least 10 times the Sun's mass; theory shows that such very massive stars evolve over less than 100 million years, becoming unstable and exploding (see STAR).

Supernovae in our Galaxy were recorded in the years 1604, 1572, 1054, and 1006. The expanding debris from the supernova of 1054 is visible as the CRAB NEBULA. Other supernovae have left expanding remnants that are strong radio emitters but generally faint optically. After about 50,000 years, however, the gas in these remnants begins to form into bright filamentary nebulae, like the Veil Nebula in Cygnus.

A supernova is expected in the Milky Way on average every 30 to 50 years. Most will appear faint, due to the INTERSTELLAR ABSORPTION of light by dust grains, but a third of them should be visible to the naked eye.

Surveyor probes

A series of American lunar soft-landers that paved the way for the first Apollo manned landings. Surveyor was a triangular-shaped craft, 10 feet (3 m) tall with three footpads of crushable aluminum honeycomb. Surveyor carried a solar panel and flat high-grain antenna at the top of a mast. As the spacecraft approached the Moon, it was decelerated by a 10,000-lb. (4,536-kg) thrust retro-rocket, which was then jettisoned. The Surveyor was braked onto the Moon by three smaller jets that switched off at an altitude of 14 feet (4.3 m), allowing the craft to

Probe	Launch date	Remarks
Surveyor 1	May 30, 1966	Landed in Oceanus Procellarum near crater Flamsteed on June 2. Returned 11,150 photographs until July 13
Surveyor 2	September 20, 1966	Impacted Moon September 23 southeast of crater Copernicus after control system failed
Surveyor 3	April 17, 1967	Landed in Oceanus Procellarum on April 20. Surface sampler dug in lunar soil. Returned 6,315 pictures until May 3. Visited by Apollo 12 astronauts in November 1969
Surveyor 4	July 14, 1967	Landed in Sinus Medii on July 17. Radio contact lost prior to touchdown
Surveyor 5	September 8, 1967	Landed by remote control from Earth in southern Mare Tranquillitatis on September 11. Carried box to analyze soil by bombardment with alpha particles. Returned 18,006 photographs until September 24
Surveyor 6	November 7, 1967	Landed in Sinus Medii on November 10. Analyzed soil with alpha-scattering device. Landing rockets refired on November 17, causing Surveyor to lift off and resettle 8 feet (2.5 m) away. Returned 30,000 photographs
Surveyor 7	January 7, 1968	Landed near crater Tycho on January 10. First highland landing. Carried sampling scoop and chemical analysis device. Returned 21,000 photographs

The U.S. Surveyor 3 Moon probe landed in the Oceanus Procellarum in 1967, where it tested the soil structure with a mechanical digger and photographed the interior of the craterlet it landed in. 2½ years later, Apollo 12 astronauts landed close by, examined the spacecraft and removed its TV camera for study back on Earth. It was found that some Earth bacteria on it had survived the hostile lunar environment.

drop to the surface at about 8 miles (5 km) per hour. Its weight on the surface was 620 lb. (280 kg). Each Surveyor carried a television camera of variable focal length which pointed at a movable mirror; tilting the mirror changed the television camera's field of view. Surveyors 3 and 7 carried mechanical scoops to dig in the lunar soil, and Surveyors 5, 6, and 7 carried small chemical-laboratory boxes that were lowered to the lunar surface to give rough soil analyses. These analyses indicated the basalt-like composition of lunar rocks, confirmed in greater detail by the Apollo missions. The Surveyor program showed that the lunar surface was safe for Apollo landings.

synchronous orbit
An orbit in which a satellite moves around a body at the same rate as the body spins on its own axis; this is also termed a GEOSTATIONARY ORBIT. The term is also applied to an object that takes as long to rotate once about its axis as it takes to complete one orbit. The Moon has such a synchronous rotation, always turning one face toward Earth; several other moons of the solar system, such as Phobos and Deimos of Mars, also have synchronous rotations. Such a synchronous spin arises because of the braking effect of the planet's gravity, and is also termed a *captured rotation.*

synchrotron radiation
Electromagnetic radiation emitted by charged particles moving near the velocity of light in a magnetic field. This emission was first observed in particle accelerators on Earth, called synchrotrons, from which the name arose. The wavelength of synchrotron radiation depends on the velocity of the particle and the strength of the magnetic field; and it is polarized (see POLARIZATION) at right angles to the lines of magnetic force.

Electrons produce synchrotron radiation more efficiently than do the heavier protons. Electrons moving in a magnetic field produce the radio emission from solar flares, from the remnants of SUPERNOVA explosions, and from RADIO GALAXIES and QUASARS. X rays produced by the synchrotron process account for some sources observed in X-RAY ASTRONOMY. The visible light from some astronomical objects, such as the CRAB NEBULA (a supernova remnant) is also synchrotron radiation.

Syncom
A series of NASA experimental communications satellites in 24-hour synchronous orbit. Their purpose was to test the suitability of synchronous orbits for

communications satellites, and to enable the difficult technique of inserting a satellite into such an orbit to be practiced. The advantage of a synchronous, or geostationary, orbit, is that the satellite hangs near-motionless over the equator, unlike the earliest forms of communications satellite which moved rapidly across the sky. However, if the satellite's orbit is inclined slightly to the equator it will move alternately north and south of the equator each day, tracing out a figure-of-eight pattern in the sky. The 86-lb. (39-kg) first Syncom was scheduled for an orbit inclined about 33° to the equator. However, the satellite went dead shortly before it reached its station in orbit, and was unusable. The identical Syncom 2 became the first successful communications satellite in synchronous orbit, although it too was inclined at 33° to the equator. Syncom 2 was stationed over Brazil and relayed telephone conversations from North and South America to Africa. Syncom 3 became the first truly geostationary satellite in September 1973, when it was maneuvered into equatorial orbit above the Pacific Ocean. It relayed television coverage of the Olympics from Japan to the United States. The Syncom satellites could carry one two-way telephone conversation, 16 teletype channels, or one black-and-white television channel. The satellites received at 7.4 GHz and transmitted at 1.8 GHz. Their success led directly to the INTELSAT series of commercial communications satellites. Syncom 1 was launched in February 1963; Syncom 2 in July 1963; and Syncom 3 in August 1964.

synodic period

The time taken for an object to return to the same position in the sky as seen from Earth. Examples of synodic periods are the intervals between two oppositions of Mars or between two full Moons. The synodic period differs from the orbital period relative to the stars (*sidereal period*), because of the Earth's motion in its own orbit around the Sun.

Taurid meteors

One of the regular annual meteor streams. They are first detectable on October 20 each year and reach a broad peak of 12 meteors per hour on November 8, declining over three weeks. The radiant, which is double, moves eastward at less than 1° a day; it lies 15° southwest of the Pleiades at maximum and reaches its greatest altitude at 1 A.M. A millennium ago, the Taurids were the strongest annual stream, but they are now very much in decline. They are closely associated with ENCKE'S COMET.

Taurus (the bull)

A major constellation of the zodiac, in the northern hemisphere of the sky. Taurus is best seen during the northern hemisphere winter. The constellation is famous for its two large star clusters, the HYADES and PLEIADES. Its brightest star is the red giant ALDEBARAN, which lies in front of the Hyades cluster but is not associated with it. Between the horns of Taurus is the famous CRAB NEBULA, the remains of an exploded star. The TAURID METEORS appear to radiate

from a point near Epsilon Tauri. Near the same point is the faint patch known as Hind's variable nebula (NGC 1554–5), discovered by the English astronomer John Russell Hind (1823–1895). This contains at its heart the star T Tauri, the prototype of a class of irregular variable stars (see T TAURI STARS). The Sun passes through Taurus from mid-May to late June.

tektites

Small, glassy objects, measuring about an inch across, found scattered in a number of localities in the world. The drop-like and button shapes of tektites and their flow structures indicate they were formed from the rapid cooling of molten material, and their extraordinarily low water content distinguishes them from the volcanic glass formed by eruptions on Earth. Tektites are, however, similar to impact glass found at meteorite crater sites, and some tektites contain metallic blobs resembling meteoritic iron.

Tektites are believed to have been formed during the impact of a meteorite, which splashed out molten rock. Some of the melted globules were projected into space, where they cooled before being further melted on reentry into the Earth's atmosphere. Tektites show no traces of encounter with cosmic rays, so that they cannot have traveled for long in space. The impact that caused them therefore occurred either on the Earth or the Moon.

Certain smaller fields of tektites are linked to known terrestrial meteorite craters. Impact glasses found at the 14-mile (22.5 km) diameter Ries crater in Germany have been dated at 14.8 million years, exactly the age of the Czechoslovakian tektites (called moldavites); both were melted from rocks about 300 million years old. Impact glass from the 6½-mile (10.5-km) diameter Lake Bosumtwi crater in Ghana has been dated at 1.3 million years, the same age as the nearby Ivory Coast tektites; both these were formed from rocks 2 billion years old. The North American (Texas and Georgia) and Far Eastern tektite fields have no known associated terrestrial meteorite crater. However, the Far Eastern field could have come from the impact which produced the crater Tycho on the Moon. Debris ejected along Tycho's major bright ray would follow a 3½-day path to Earth and fall in exactly the same pattern as the Far Eastern tektite field.

telescope

A device for collecting and magnifying light, which has made possible the development of modern astronomy. The first telescope was made in 1608 by Hans Lippershey, a Dutch lens maker; telescopes were first turned to the sky by the Italian physicist GALILEO. In 1609 Galileo constructed an instrument magnifying 30 times, with which he discovered the four bright satellites of Jupiter, the phases of Venus, and the mountains of the Moon.

A telescope is superior to the naked eye for two principal reasons: it collects more light, thus revealing fainter objects; and it can be made to magnify the images it receives. Some work in astronomy, involving the discovery of fine planetary detail or the separation of two close stars, requires high magnification. On the other hand, the detection of very faint and nebulous objects, such as remote galaxies, is aided much more by a telescope's light-gathering power than by its magnification.

Telescopium

The earliest instruments were REFRACTING TELESCOPES, using a large lens or OBJECT GLASS to collect and focus the light. The drawback of a simple lens system is that it acts like a prism, causing light to spread out and form a colored image (CHROMATIC ABERRATION). The only way then known of combating this was to make the FOCAL LENGTH of the lens very long in relation to its aperture. In the later years of the 17th century, telescopes 150 feet (46 m) or more in length were being built, but with apertures of only a few inches. These *aerial telescopes,* though very cumbersome, were used to good effect by such observers as Christiaan HUYGENS and Giovanni CASSINI to discover objects that included the rings of Saturn and several of its satellites. However, a great advance was achieved in 1721, when the first truly workable REFLECTING TELESCOPE was constructed by the English instrument maker John Hadley (1682–1744); although reflectors had been made previously, they were of poor optical quality and unsuitable for astronomy. Using a concave mirror instead of a lens to form the image, Hadley's reflector was less then 6 feet (1.8 m) long, yet had an aperture of 6 inches (15 cm). This was possible because a mirror produces no chromatic aberration. Reflecting telescopes, in their various forms, have been used for astronomical research ever since. Most have been either of the *Newtonian* type introduced by Hadley, with a small plane mirror diverting the light through the side of the tube, or of the CASSEGRAIN configuration, which effectively folds a telescope of long focal length into a short tube.

The refracting telescope did not regain its position until the middle of the 18th century, when the London lens maker John DOLLOND marketed an achromatic object glass, consisting of two lenses, each of which compensated for the other's chromatic aberration. Achromatic refractors soon became popular because of their highly efficient light-transmission capabilities. After Dollond, the great makers of refracting telescopes included Joseph FRAUNHOFER, and the firm of Alvan CLARK, which made the two largest refracting telescopes in the world, the 36-inch (91-cm) LICK OBSERVATORY instrument, and the 40-inch (102-cm) at YERKES OBSERVATORY. However, mirrors can now be made much larger than lenses, which gives reflectors a major advantage in light-gathering. The age of huge reflectors began with William HERSCHEL, who built a 48-inch (122-cm) reflector in 1789 at Slough, England. This was followed by Lord ROSSE's 72-inch (183-cm) instrument of 1845.

All early reflectors used a shiny alloy called *speculum metal* for their mirrors. This did not reflect light very efficiently, and tarnished quickly. The most important development in the reflecting telescope came in the late 19th century with the introduction of mirrors made from glass, coated with a highly reflective layer of silver or (much more commonly) aluminum. The first of the modern giants was the 60-inch (152-cm) at Mount Wilson Observatory, California, in 1908. A 100-inch (254-cm) was installed at the same observatory in 1917, and the giant 200-inch (508-cm) at Mount Palomar, California, was completed in 1948. At present several instruments approaching this aperture are in existence or being completed (see REFLECTING TELESCOPE). In the few remaining professional fields where visual work,

The first reflecting telescope, made by Isaac Newton in 1671. Its mirror, shown beside it, is 2 inches (5 cm) in diameter and 8 inches (20 cm) focal length. Like all early mirrors, it is made of metal. The telescope tube is made of oiled parchment. A trial eyepiece hole, covered over and waxed by Newton, is still visible.

rather than photographic recording, is still undertaken, refracting telescopes are generally preferred because of their more critical definition and the relative permanence of their adjustments. The same instrument may be used continuously for years or even decades. (See also CATADIOPTRIC TELESCOPE; MIRROR; SCHMIDT TELESCOPE.)

Telescopium (the telescope)
A faint constellation in the southern hemisphere of the sky, introduced by Nicolas Louis de Lacaille. Telescopium lies below Sagittarius. Its brightest star is of only the fourth magnitude.

Telstar
Two American communications satellites, built by the American Telephone and Telegraph Company for experiments in long-distance television, telephone, and telegraph transmissions. Telstar 1 carried the first live transatlantic television broadcast, beamed from a ground station at Andover, Maine, to Europe. The Telstar satellites were 32-inch (81-cm) spheres, studded with solar cells. They received transmissions at frequencies of 6.39 GHz and retransmitted at 4.18 GHz; they could carry 600 telephone circuits or one television channel. The Telstars orbited at medium heights, so that they moved across the sky and had to be continuously tracked by ground-station antennae. The first Telstar was damaged by radiation

from the Earth's Van Allen belts, and the second Telstar was inserted into a slightly higher orbit. Their immediate success established the potential of commercial communications satellites.

Satellite	Launch date	Remarks
Telstar 1	July 10, 1962	Orbited between 593 and 3,503 miles (954 and 5,638 km) every 158 minutes. Weight 170 lb. (77 kg)
Telstar 2	May 7, 1963	Orbited between 604 and 6,713 miles (972 and 10,803 km) every 225 minutes. Weight 175 lb. (79 kg)

Tereshkova, Valentina Vladimirovna (b. 1937)

Soviet cosmonaut, the first woman to fly in space. She piloted the Vostok 6 spacecraft in June 1963, making 48 orbits of the Earth over a period of 70.8 hours, while Valery Bykovsky was orbiting separately in Vostok 5. Although not a pilot, Tereshkova was an avid parachutist, and learned to fly during her cosmonaut training. She is said to have volunteered for spaceflight in a letter following the flight of Vostok 2. Some reports suggest she was actually the back-up pilot for Vostok 6, and was substituted at the last moment when the woman chosen to make the flight became indisposed. Tereshkova married cosmonaut Andrian NIKOLAYEV on November 3, 1963.

terminator

The sunrise–sunset line dividing the lit and unlit sides of a planet, moon, or asteroid.

Tethys

The fourth satellite of Saturn in order of distance from the planet, probably only about 600 miles (1,000 km) in diameter. Tethys orbits Saturn every 45 hours 18½ minutes in a circular path 183,150 miles (294,750 km) above the planet's surface. It has a density about 1.1 times that of water, indicating that it is made of loosely-compacted frozen gas. Tethys was discovered in 1684 by G. D. CASSINI.

Thales of Miletus (c.624 B.C.–c.546 B.C.)

Greek philosopher, regarded as one of the founders of physical science. Thales believed that water is the primary substance of the Universe, and he visualized the Earth as a flat floating disk. Thales plotted the course of the Sun around the sky, and predicted the eclipse of May 28, 585 B.C. The eclipse was said to have stopped a battle between the Lydians and the Medes. Thales is known to have written on the movements of the Sun, and to have made suggestions for celestial navigation. One of his pupils was ANAXIMANDER.

Thor rocket

An American space launcher, based on the Thor intermediate-range ballistic missile. Thor was the first American intermediate-range missile, test fired in 1957. The basic Thor was 65 feet (19.8 m) tall and 8 feet (2.4 m) in diameter, tapering toward the top. It had a single engine similar to the two booster engines used in the ATLAS ROCKET, of 150,000 lb. (70,000 kg) thrust, burning liquid oxygen and kerosene.

By the time Thor was withdrawn from military use in 1963, it had already been modified into a reliable space launcher. With upper stages such as the Able and AbleStar (modified from the VANGUARD rocket) and the AGENA, Thors were used to launch many early U.S. satellites and probes. These included several Pioneers and Explorers, together with the Discoverer series. In 1963 the Thrust-Augmented Thor (TAT) was introduced, with lift-off power augmented to 330,000 lb. (150,000 kg) by three solid-fuel strap-on motors. In 1966 a long-tank Thor, known as Thorad, was developed; this was 71 feet (21.5 m) long and gave Thor a 20 percent increase in payload capacity. With strap-ons, Thorad became known as the Long-Tank Thrust Augmented Thor (LTTAT). Thor is also used as part of the DELTA launch rocket.

tides

The rhythmic rise and fall of the surface of the sea that occurs twice each day and is due to the gravitational attraction of the Moon and Sun. Because the Moon is so much nearer, it has about twice the effect on the tides as the Sun. The Moon's gravity pulls the Earth's water surface into two bulges, one on the side facing the Moon and the other on the opposite side. Similar effects are caused by the Sun. Twice a month, at new and full Moon, the Sun and Moon are pulling in line, and their tidal effects combine to produce tides higher than normal (called "spring" tides because they spring up). When the Moon is at first or last quarter, it is pulling at right angles to the Sun, and the tidal effects work against each other. This produces tides of a small range (called "neap" tides, meaning scanty). As the Earth rotates under these two tidal bulges, points on its surface experience a cycle of low tides twice each day. The actual height of the tide is partly determined by the shape of the coastline and the depth of water.

Tikhonravov, Mikhail Klavdiyevich (1900–1974)

Soviet rocket engineer, designer of early Russian rockets and later the anonymous "Chief Theoretician of Cosmonautics." During the 1920s Tikhonravov began research into aerodynamics, and in about 1931 joined the Moscow Group for the Study of Reactive Propulsion (MosGRID). He and Sergei KOROLEV tested their first successful rocket, the GIRD 09, on August 17, 1933; this used a solidified form of gasoline as fuel. Later, engines built by Valentin Glushko were used to power the rockets Tikhonravov had designed; one reached an altitude of over 6 miles (9.5 km), and was apparently being developed as an antiaircraft missile. In 1939 and 1940 Tikhonravov worked on a long-range ballistic missile, a forerunner of the later Soviet space launchers. Tikhonravov eventually became responsible for calculating trajectories and flight plans for Russian space rockets, satellites, and probes.

time

The measurement of time is one of the basic functions of astronomy. Early communities measured time by the obvious alternation of day and night, the Moon's cycle of phases, and the progress of the seasons. The day, the month, and the year are the fundamental astronomical divisions of time on which the CALENDAR is based. Hours, minutes, and seconds are only arbitrary units of convenience.

Primitive clocks, which worked by measuring sand running through an orifice, or water dripping into a receptacle, had to be checked and regulated by observations of the rotation of the Earth. These observations were made by noting the moment at which the Sun or a star returned to its original position (usually due south) in the sky. Such crude methods could produce huge errors. Serious attempts at more precise timekeeping came when long sea voyages were first attempted in the 16th and 17th centuries. Only through having an accurate knowledge of the time could a navigator calculate his longitude from observations of the Sun and stars. Eventually, observatory clocks and marine chronometers became essential equipment for any maritime nation. As ordinary life became more regulated, domestic timepieces also came into demand.

Until the 1920s, it was assumed that the rotation of the Earth must be the prime standard for timekeeping; no clock could be as steady and unvarying. The introduction of the Synchronome-Shortt free pendulum clock at the Royal Greenwich Observatory in 1925 destroyed this assumption. The clock was so accurate that the tiny effects on the Earth's rotation caused by NUTATION were revealed. Nutation can produce a change of up to 0.003 second per day. Even more accurate clocks soon followed, devices measuring the rate of vibration of quartz crystals (see QUARTZ-CRYSTAL CLOCK) and, more recently, cesium atoms (see ATOMIC CLOCK). By comparing the time derived from such equipment with direct observations of the rotation of the Earth, not only nutation but a gradual slowing-down of the Earth's rotation, amounting to about one second per year, can be demonstrated. The most accurate measurements of the Earth's rotation are now made with a photographic zenith tube (PZT), a telescopic camera aimed directly at the zenith. This photographs stars as they pass overhead, from which the Earth's rotation relative to the stars can be obtained to an accuracy of a few thousandths of a second. Since nutation is a cyclic phenomenon, with a main period of 18.6 years and a range of ± 1.2 seconds, it is ignored for all civil purposes. But the slowing of the Earth's spin, which involves a cumulative error, is allowed for by the occasional inclusion of an extra leap second in time signals. Astronomical computing practice, which demands a constant time interval for its basis, ignores these effects and uses ephemeris time, which assumes a constant rotational speed for the Earth.

Tiros satellites

A series of American weather satellites; the name TIROS stands for Television and Infra-Red Observation Satellite. The Tiros series returned photographs of the Earth's cloud cover and monitored the flow of heat from the Earth into space; the "heat balance" of the Earth had never before been measured. Tiros satellites were shaped like hatboxes, 42 inches (107 cm) across and 22 inches (56 cm) high, with two television cameras pointed at Earth. The weight of the satellites increased from 263 lb. (119 kg) at the start of the program to 305 lb. (138 kg) at the end. The Tiros satellites orbited Earth about every 100 minutes at altitudes of around 450 miles (725 km). The first

four were in orbits inclined at 48° to the equator, the next four in 58° orbits, and the last two in polar orbits. The Tiros series returned a total of over 500,000 photographs, establishing the value of satellites for weather monitoring and forecasting. They were succeeded by the Tiros Operational System (TOS) satellites, which were designated ESSA SATELLITES once in orbit. An Improved Tiros Operational System (ITOS) has since come into operation; these satellites are given a NOAA designation in orbit (see WEATHER SATELLITES).

Satellite	Launch date	Remarks
Tiros 1	April 1, 1960	First weather satellite. Sent 22,952 photographs until June 17
Tiros 2	November 23, 1960	Sent 36,156 photographs until December 4, 1961
Tiros 3	July 12, 1961	Returned 35,033 photographs until February 27, 1962
Tiros 4	February 8, 1962	Returned 32,593 photographs until June 10
Tiros 5	June 19, 1962	Returned 58,226 photographs until May 4, 1963
Tiros 6	September 18, 1962	Returned 66,674 photographs until October 11, 1963
Tiros 7	June 19, 1963	Over 125,000 photographs returned until February 3, 1966
Tiros 8	December 21, 1963	Over 100,000 photographs returned until July 1, 1967
Tiros 9	January 22, 1965	In near-polar orbit; gave first photographs of Earth's entire cloud cover until February 15, 1967
Tiros 10	July 2, 1965	Final Tiros. In polar orbit; abandoned July 3, 1967

Titan

The largest satellite of Saturn, and the only moon in the solar system known to have a substantial atmosphere. Titan was discovered in 1655 by Christiaan HUYGENS. It orbits Saturn every 15 days 22 hours $41\frac{1}{2}$ minutes at an average distance of 759,080 miles (1,221,620 km). In 1944 the American planetary astronomer Gerard KUIPER detected an atmosphere of methane gas around Titan. More recent observations suggest that Titan may have a cloudy atmosphere with between one and ten percent the Earth's atmospheric pressure, composed not only of methane but also of hydrogen. These gases seem to produce a mild GREENHOUSE EFFECT on Titan; however, its surface temperature is probably only about 115°K. Titan's clouds are reddish in color, like the clouds of Jupiter and Saturn. Accurate measurements place the visible diameter of Titan at 3,600 miles (5,800 km). But since the clouds are probably 100 miles (160 km) deep, the actual diameter of Titan must be about 3,400 miles (5,500 km). Titan therefore is larger than Mercury and has a denser atmosphere than Mars. It is bigger than any of Jupiter's moons, but Neptune's satellite TRITON may be larger still. Titan has a low density of about 1.6 times that of water. It is probably made mostly of frozen gases; these would be slushy near the surface, releasing clouds of vapor to enrich the atmosphere. Some of the lightest gases probably escape from the atmosphere of Titan, spreading out into a gas ring around Saturn.

Titan rocket

A family of American space launchers, developed from the Titan intercontinental ballistic missile. The two-stage Titan I missile entered service in 1962. Titan I was fueled by liquid oxygen and kerosene, which had to be pumped in at the last minute, and this could delay launches in an emergency. The improved Titan II used storable HYPERGOLIC propellants. Titan II was modified to launch the Gemini series of manned spacecraft. The first stage of this modified version was 63 feet (19.2 m) long and its second stage 27 feet (8.2 m) long; first-stage thrust was 430,000 lb. (195,000 kg), and second-stage thrust was 100,000 lb. (45,000 kg). The Air Force, which developed Titan, produced the more powerful Titan III specifically for space launches. Titan III has several versions. Titan IIIA was a three-stage extension of the Titan II; Titan IIIB is a Titan II with an AGENA upper stage. Titan IIIC is based on the Titan IIIA, with the addition of two solid-fuel strap-on boosters. Each of these 10-foot(3-m)-diameter boosters is built in segments. The five-segment version, 85 feet (25.9 m) long, produces a thrust of 1,200,000 lb. (540,000 kg). An advanced vehicle, the Titan IIIM, using seven-segment strap-on boosters each of 1,400,000 lb. (640,000 kg) thrust, was planned to launch the Air Force's Manned Orbiting Laboratory space station, but the project was canceled in 1969.

Titan III operations. Titan III rockets are assembled vertically in special buildings and rolled out to the launchpad like a Saturn V. The Air Force operates two Titan III launchpads, complexes 40 and 41 at Cape Canaveral. Similar facilities exist at Vandenberg Air Force Base, California (see WESTERN TEST RANGE). Titan IIIC orbits military satellites, particularly communications satellites which it carries eight at a time. A modified version, Titan IIID, launches the Air Force's Big Bird reconnaissance satellites. The most advanced member of the series is the Titan IIIE, which has a CENTAUR upper stage. This combination launched the VIKING probes to Mars, and the HELIOS Sun probes. The Titan IIIE is boosted at launch by two solid-fuel strap-ons, like those on the Titan IIIC, which drop away after two minutes. The central core's 520,000-lb. (236,000-kg) thrust first stage, 72.9 feet (22.2 m) long, then takes over and burns for 2½ minutes. After this the 23.3-foot (7.1-m) second stage of 101,000 lb. (46,000 kg) thrust ignites. Finally, the Centaur top stage fires to place the payload in orbit. It is later restarted to carry the probe onto the planets. Overall height of Titan IIIE on the launchpad, including payload and its large shroud, is 160 feet (48.8 m); lift-off weight is over 1,400,000 lb. (640,000 kg). Titan IIIE can place up to 38,000 lb. (17,000 kg) in low Earth orbit, or send 8,000 lb. (3,600 kg) to the planets.

Titania

The fourth satellite of Uranus in order of distance from the planet, discovered in 1787 by William Herschel. Titania orbits Uranus every 8 days 16 hours 56½ minutes at an average distance of 272,390 miles (438,370 km). Its diameter is about 1,100 miles (1,800 km).

Titius-Bode law

See BODE'S LAW.

Titov, Gherman Stepanovich (b. 1935)

Soviet cosmonaut, the first man to spend a full day in space. Titov made a 17-orbit flight lasting 25.3 hours in Vostok 2 in August 1961. He was the second Russian in space; he had already been backup to Yuri Gagarin, who made the first manned spaceflight. Early in his flight, Titov reported feelings of nausea, or space sickness, and he suffered from an inner-ear disorder for some time after his return to Earth. Titov was made commander of the cosmonaut corps in 1964, and he is believed to be involved in the development of a number of important future space projects.

Tombaugh, Clyde William (b. 1906)

American astronomer, who discovered the planet Pluto on February 18, 1930, from photographs taken at the Lowell Observatory, Arizona. Tombaugh, not a professionally trained astronomer, had been hired by the observatory to photograph the sky in search of a new planet beyond Neptune. Tombaugh compared two plates in a device called a BLINK MICROSCOPE; Pluto revealed itself because it had moved in position during the few days between each exposure. Tombaugh's studies also revealed many previously unknown asteroids and variable stars. After Pluto had been discovered, the planet search continued, and showed that no other bodies of any size existed at the edge of the solar system. Tombaugh later made a similar search for possible small natural satellites of the Earth, finding that no bodies bright enough to be photographed were orbiting Earth except our one Moon. Tombaugh has since become professor of astronomy at New Mexico State University, where he continues planetary studies.

total eclipse

See ECLIPSE.

transit

The moment when a celestial body crosses the MERIDIAN, the north–south line passing overhead in the sky. The meridian is the normal reference line for observing stars either to check their position or, if this is already known, to measure the rotation of the Earth. The precise observation of transits has been vital for accurate timekeeping.

Transits are observed by a *transit instrument,* a refracting telescope mounted on a horizontal shaft so that it can be pointed only along the meridian. Cross-wires run vertically through the field of view, sharply in focus with the star, and the passage across the meridian can be timed to an accuracy of $\pm .05$ second or less. Currently, however, the most precise timing observations are made with a prismatic astrolabe (see ASTROLABE) or with a photographic zenith tube (see TIME). The word *transit* also applies to the passage of a planet (such as Mercury or Venus) across the face of the Sun, or of a satellite or a surface feature across the disk of a planet.

Transit satellites

A series of U.S. Navy NAVIGATION SATELLITES, initially intended to provide accurate position-fixing for Polaris submarines, but later used by merchant shipping and airliners. Transit 4A was the first satellite to have a nuclear power source, the small radioisotope generator known as SNAP-3 (Systems for Nuclear Auxiliary Power). The Transit series has been superseded by the polar-orbiting Defense Navigation Satellite System (DNSS), also nuclear powered. Transit 4A was launched on June 29, 1961.

Triangulum (the triangle)

A small constellation in the northern hemisphere of the sky between Andromeda and Perseus, known since ancient times. None of its stars is of particular interest, but it contains the spiral galaxy M33 (NGC 598), a prominent member of our LOCAL GROUP about 2 million light-years away.

Triangulum Australe (the southern triangle)

A small but prominent constellation of the southern hemisphere of the sky, introduced on the 1603 star map of Johann BAYER. It lies on the edge of the Milky Way. Its brightest star, Alpha Trianguli Australis, 99 light-years away, is of magnitude 1.88. Beta Trianguli is of magnitude 3.04, 38 light-years away. The triangle is completed by Gamma, magnitude 3.06.

Triton

The largest satellite of Neptune, and possibly the largest satellite in the solar system; one recent estimate gives it a diameter of roughly 3,700 miles (6,000 km). Triton is also remarkable because it orbits Neptune from east to west, a motion termed RETROGRADE; this is opposite to the general direction of motion in the solar system. A possible explanation suggests that Triton's orbit was disturbed by undergoing a near-collision with another Neptunian moon, which was ejected to become the planet PLUTO. Triton moves around Neptune every 5 days 21 hours 2 minutes in a circular orbit at 220,740 miles (355,250 km). Triton may have an atmosphere.

Trojans

Two groups of asteroids moving in the same orbit as the planet Jupiter. They occupy two of the LAGRANGIAN POINTS, and form equilateral triangles with Jupiter and the Sun. The orbit of Jupiter is, however, elliptical, and the Trojans are perturbed by other planets, so that they wander considerably either side of the Lagrangian points known as L_4 and L_5. The first Trojan, minor planet 588 Achilles, was found in 1906. The Trojans are named for the legendary heroes of the Trojan Wars. They are large, dark, and slightly reddish bodies; their generally elongated shapes suggest they are the debris from collisions of larger asteroids. One Trojan particularly unusual in shape is 624 Hektor (see ASTEROID). There are only 15 numbered Trojans, but a recent survey suggests there may be a total of as many as 700 brighter than magnitude +21 at the two Lagrangian points.

Tsander, Friedrich Arturovich (1887–1933)

Russian rocket pioneer, who developed early Soviet liquid-fueled rocket engines. Tsander, a disciple of Konstantin TSIOLKOVSKY, published in 1924 a design for a hybrid vehicle that would fly like an airplane to the edge of the atmosphere, then blast into space with its rocket motors. The plane's wings would be taken off and melted down to reduce weight and provide additional fuel. During the 1920s Tsander worked on designs for liquid-fuel rocket engines, and in 1930 built his first rocket motor, the OR-1, which ran on gasoline and compressed air. The following year he became head of engine development in the Moscow Group for the Study of Rocket Propulsion (MosGIRD), where with younger colleagues such as Sergei KOROLEV he produced the more powerful OR-2, which used liquid oxygen instead of compressed air. This engine was test fired in 1933, shortly before Tsander's premature death. His followers later used a modified OR-2 to launch a Tsander-designed rocket, the GIRD-10. This was the first fully liquid-fueled Soviet rocket; it flew on November 25, 1933, reaching an altitude of over 3 miles (5 km).

Tsiolkovsky, Konstantin Eduardovich (1857–1935)

Russian spaceflight pioneer, called the Father of Astronautics. Tsiolkovsky became partly deaf at the age of 10; he was introspective and studious by nature. He qualified to become a teacher, but he dreamed of flight, and in his spare time he developed plans for a gas-filled metal dirigible, and built the first Russian wind tunnel to carry out research in aerodynamics. His studies in aeronautics led him to even bolder speculations, long before the Wright Brothers had made even their first flights. His earliest article on spaceflight, "Free Space" (1883), accurately described the weightless conditions of space; he later wrote a science-fiction story about a trip to the Moon. In his book *Dreams of Earth and Sky* (1895), he wrote of an artificial Earth satellite, orbiting at a height of 200 miles (320 km).

By 1898 he had worked out the simple theory of rocket propulsion, relating the rocket's final speed to its exhaust velocity and showing the amount of propellant needed for a rocket of given mass. In 1903 he published an article titled "Exploration of Space by Reactive Devices;" this was the first theoretical demonstration that space travel by rocket was possible. It also contained a design for a rocket that would work on liquid hydrogen and liquid oxygen— the fuels of many modern rockets. Tsiolkovsky realized that single-stage rockets were not powerful enough to escape from Earth on their own, and that a multistage rocket would be needed. He termed this a *rocket train*. His book *Cosmic Rocket Trains* (1924) described a multistage rocket with the various sections stacked on top of each other—what is termed series staging. But in 1935, shortly before his death, he wrote of a rocket squadron, consisting of boosters arranged side-by-side, similar to the parallel staging technique actually adopted by Russian space boosters. Tsiolkovsky established the theoretical basis of spaceflight, and his achievement acted as an inspiration to rocketeers such as Friedrich TSANDER, culminating in the first Soviet Earth satellite.

T Tauri stars

Irregular variable stars, often found in groups (*T-associations*) together with large amounts of dust and gas. They are named after the prototype star in the constellation Taurus. Many T Tauri stars show high outputs of infrared radiation thought to come from warm dust clouds that surround them. The variability of these stars is possibly due to the obscuring effect of the cloud as it swirls around the star. T Tauri stars are PROTOSTARS, settling down into a stable existence as dwarfs. As they contract from a large cloud of gas and dust they spin progressively faster, throwing off material to form a disk from which planets may form, as happened early in the history of our own solar system (see PLANETS). The class of T Tauri variables was first recognized in 1945 by the American astronomer Alfred Harrison Joy (1882–1973).

Tucana (the toucan)

A constellation near the south pole of the sky, containing the smaller of the two MAGELLANIC CLOUDS. The constellation also contains one of the brightest globular star clusters, 47 Tucanae (NGC 104), of magnitude 3, about 19,000 light-years away. Beta Tucanae is a multiple star; Beta[1] Beta[2] are a binary pair, of magnitudes 4.52 and 4.48, 148 light-years away; however, Beta[1] itself has a magnitude 14 companion, and Beta[2] is also binary, made up of components of magnitudes 4.9 and 5.7 orbiting each other every 43 years. Beta[3] is an optical companion, only 93 light-years away; however, it too is double, with components of magnitudes 5.7 and 6.1.

Tycho

See BRAHE, TYCHO.

Tyuratam

The main Russian space launch site, near the town of the same name northeast of the Aral Sea. The Tyuratam cosmodrome is the Soviet equivalent of Cape Canaveral; confusingly, the Russians refer to it as Baikonur, which is actually a place 173 miles (278 km) to the northeast. A new city, Leninsk, has grown up 15 miles (24 km) north of Tyuratam to service the space center; it is not marked on official maps, but has a population of some 50,000. The cosmodrome has an area five times that of Cape Canaveral. It contains an estimated 80 to 85 launchpads for firing military test missiles, Earth satellites, manned rockets such as Soyuz, and unmanned PROTON rockets, together with two additional pads for a giant superrocket that will be larger than the American Saturn V. The first Sputniks, and all Soviet manned launches and space probes, have been launched from Tyuratam. It was first used as a launch site in the 1950s, during development of missiles and sounding rockets. Tyuratam remains the site from which the most important space missions are launched, but it is now less busy a place than PLESETSK, used mainly for military purposes.

U

UBV system

The measurement of the apparent brightness of a star at three specific wavelengths in its spectrum. In a photoelectric PHOTOMETER, the required wavelength region is selected by a colored glass filter, and the APPARENT MAGNITUDE is measured by a PHOTOELECTRIC CELL. In 1951 Harold Lester Johnson (b. 1921) and W. W. MORGAN devised the most commonly used set of three filters: U (ultraviolet), B (blue) and V (visual, or yellow). More recently other filters, passing red and infrared bands have been added to the system.

The difference in magnitude between two wavelengths is known as a COLOR INDEX. The color index B-V, the ratio of blue to yellow light, is determined by the temperature of the star. The second color index, U-B, the ratio of ultraviolet to blue, is determined by the surface gravity of a star, and distinguishes dwarf, giant, and supergiant stars.

All color indices can be affected by INTERSTELLAR ABSORPTION, which must be taken into account.

UFO

Unidentified Flying Objects, or UFOs, are regularly reported by both the general public and by trained observers. Many of these sightings are found on investigation to be misidentifications of aircraft, bright stars or planets, meteors, balloons and satellites, birds, or atmospheric and electrical phenomena. Only when a sighting cannot be readily explained in known terms does it become a genuine UFO. UFOs are often described as cigar-shaped or disk-shaped. Some appear as glowing objects at night, others as opaque disks in the daytime sky, and some have been tracked on radar. Occasional reports speak of apparent "landings" by these objects, and even encounters with supposed "occupants." Some or all of these cases may be hoaxes or hallucinations. Some scientists believe that with sufficient investigation, all UFO cases could be explained in known terms. Others believe that among the most puzzling UFO reports may be genuine phenomena new to science. Certain individuals, however, will always regard the classification of an object as a UFO as a sufficient explanation in itself. These people equate UFOs with the mythical "flying saucers," supposed flying craft reputedly under the control of other beings. Although many scientists believe we may not be alone in space (see LIFE IN THE UNIVERSE), there is no scientifically accepted evidence that we have been visited by alien beings, either in the distant past or more recently.

ultraviolet astronomy

The study of the Universe in the ultraviolet part of the electromagnetic spectrum, which lies between visible light and X rays, from a wavelength of 3000 Å down to 300 Å. Ultraviolet radiation is absorbed by the Earth's atmosphere, and detectors must be taken above it by rocket or satellite.

Copernicus satellite. The Orbiting Astronomical Observatory OAO-II, launched in 1968, surveyed the ultraviolet sky for the first time. Its successor, the $2\frac{1}{2}$-ton Copernicus satellite (OAO-III), launched in 1972, carries the largest reflecting telescope yet sent into orbit, a 32-inch (81-cm) mirror with associated sensors. It is designed to detect and measure the ultraviolet spectra of stars and interstellar molecules with better resolution than ever before. Copernicus also carries three X-ray telescopes, and can study objects across a very broad band of the electromagnetic spectrum. The guidance system of

Copernicus, which points the orbiting telescope toward a star, is precise enough to take full advantage of its high-resolution telescope; such a telescope on earlier satellites would have been wasted because it could not have been "locked on" the star with sufficient accuracy.

One of the most important Copernicus research projects is the study of hydrogen molecules in space, by the ultraviolet absorption lines they produce in the spectra of background stars. Possibly half the interstellar gas in our Galaxy is in this form, and it can only be detected at ultraviolet wavelengths. Hydrogen atoms are also easily detected at short ultraviolet wavelengths, and their distribution near the Sun will be determined by future satellites.

The Sun. Long before the advent of such long-lived, accurately pointing space platforms, many ultraviolet observations of the SUN were made using sounding rockets. The first of these marked the beginning of space-age astronomy: in October 1946 a U.S. Naval Research Laboratory experiment photographed the solar spectrum down to a wavelength of 2200 Å from a V-2 ROCKET at an altitude of 50 miles (80 km). As well as many other sounding rocket experiments, a series of Orbiting Solar Observatory (OSO) satellites was launched by NASA from 1962 onward, each carrying about 70 lb. (21.5 kg) of instruments. These could be pointed with an accuracy of about 1 arc minute, and scanned about 1/30 of the Sun's diameter at a time.

Many spectral lines of highly ionized silicon, oxygen, iron, and other elements have been studied at ultraviolet wavelengths in the solar spectrum, providing valuable information about the region of rapidly changing temperature that forms the boundary between the CHROMOSPHERE and the CORONA. Ultraviolet observations are also providing new insights into solar FLARE activity. Observations from the OAO satellites have now shown that, as expected, there are lines of similar ionized elements in the spectra of other stars. This implies that a high-temperature corona is a common feature among stars, and could provide a means of monitoring "starspot" activity. The SOLAR CYCLE of sunspot activity is roughly 11 years long, and it affects the intensity of the ultraviolet lines with the same 11-year period. Although it is not possible to observe equivalent spots on other stars directly, a similar regular variation in the strength of the ultraviolet lines would show that they, too, go through similar cycles of activity. A period of 11 years would probably not be expected, but if such variations were found after several years of monitoring, they could provide a valuable insight into the workings of all stars, including the Sun.

Galaxies and quasars. Ultraviolet observations of galaxies indicate what proportion of their stars are young blue stars, which radiate strongly in the ultraviolet. The first ultraviolet observations of the ANDROMEDA GALAXY showed an unexpected amount of radiation at wavelengths less than 2500 Å. If the radiation from distant galaxies is similar to that from Andromeda, this ultraviolet "peak" in their spectra might be moved into the visible region by the cosmological RED SHIFT. Galaxies farther from us would then appear brighter and bluer than they really are, and this effect could have important implications for the entire field of cosmology, which uses observations of galaxy luminosities.

The same red shift moves ultraviolet lines in the spectra of QUASARS into the visible part of the spectrum, and in many cases it is the identification of these lines that makes measurement of the red shift possible. In the mid-1960s, when the first quasar red shifts were being measured, observers drew on calculations of strong ultraviolet emission lines made originally to help in the study of gaseous nebulae within our Galaxy. The ionization of these H II REGIONS is caused primarily by ultraviolet radiation from very hot stars, and the successful application of this work to quasar observations extends "ultraviolet astronomy" to include optical observations under these special conditions of high red shift.

umbra

The darkest part of a shadow. In the solar system, the umbra is that part of an object's shadow in which light from the Sun is totally cut off; therefore a body entering another's umbra is totally eclipsed. The umbra is surrounded by a much larger, partially shaded area termed the penumbra. The dark central portion of a sunspot is also referred to as the umbra.

Umbriel

The third satellite in order of distance from Uranus, discovered in 1851 by William LASSELL. It orbits Uranus every 4 days 3 hours $27\frac{1}{2}$ minutes at an average distance of 166,020 miles (267,180 km). Umbriel is about 600 miles (1,000 km) in diameter.

Universe

The Universe contains everything that exists—all of space, time, and matter. Most of the Universe is near-empty space. The matter that we see is clustered into vast systems called galaxies, which may be a billion billion miles across. Much of this matter is in the form of luminous stars. Our own Sun is an undistinguished star belonging to a typical galaxy called the Milky Way, which contains about 100 billion other stars, together with some gas and dust. Many millions of other galaxies are visible through large telescopes. Some of these are so distant that the light we see left them before the Earth was born. By looking deep into space we therefore see the Universe as it appeared in the remote past. The study of the structure and evolution of the Universe as a whole is the subject of COSMOLOGY.

Early ideas about the Universe. To ancient peoples, the Earth seemed the center of a small Universe, consisting of spheres on which the celestial bodies were fixed (see ASTRONOMY, HISTORY OF). In the 16th and 17th centuries the Earth was removed from the center of the Universe, to be replaced by the Sun. But scientists' ideas about the total extent of the Universe were little changed until they slowly began to realize that the stars are separate Suns, much more distant than the planets; the astronomer James GREGORY is credited with one estimate of their distance. Only toward the end of the 19th century, with the development of ASTROPHYSICS, was the gaseous nature of stars fully verified. In 1917 the American astronomer Harlow SHAPLEY transformed the known scale of the Universe by showing that our Galaxy was

The vastness of the Universe is vividly demonstrated by this view of an irregular cluster of galaxies in Hercules. Each galaxy here is as big as our Milky Way; there are thousands of galaxies in this cluster, and tens of thousands of similar clusters throng the Universe to the limits our telescopes can reach.

much larger than had previously been supposed. But most astronomers still assumed that the limits of our own Galaxy were the limits of the entire Universe, until in the 1920s Edwin HUBBLE proved the existence of other galaxies of stars, stretching far into space. Since then, progressively more powerful telescopes reaching ever-deeper into the Universe at optical and radio wavelengths have attempted to trace the structure of the Universe to its visible edge.

Modern views of the Universe. Hubble laid the foundation for modern theories of cosmology with his discovery that the Universe is apparently expanding. The concept of an expanding Universe suggested that all the galaxies were once compressed close together, and that the Universe originated in a BIG-BANG explosion; this theory was given support by the discovery in 1965 of a weak BACKGROUND RADIATION in the Universe, apparently caused by heat left over from the big bang itself. According to

current estimates, the Universe originated about 18 billion years ago.

As astronomers probe deep into the Universe, ordinary galaxies become increasingly faint and difficult to see. At optical and radio wavelengths the super-bright RADIO GALAXIES and QUASARS dominate the picture. Observations suggest that the appearance of objects in the Universe has changed over time, as the Universe has evolved. By observing the rate of expansion at great distances, we can determine whether the expansion will continue for ever, or whether it will slow down and possibly reverse (see OSCILLATING UNIVERSE).

An alternative view of the Universe, called the STEADY-STATE THEORY, held that the Universe has no origin and has always appeared the same. Unfortunately, there seems to be no way of fitting either background radiation or the apparent evolution of the Universe into the steady-state theory, and most astronomers have therefore abandoned it.

But it is conceivable that we see only a small section of a much larger Universe, and that our restricted view prevents us from perceiving the true nature of the cosmos. Our present models of the Universe may seem as naive to future astronomers as do the ideas of the Greeks to us today.

215

Uranus

The seventh planet from the Sun, and the first to be discovered telescopically. Uranus is a large planet, about four times the diameter of the Earth, with a thick atmosphere. Although at its brightest it is just visible to the naked eye, Uranus was not discovered until March 13, 1781, when it was found by Sir William Herschel during his systematic search of the sky. Uranus had in fact been sighted but not recognized as a planet by earlier observers; listings of it as a star in a number of catalogs stretched back nearly a century. These prediscovery observations allowed the orbital motion of Uranus to be calculated, and revealed irregularities which later led to the discovery of Neptune.

Physical properties. Uranus has a mean distance from the Sun of 19.18 astronomical units (1,783,000,000 miles; 2,869,600,000 km). The orbit is inclined by only 0° 46′ to the ecliptic and has an eccentricity of 0.047; the actual distance from Uranus to the Sun varies by 167,000,000 miles (269,000,000 km). Uranus orbits the Sun every 84.01 years. As a result of its varying distance from the Earth, Uranus changes in brightness by 26 percent between conjunction and opposition. It also varies regularly in brightness as it turns on its axis. From this regular variation, together with the Doppler shift in its spectrum, it has been found that Uranus rotates once every 10 hours 49 minutes. The most remarkable feature of Uranus is that its axis is nearly in the plane of its orbit; the north pole is inclined at 98° to the vertical, so that the planet appears to rotate about its axis in a retrograde direction. The orbital motion is, of course, direct. A peculiar result of its highly tilted axis is that Uranus has extreme seasons, each pole experiencing a 42-year "summer" and a 42-year "winter."

Uranus appears through a telescope as a greenish disk, on which some observers have reported seeing faint bands. However, high-quality photographs taken from balloon-borne telescopes have detected no markings of any kind on the planet. These photographs reveal that the equatorial diameter of Uranus is 32,200 miles (51,800 km). The planet is slightly oblate (squashed) in shape; its polar diameter is about 2,000 miles (3,000 km) smaller. The mass of Uranus, determined from the motions of its satellites and the perturbations that it produces on the motion of Saturn, is 14.6 times the Earth's. Uranus has a density of 1.2 times that of water.

Structure of Uranus. Uranus and Neptune are smaller and colder than their companion giant planets Jupiter and Saturn, and they form a separate family of planets. They are composed of a smaller proportion of hydrogen and helium than are Jupiter and Saturn, and a larger proportion of heavy elements. While Jupiter and Saturn are similar to the Sun in composition, Uranus and Neptune are more like the comets.

The atmosphere of Uranus shows a strong presence of methane, which gives the planet its greenish tinge. Ammonia probably also exists, but because of the low temperature (about −210°C) at the top of the atmosphere, it has condensed into clouds at a lower level. Hydrogen is the major constituent of the atmosphere, and helium is also thought to be present. Cloud layers of methane and ammonia shield the deeper levels from sunlight.

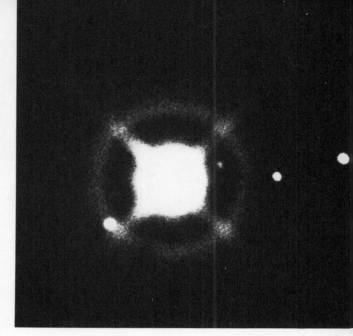

All five of Uranus' satellites are visible on this long exposure photograph taken with the McDonald 82-inch reflector; from left to right they are Oberon, Ariel, Miranda, Umbriel, and Titania. The planet itself is highly over-exposed, and a combination of photographic and telescopic effects causes the apparent cross and ring around it.

Beneath its atmosphere Uranus is believed to have a rocky core about 10,000 miles (16,000 km) in diameter, coated with a layer of ice 5,000 miles (8,000 km) thick. The pressure at the center is over 2 million atmospheres and the temperature about 4,000°C. The rocky core and the layer of ice together account for about four-fifths of the planet's mass, but the overall low density of Uranus is due to its extensive gaseous atmosphere.

Satellites. Uranus has five known moons: MIRANDA, ARIEL, UMBRIEL, TITANIA, and OBERON. The moons orbit in the plane of the planet's highly tilted equator, moving in the same direction as the planet's rotation. It is not known how the satellites could have so exactly taken up the planet's extreme axial tilt. One suggestion is that the formation of the satellites was somehow part of the same event that tilted the planet's axis. The system of moons presents itself edge-on every 42 years, as, for example, in 1882, 1924, and 1966, while in 1861, 1903, and 1945, the orbital paths appeared to observers as circles.

Satellites of Uranus

	Discoverer	Diameter (km)	Orbit radius (10³ km)	Period (days)	Eccentricity
Miranda	Kuiper (1948)	550	130	1.4135	0.017
Ariel	Lassell (1851)	1,500	192	2.5204	0.0028
Umbriel	Lassell (1851)	1,000	267	4.1442	0.0035
Titania	Herschel (1787)	1,800	438	8.7059	0.0024
Oberon	Herschel (1787)	1,600	586	13.463	0.0007

Urey, Harold Clayton (b. 1893)
American chemist and Nobel Prize winner, notable in the field of astronomy for computing the table of the relative cosmic abundances of the elements, and for championing the view that the planets were formed by the coalescence of asteroid-sized bodies. He proposed that the early atmosphere of the Earth was rich in hydrogen, methane, and ammonia, and suggested it was from this mixture that the first building blocks of life formed. A student of his, Stanley Lloyd Miller (b. 1930), confirmed this in 1952 when he passed an electric spark through such a mixture of gases, simulating the effect of lightning in the atmosphere of the early Earth, and produced a host of amino acids, the basic components of protein. This work suggested that there may be many other planets in space on which life has evolved (see also LIFE IN THE UNIVERSE).

Ursa Major (the great bear)
A major constellation of the northern hemisphere of the sky, circumpolar from mid-northern latitudes but best seen during the northern hemisphere winter. Its most conspicuous part is a pot-shaped group of seven bright stars, forming what is commonly termed the Big Dipper or Plow. The star Beta, also called Merak, of magnitude 2.44, and Dubhe, Alpha Ursae Majoris, magnitude 1.95, form the Pointers, so called because a line drawn through them leads to POLARIS, the Pole Star. Dubhe is actually double, with components of magnitude 2.0 and 4.8 orbiting every 44 years. Alioth, Epsilon Ursae Majoris, is a spectroscopic binary varying between magnitudes 1.68 and 1.83 every five days. Among other double and multiple stars in Ursa Major is MIZAR, Zeta Ursae Majoris. Gamma Ursae Majoris, Phekda, is of magnitude 2.54, and Eta Ursae Majoris, Benetnash, of magnitude 1.91. The faintest star is Megrez, Delta Ursae Majoris, of magnitude 3.44. The stars of the Dipper, except Dubhe and Benetnash, form an associated group, moving through space together. There are numerous galaxies in the region of Ursa Major, including the beautiful spiral M81 (NGC 3031), 12 million light-years away, and M101 (NGC 5457), 23.5 million light-years

distant. M82 (NGC 3034) is a well-known exploding galaxy, and M97 (NGC 3587) is a planetary nebula known as the Owl, 2,600 light-years distant. Lalande 21185, a magnitude 7.48 red dwarf, lies near coordinates 11^h, $+36°$. It is the fourth-closest star to the Sun, 8.1 light-years away, and may possess a planetary system. Xi Ursae Majoris, 26 light-years away, was the first double star to have its orbit computed. It is composed of stars with masses 1.28 and 0.99 that of the Sun, orbiting each other every 59.84 years; they are of magnitudes 4.41 and 4.87. Both components are also spectroscopic binaries, of periods 669 and 3.98 days.

Ursa Minor (the little bear)
A constellation at the north pole of the sky, containing the north pole star, POLARIS, Alpha Ursae Minoris. The constellation's second-brightest star is Beta, called Kochab, of magnitude 2.24; this was the north pole star between about 1500 B.C. and 300 A.D. The constellation is also known as the little dipper.

U.S. Naval Observatory
Astronomical facility of the U.S. government, with headquarters in Washington D.C. It originated in 1830 as a part of the Navy's department of charts and instruments, and was established under its current name in 1844. It moved to its present site in northwest Washington in 1893, when it also absorbed the U.S. Navy's *Nautical Almanac* Office. The Naval Observatory is responsible for the nation's time service, and for the production of almanacs for astronomers, navigators, and surveyors; these publications are prepared in collaboration with the *Nautical Almanac* Office of the Royal Greenwich Observatory in England. The Naval Observatory operates a time service substation in Richmond, Florida, near Miami. Among the observatory's instruments at Washington is the famous 26-inch (66-cm) refractor, installed in 1873, with which Asaph HALL discovered the two moons of

A flare on UV Ceti observed simultaneously by Russian astronomers at Odessa and by the Jodrell Bank 250-foot radio telescope.

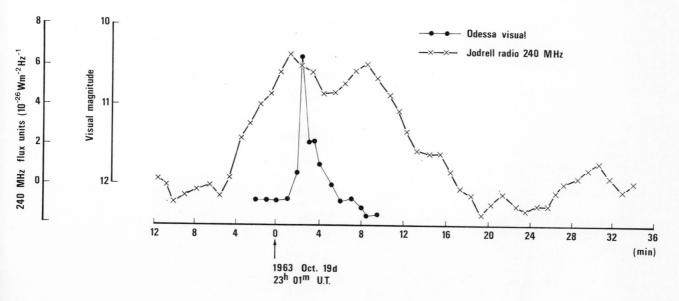

Mars. This telescope was remounted when the observatory moved in 1893, and was completely modernized in 1960. The observatory also operates a 24-inch (61-cm) reflector and 15-inch (38-cm) astrographic telescope at its Washington headquarters. In 1955 the U.S. Naval Observatory established an observing station near Flagstaff, Arizona, at an elevation of 7,579 feet (2,310 m). This houses a 61-inch (155-cm) astrometric reflector with a 50-foot (15.2 m) focus, installed in 1963 for accurate measurements of star positions. Another telescope at the Flagstaff station, a 40-inch (102-cm) reflector, is used for observations of comets and asteroids.

UT

Abbreviation for Universal Time, the standard time reference of the world. UT is equivalent to Greenwich Mean Time (GMT), and the name Universal Time was introduced at the 1928 meeting of the International Astronomical Union; GMT has been the world's standard time system since 1884. Universal Time is based on the rotation of the Earth, and is obtained by observations of star positions as the Earth spins; because the rate of the Earth's rotation is not quite constant, UT differs from the more regular scale of ATOMIC TIME.

UV Ceti stars

Variable stars which suddenly flare in brightness by up to 250 times (six magnitudes). The flare-up takes only a few seconds, and is followed by a steady decline to normal brightness over a few minutes. UV Ceti stars are small, cool dwarfs, intrinsically very faint. Only about 30 are known, because they are normally so faint, but they may actually be the commonest type of variable star in the Galaxy. Sir Bernard LOVELL has observed bursts of radio waves from UV Ceti—the prototype star—and others of its type, occurring simultaneously with the optical flares. UV Ceti stars are apparently rapidly rotating protostars, not yet having reached the mature stable state of a dwarf (see also FLARE STAR).

V

V-2 rocket

The forerunner of modern ballistic missiles and space rockets, developed for the German army during World War II by Wernher VON BRAUN. The V-2, first successfully fired on October 3, 1942, was the first rocket to exceed the speed of sound. The V-2 was 46 feet (14 m) long and 65 inches (1.65 m) in diameter; it weighed about 28,000 lb. (12,700 kg), had a thrust of 56,000 lb. (25,400 kg), and could carry a 1-ton payload 200 miles (320 km). The first V-2 was launched against an enemy target on September 6, 1944. (The earlier V-1 was a subsonic pulsejet, not developed by the von Braun group.) After the war, remaining V-2 rockets were used by the American government for upper-atmosphere research, and from them developed the American REDSTONE rocket.

Van Allen, James Alfred (b. 1914)

American physicist, who supervised the use of V-2 rockets for upper-atmosphere research after World War II, and developed the Aerobee sounding rocket.

The ancestor of the giant rockets of the U.S. space program, the V2 rocket was developed in Germany during World War 2 and used to attack London. The firing shown here was made by Allied scientists after the capture of the rockets and launch sites.

Van Allen was among those who proposed the 1957–1958 International Geophysical Year and devised the experiments sent aboard the first U.S. satellite, Explorer 1, launched as part of the IGY program. Explorer 1 carried a geiger counter to measure cosmic ray particles in the upper atmosphere. Its results, confirmed by those of its successors, revealed two doughnut-shaped belts of charged particles around the Earth, known as the VAN ALLEN BELTS. The discovery provided new knowledge about the Earth's magnetic shell, called the MAGNETOSPHERE.

Van Allen belts

Two zones surrounding the Earth in which charged particles are concentrated and trapped by the Earth's magnetic field. James VAN ALLEN deduced their existence in 1958 from measurements made by the early Explorer satellites.

The two belts are both toroidal (doughnut-shaped), and lie at heights of 1,900 and 14,000 miles (3,000 and 22,000 km) above the Earth's equator. High-energy electrons fill both belts, but the principal constituents of the inner belt are the heavier protons (hydrogen nuclei). Both types of particles have probably been ejected from the Sun as part of the SOLAR WIND, and captured by the Earth's magnetic field.

Space probes have shown that Venus and Mars do not possess Van Allen belts, and therefore cannot have magnetic fields comparable to that of the Earth. But Jupiter has a very strong magnetic field, and with it, associated Van Allen belts 10,000 times as intense as the Earth's.

Van De Kamp, Peter (b. 1901)

Dutch-born American astronomer who discovered the nearest planetary system to our own, orbiting the red dwarf known as BARNARD'S STAR, the second-closest star to the Sun. At Sproul Observatory in Pennsylvania, Van De Kamp began in 1937 to measure the positions of nearby stars in the hope of finding evidence for planets around them. A planetary system would cause a slight wobble in the PROPER MOTION of a star across the sky. Van De Kamp first noted such a wobble in the motion of Barnard's star in 1956, and in 1963 he proposed that a planet 1.6 times the mass of Jupiter orbited the star. From further observations he concluded in 1969 that two planets orbited Barnard's star. According to an analysis published in 1975, these planets have masses 1.0 and 0.4 times that of Jupiter, and orbit every 11.5 and 22 years at a distance of 2.71 and 4.17 astronomical units. There may be other, smaller planets whose effects are undetectable. In 1974 Van De Kamp announced the presence of a planetary companion about six times Jupiter's mass orbiting the star Epsilon Eridani.

Vandenberg Air Force Base

Headquarters at Lompoc, California, of the WESTERN TEST RANGE for rockets and missiles.

Vanguard project

A U.S. Navy project begun in 1955, intended to launch the first American satellite. The Vanguard launcher was based on a sounding rocket called Viking, with a modified Aerobee sounding rocket as a second stage, and a new solid-propellant third stage. The Vanguard rocket stood 72 feet (22 m) tall overall, and had a maximum diameter of 45 inches (114 cm); first-stage thrust was 27,000 lb. (12,250 kg), second stage thrust 7,500 lb. (3,400 kg), and third-stage thrust 3,100 lb. (1,400 kg). Vanguard was beaten into orbit by Sputnik; the earliest Vanguard launch attempts failed, and the first American satellite was Explorer 1, launched by a JUNO ROCKET.

variable stars

Stars whose light reaching the Earth varies in brightness. The star's actual light output may change, or part of its light may be temporarily blocked by another star or a dust cloud. In 1975 the *Moscow General Star Catalog* listed 25,140 variable stars, and they are being discovered at an ever-increasing rate. Variability down to about a hundredth of a magnitude can be measured with a photoelectric photometer, which converts the light received from a star into an easily measured electric current. This technique is time-consuming, and can only be applied to one star at a time. A more usual method is to take two photographs of a star field at different times and examine them in a blink microscope. The operator can spot stars which vary from one plate to another by as little as 0.2 magnitude.

Except for those stars already named when their variability was discovered, such as Delta Cephei, variable stars are noted with a letter or pair of letters plus the constellation (RR Lyrae, for example). Where even these possible permutations are insufficient, variable stars are numbered with a prefix V.

Variable stars are normally classified into three

Vanguard project

Satellite	Launch date	Remarks
Vanguard TV3 (TV = test vehicle)	December 6, 1957	Lost thrust after 2 seconds; launch failure
Vanguard TV3 backup	February 5, 1958	Broke up in flight due to control system malfunction
Vanguard 1 (Vanguard TV4)	March 17, 1958	Second U.S. satellite; 3¼-lb. (1.5-kg) sphere orbited between 405 and 2,462 miles (652 and 3,962 km) every 134.3 minutes; tracking revealed Earth slightly pear-shaped
Vanguard TV5	April 28, 1958	Third-stage failure
Vanguard SLV1 (SLV = satellite launch vehicle)	May 27, 1958	Second-stage malfunction
Vanguard SLV2	June 26, 1958	Premature second-stage cutoff
Vanguard SLV3	September 26, 1958	Insufficient second-stage thrust
Vanguard 2	February 17, 1959	22-lb. (10-kg) satellite orbited between 347 and 2,064 miles (558 and 3,322 km) every 125.9 minutes; attempted TV pictures of cloud cover spoiled by satellite wobble
Vanguard SLV5	April 13, 1959	Second-stage failure
Vanguard SLV6	June 22, 1959	Second-stage malfunction
Vanguard 3	September 18, 1959	100-lb. (45.4 kg) satellite orbited between 317 and 2,329 miles (510 and 3,748 km) every 130.2 minutes; mapped Earth's magnetic field, recorded micrometeorites, and monitored solar radiation. Final Vanguard

groups, although there are some which do not fit well into any of them. The groups are pulsating, eruptive, and eclipsing stars.

Pulsating variable stars throb like a heart beat, some regularly (like CEPHEIDS) and some irregularly (like many red giants). Pulsating stars change in size, and also in surface temperature, both of which affect the star's brightness. The temperature change is usually most important, so that the star is brightest when hottest; however, maximum temperature does not always coincide with maximum (or minimum) size. The surface of the star rises and falls as it pulsates, but this motion does not penetrate deeply, and the central core containing 95 percent of the star's mass remains unaffected. The pulsation is caused by the ionization of helium and hydrogen.

Pulsating variable stars include CEPHEIDS and RR LYRAE stars. These, together with the very short period stars called Delta Scuti stars (periods less than 5 hours, range of variation about 0.1 magnitude), occupy only a small area of the HERTZSPRUNG-RUSSELL DIAGRAM, called the *instability strip*. Nearly all stars pass through this area sometime in their evolution; presumably all are pulsators at that time. Other pulsators include the *Beta Cephei stars* (also called *Beta Canis Majoris stars,* after another well-

Name	prototype	Pulsating variable stars period (days)	variation (magnitudes)	Kind of star	Where found
cepheids	Delta Cephei	1 to 70	0.1 to 2	yellow supergiants	spiral arms, galactic clusters
Type II cepheids	W Virginis	1 to 70	0.1 to 2	yellow supergiants	halo, galactic center
long-period variables	Mira	80 to 1000	2.5 to 5	red giants	everywhere
semi-regular variables	Z Aquarii	about 100	0.5 to 2.5	red giants	everywhere
irregular variables	Betelgeuse	irregular	about 1	red supergiants	spiral arms
RV Tauri stars	RV Tauri	30 to 150	1 to 3	yellow supergiants	halo, galactic center
RR Lyrae stars	RR Lyrae	0.05 to 1.2	0.1 to 2	blue giants	halo, globular clusters
Beta Cephei stars	Beta Cephei	0.1 to 0.6	0.1	blue giants	spiral arms
Delta Scuti stars	Delta Scuti	less than 0.2	0.1	blue/yellow giants	disk, galactic clusters

studied example), which are periodically varying blue giants. Red giants also vary more or less regularly and are known as *Mira-type, semi-regular variables,* and *irregular variables,* in decreasing order of regularity. Being large stars with low densities, their periods are long (typically 100 days).

Eruptive variable stars are stars on whose surfaces occur explosions (flares) or which themselves explode (like *novae*). These unpredictable stars are the most spectacular variables, increasing in brightness up to 100 million times (SUPERNOVAE). Eruptive variables include FLARE STARS, like T TAURI STARS and UV CETI STARS, in which the flares are finished in a few minutes, as well as U Geminorum and Z Camelopardis variables, which show slow flares lasting a day or two. These latter are close double stars which contain a WHITE DWARF star, surrounded by a disk of gas. Gas streams from the companion star to the white dwarf, colliding with the disk to form a hot spot which comes and goes as the binary system rotates. These variable stars show a fascinating range of light variations.

In their gross light output, the *R Coronae Borealis stars* are the reverse of flare stars. These enigmatic stars from time to time suddenly *drop* in brightness, staying faint for weeks or months, and then gradually recover their former brightness. These stars have an excess of carbon in their atmosphere, and possibly they periodically puff a cloud of atmosphere into space. Away from the star the carbon cools to form graphite flakes, which temporarily obscure the star. It recovers its former brightness as the puff of graphite disperses into space.

Eclipsing variable stars are DOUBLE STAR systems in which one star periodically passes in front of the other, blocking some of the light from Earth (see ECLIPSING BINARY). Even when the orbit is angled slightly, so the two stars do not eclipse, the total light may still vary during each revolution because the gravitational pull of each may be strong enough to distort the other into a non-spherical shape. When seen side-on, an ellipsoid, a shape like a football, presents a larger surface area than when seen end-on (along its long axis); for this reason ellipsoidal stars vary as they rotate. The closer together two stars are in the double star system, the more pronounced is their distortion and the greater the variation. The star Beta Lyrae is a typical close eclipsing variable star,

having a period of 12.9 days and a range of brightness of 0.7 magnitude.

Although not eclipsing variables, MAGNETIC STARS also vary during their rotation, as large spots are carried across their face. They are known after their typical example as *Alpha² Canum Venaticorum variables.*

Vega
The brightest star in the constellation Lyra, and the fifth-brightest in the sky. Vega, of magnitude 0.04, is a white star 26 light-years from Earth. Its diameter is three times that of the Sun. Vega, also known as Alpha Lyrae, will be the pole star in about 14000 A.D.

Vela (the sails)
A constellation of the southern skies, once part of the larger constellation Argo Navis until made separate by Nicolas Louis de LACAILLE. It lies in a bright part of the Milky Way next to Centaurus, and contains several interesting star clusters. Its brightest star, Gamma, is a visual double of magnitudes 2.22 and 4.79; the brighter component is the brightest WOLF-RAYET STAR known. Both components are also spectroscopic binaries, making Gamma Velorum a quadruple star. Vela also contains a famous PULSAR.

Venus
The second planet from the Sun. Venus is a rocky planet with an extremely hot, dry surface enveloped by a dense, smog-ridden atmosphere. Although the Earth's twin in size, Venus has evolved in a quite different way, producing a planet where there is little chance of finding even the simplest forms of life.

Physical nature. Venus is 7,521 miles (12,104 km) in diameter, slightly smaller than the Earth, with a mass of 4.87×10^{21} tons and a mean density of 5.25 times that of water, compared with the Earth's 5.52. The planet takes 243 days to rotate on its axis, and it does so in a retrograde direction, opposite to that of all the other planets except Uranus. Its axis is inclined at only 3°. Venus moves at about 22 miles (35 km) per second in an almost circular orbit some 67,238,000 miles (108,210,000 km) from the Sun. The orbit is inclined at 3° 24′ to the ecliptic, and Venus completes one circuit every 224.7 days, showing phases in the same way as Mercury.

Venus is the morning and evening star, the

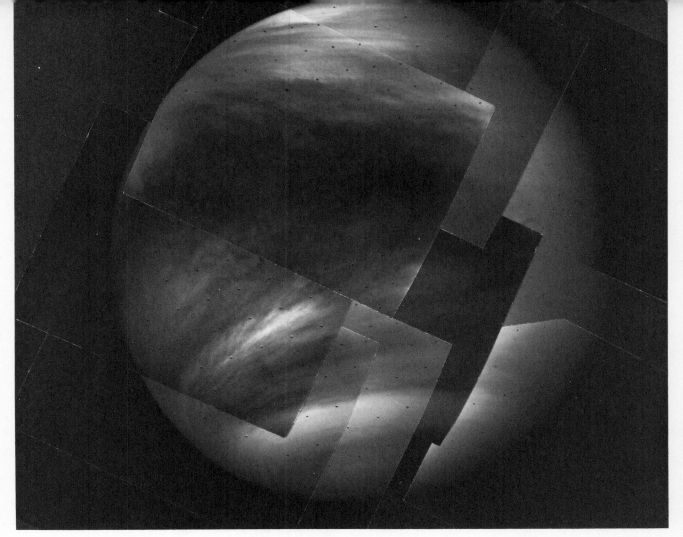

A mosaic of photographs of Venus taken by Mariner 10 from 525,000 miles. This ultraviolet view reveals details of the top of the cloud layer, in particular the high-speed winds blowing parallel to the equator, and the undisturbed cloud near the south pole, **lower center**.

Phosphorus and Hesperus of antiquity, which were long thought to be different bodies. It is the most brilliant object in the sky after the Sun and Moon, reaching magnitude −4.3. It is often visible in daylight and can cast shadows at night. Venus passes closer to the Earth than does any other major planet, but its position inside the Earth's orbit means that the planet's dark side then faces us.

Venus stays close to the Sun in the sky, reaching a maximum separation (elongation) of 47°. At maximum evening elongation, the planet appears as a half-illuminated disk. It then rapidly moves sunward, increasing in size with the phase changing into a slender crescent. On rare occasions Venus crosses the solar disk, passing directly between the Earth and Sun, and it is then visible as a slow-moving dark spot. The last such transits occurred in 1874 and 1882, and the next will occur on June 7, 2004 and June 5, 2012.

Markings and rotation. Venus is unrewarding to the telescopic observer because its surface is always obscured by an unbroken layer of opaque yellowish clouds, showing only occasional dusky markings.

When Venus is close to the Sun in the sky at inferior conjunction, it appears as a fine crescent. Frequently the whole outline of the planet is then visible by light scattered in the planet's dense cloudy atmosphere. The first clear glimpse of markings on Venus came in 1928, when Frank E. Ross took photographs of the planet in ultraviolet light. These showed dark patches and streaks near the equatorial and temperate regions, while the poles always appeared bright. The markings changed rapidly from day to day and were clearly cloud formations, as confirmed by photographs from the Mariner 10 space probe in 1974, which showed clouds spiraling outward from the equator to the poles. As first found by Earth-based observers in the 1960s, the clouds of Venus rotate in a retrograde direction (from east to west) around the planet every four days.

The rotation period of the solid body of Venus remained a mystery until the study of radar echoes from the planet in the early 1960s showed that Venus has a very slow east-to-west rotation of 243.0 days. This may have been caused by the impact on Venus of a former moon. The planet has no satellites today.

Surface conditions. In the mid-1950s, radio astronomers detected radiation from Venus, which suggested that the planet's surface was very hot. The high surface temperature has since been confirmed by space probes, which have also measured temperatures throughout the atmosphere. The temperature is about −40°C at the cloud-top level of 55 miles (99 km), but rises sharply deeper into the atmosphere. It reaches 70°C at 30 miles (50 km) in altitude, 200°C at 20 miles (32 km), 350°C at 10 miles (16 km), and a blistering 475°C on the planet's surface. Heat and light absorbed by the planet are re-radiated as infrared radiation, which is trapped by the carbon dioxide in the atmosphere in what is termed the GREENHOUSE EFFECT, and builds up the high observed temperatures. The surface of Venus reflects radio waves much as do the surfaces of the Moon and Mercury, although its higher radio reflectivity of 12 percent suggests that it is smoother than either of these. Computer analysis of complex radar echoes has enabled scientists to map Venus, showing mountainous areas and craters up to several hundred miles across.

The first direct information about the nature of the planet's surface came in October 1975, when the Soviet spacecraft Venus 9 and 10 each made a soft landing and sent back a panoramic photograph. The photographs were surprising, showing clear landscapes with large and small boulders that cast clear shadows. The rocks are only slightly rounded by wind erosion during dust storms, and the landscapes seem to be relatively young. The deep layer of dust and debris, or regolith, that characterizes the ancient surfaces of the Moon and Mercury, is absent on Venus. This suggests that geological forces may still be active on Venus, as on Earth. The considerable amount of light that penetrated the planet's thick atmosphere was also unexpected. Internally, Venus is believed to be much like the Earth, with a liquid core probably smaller than the Earth's, overlain by a mantle and a granite-like crust.

The atmosphere. The dense blanketing atmosphere of Venus keeps the surface temperature constant at about 475°C from the equator to the poles, both midday and midnight. The atmospheric pressure is 91 times that at the Earth's surface. The atmosphere is over 90 percent carbon dioxide, with some hydrogen, oxygen, helium, atomic carbon, carbon monoxide, water, hydrogen chloride, hydrogen fluoride, and sulfuric acid. The clouds on Venus are quite unlike those on Earth, more closely resembling industrial smogs or natural fogs than familiar terrestrial clouds. The main cloud region on Venus occurs at about 45 to 55 miles (70–90 km) in altitude, with several layers of less conspicuous haze extending up to 65 miles (100 km). This is in striking contrast to the Earth, where normal clouds rarely extend to heights above 8 miles (13 km). Initially, the clouds were thought to be water clouds topped with ice crystals, but the water content of the atmosphere is now known to be very slight. It is now thought that the clouds are composed of droplets of sulfuric acid with some hydrochloric acid and hydrofluoric acid, together with other minor constituents forming individual haze layers.

The weather in the bottom few miles of the atmosphere is fairly quiet, with only gentle breezes of a few miles an hour most of the time. At greater altitudes, however, the atmosphere becomes much more turbulent with quite violent winds. The Mariner 10 photographs of the planet show that the atmosphere is heated strongly by the Sun and jet streams blow away toward the poles. Great belts of cloud cover the equatorial and temperate latitudes, while the poles lie beneath permanent smooth cloud caps. Above this activity is a fierce 220 mile (350 km) per hour gale moving westward, which seems to be a permanent feature of the visible cloud layer. This motion causes the four-day cloud rotation period clearly visible in ultraviolet photographs taken from Earth.

The small amount of hydrogen present in the atmosphere is carried to Venus by the solar wind. Unlike the Earth, which has a strong magnetic field and is protected by an extensive magnetosphere, Venus apparently lacks such a protective field. The planet's slow rotation can hardly generate a significant field through dynamo action even though Venus probably has an iron core. The planet's only protection is its ionosphere.

The evolution of Venus and Earth. Although the Earth and Venus must have begun their evolution as almost identical bodies, they have clearly evolved in quite different ways. On the Earth, the temperature was low enough for water eventually to condense to form oceans, dissolving carbon dioxide and enabling it to be incorporated into carbonate rocks. This left an atmosphere of nitrogen, oxygen, and lesser trace gases. On Venus, however, the temperature never permitted water to condense but led to a permanent cloud layer which promoted the greenhouse effect, kept carbon dioxide as an atmospheric constituent, and allowed water molecules to be broken up and lost in the high atmosphere.

Venus has no moons. Scrutiny of spacecraft photographs of the planet that would have revealed any bodies down to a size of a few hundred yards has been fruitless.

The surface of Venus photographed by the Russian craft Venus 9. The light curved area in the foreground and the density meter, **arrowed**, are part of the spacecraft; the vertical stripes result from breaks in the transmission. In this distorted view, the horizon is seen crossing the top right corner, and rocks of all shapes and sizes are evident.

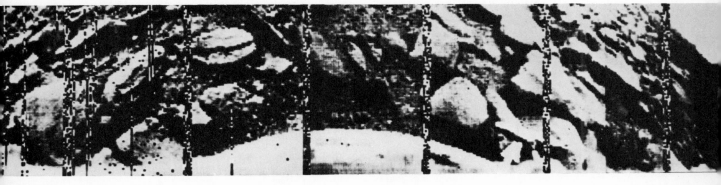

Venus probes

A series of Soviet space probes (sometimes also called Venera) to the planet Venus; for American Venus probes see MARINER. Venus 1 was the world's first planetary probe, though it failed in its mission, as did its two successors. Venus 2 was intended to fly past Venus and scan its clouds; Venus 3 contained a reinforced capsule designed to descend to the surface. Venus 4 began a new series of probes that ejected lander capsules as they sped past the planet. The ball-shaped capsule, 3.3 feet (1 m) in diameter, descended under parachutes, sending back data on temperature, pressure, and atmospheric composition. The first three capsules were all crushed by the enormous atmospheric pressure on Venus before they reached the surface. The improved descent capsule of Venus 7 was the first object to land successfully on the planet. All these capsules had entered the night side of Venus, the side which is turned toward us when Venus is at its closest. The Venus 8 lander touched down on the daylight side of the planet, and showed that conditions there are very similar. The heavier Venus 9 and 10 capsules each returned one photographic panorama of the surface after soft landing, while their parent probes scanned the planet's atmosphere from orbit.

Probe	Launch date	Remarks
Venus 1	February 12, 1961	Contact lost at 4.7 million miles (7.5 million km). Bypassed planet at 60,000 miles (100,000 km)
Venus 2	November 12, 1965	Passed Venus at 15,000 miles (24,000 km) on February 27, 1966, but failed to return data
Venus 3	November 16, 1965	Impacted Venus March 1, 1966, but failed to return data
Venus 4	June 12, 1967	Ejected capsule into Venus atmosphere on October 18, transmitted for 94 minutes during descent
Venus 5	January 5, 1969	Ejected capsule into Venus atmosphere on May 16, transmitted data for 53 minutes during descent
Venus 6	January 10, 1969	Ejected capsule into Venus atmosphere on May 17, data returned for 51 minutes during descent
Venus 7	August 17, 1970	Ejected capsule into Venus atmosphere December 15, which transmitted data from surface for 23 minutes
Venus 8	March 27, 1972	Ejected capsule into Venus atmosphere on July 22, which soft-landed and returned data from the surface for 50 minutes
Venus 9	June 8, 1975	Lander capsule descended on October 22, returning panoramic photograph and other data from surface for 53 minutes. Orbiter section continued in orbit around Venus
Venus 10	June 14, 1975	Lander capsule descended on October 25, returning panoramic photograph and other data for 65 minutes. Orbiter section continued around Venus

vernal equinox

The moment when the Sun, moving north, lies on the celestial equator, it is also termed the spring equinox. The vernal equinox indicates the end of winter in the northern hemisphere, and the end of summer in the southern, being the commencement of the six-month period when the Sun lies north of the celestial equator. The vernal equinox is the zero point for the reckoning of the celestial coordinate called RIGHT ASCENSION, the equivalent of longitude on Earth. The vernal point is moving slowly westward around the sky because of the slow wobbling or PRECESSION of the Earth's axis; when first established by the Greek astronomers it lay in the constellation Aries and was referred to as the *First Point of Aries,* a name it still bears even though precession has now carried it into Pisces.

Very Large Array (VLA)

The world's largest and most sensitive radio telescope, located about 40 miles (64 km) west of Socorro, New Mexico, and operated by the National Radio Astronomy Observatory. Construction began in 1973, and is planned for completion in 1981; limited observations started in 1976. When finished, the VLA will consist of 27 movable antennae, each 82 feet (25 m) in diameter, arranged in a Y-shaped pattern along arms 13 miles (21 km) long. The instrument will work at wavelengths down to 1 centimeter, and will use the principle of APERTURE SYNTHESIS to give the resolution of a single dish 17 miles (27 km) in diameter.

Vesta

The fourth asteroid to be discovered, also known as minor planet 4, found by Wilhelm OLBERS on March 29, 1807. Vesta moves in a slightly elliptical orbit between 2.15 and 2.57 a.u. from the Sun, in the inner regions of the asteroid belt. The orbit has a 3.63-year period and is inclined at only $7°$ to the planetary system. Vesta is roughly spherical and rotates in $10\frac{1}{2}$ hours. It has a diameter of $312\frac{1}{2}$ miles (503 km) and a mass of about 3×10^{17} tons. Vesta's basalt-like surface gives it a relatively high reflectivity of 26.4 percent. This, combined with its position on the inner edge of the asteroid belt, means that Vesta can become brighter than magnitude 6, making it the brightest of the main-belt asteroids, and just visible to the naked eye in clear dark skies.

Viking spacecraft

Two American probes designed to look for life on Mars. Each Viking consisted of two halves: the orbiter craft to survey Mars from orbit, which photographs its surface and returns data on its atmosphere; and the lander, which touches down on Mars to sample its soil and return photographs. After the orbiter's cameras confirm safe landing sites, the lander parachutes through the thin Martian atmosphere, final touchdown being cushioned by retro-rockets. The three-legged lander, weighing 1,320 lb. (599 kg), is 7 feet (2.1 m) tall to the top of its 30-inch (76-cm) dish antenna. This relays data direct to Earth, or more efficiently via the orbiter when above the horizon. Each lander carries two television cameras to take panoramic photographs in color. On-board instruments are designed to analyze the

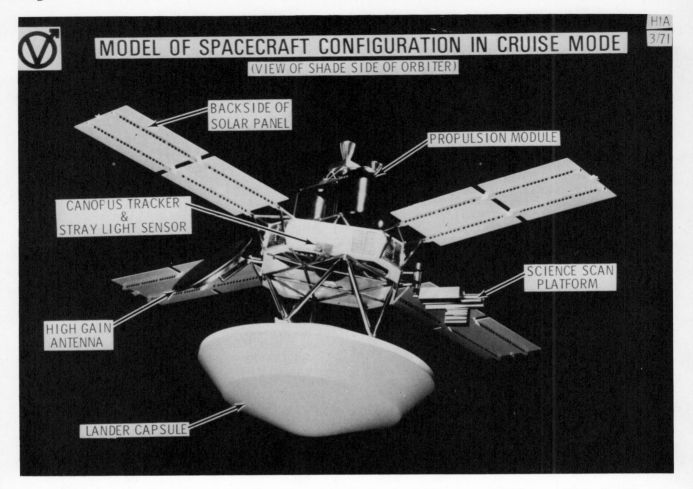

MODEL OF SPACECRAFT CONFIGURATION IN CRUISE MODE
(VIEW OF SHADE SIDE OF ORBITER)

BACKSIDE OF SOLAR PANEL

PROPULSION MODULE

CANOPUS TRACKER & STRAY LIGHT SENSOR

SCIENCE SCAN PLATFORM

HIGH GAIN ANTENNA

LANDER CAPSULE

The Viking spacecraft in its cruise configuration, during transit between Earth and Mars. The upper half acts as an orbiting "mother ship" relaying transmissions to Earth, while the lander housed in the capsule below carries out experiments on the Martian surface.

Martian atmosphere and detect ground tremors. The major part of the lander's program is analysis of soil samples dug up by the craft's 10-foot (3-m) sampler arm. Two experiments examine the overall composition of the Martian soil, while three experiments test for the presence of life. Each lander has a planned working life of three months on the Martian surface.

Probe	Launch date	Remarks
Viking 1	August 20, 1975	Mars encounter scheduled for June 19, 1976
Viking 2	September 9, 1975	Mars encounter scheduled for August 7, 1976

Virgo (the virgin)
A constellation of the equatorial region of the sky and one of the constellations of the zodiac, best seen during the northern hemisphere spring. The Sun passes through Virgo from mid-September to early November; the autumnal equinox lies in Virgo. The constellation's brightest star is SPICA. Gamma Virginis, called Porrima, is a double star, of magnitudes 3.63

and 3.6, orbiting every 171.76 years. Virgo is well known for the thousands of distant galaxies that lie within its boundaries. A nearer galaxy, M104 (NGC 4594), is called the Sombrero galaxy because of its distinctive shape.

Volans (the flying fish)
A constellation of the southern hemisphere of the sky, introduced on the 1603 star map of Johann BAYER. None of its stars is brighter than magnitude 3.5.

von Braun, Wernher (b. 1912)
German-born American rocket engineer, designer of the world's largest rocket, the Saturn V Moon launcher. While still in his teens von Braun assisted the German rocket pioneer Hermann OBERTH with engineering experiments and test flights in the German Society for Space Travel. Other experimenters were Klaus Riedel (1910–1944), Rudolf Nebel (b. 1897), and Willy Ley (1906–1969). The German army soon took an interest, and von Braun's Ph.D. thesis on rocket engine development was completed under the army's auspices. This work laid the foundation for the V-2— and, ultimately, the space age itself.
In 1937 the German army moved its rocket development center from Kummersdorf, near Berlin, to Peenemünde on the Baltic coast. There, under the direction of the army general Walter Robert Dornberger (b. 1895), the world's first missiles for long-range bombardment took shape, as well as

radically new shorter-range missiles and rocket-propelled aircraft.

Foremost of von Braun's developments was the V-2 ROCKET, the world's first successful large missile. The V-2 was first successfully fired on October 3, 1942, reaching a maximum altitude of 53 miles (85 km); by the end of the war over 5,000 V-2s had been built. In 1944 von Braun was arrested by the Gestapo and charged with being more interested in space flight than in military rockets. In part, this was true, for among designs by the von Braun team at Peenemünde were three-stage rockets for putting men and spacecraft into orbit. However, Dornberger pointed out that without von Braun there would be no V-2, and he was released.

At the end of the war von Braun and most of his Peenemünde colleagues fled south to surrender to the American army, which allowed them to continue V-2 development at White Sands Proving Ground in New Mexico. In 1950 von Braun was moved to missile development in Huntsville, Alabama, where he designed the REDSTONE ROCKET, which was eventually used for the first U.S. manned suborbital launches, and which provided the basis for the Jupiter and Juno rockets that launched early American satellites.

After the setting up of NASA, von Braun in 1960 was placed in charge of the George C. Marshall Space Flight Center at Huntsville. Von Braun had already begun to design powerful rockets made up from clusters of existing engines, and this approach led to the family of SATURN ROCKETS used in the Apollo program to put men on the Moon. With the run-down of the American space program, von Braun retired from NASA in 1972 to enter private industry.

von Kármán, Theodore (1881–1963)
Hungarian-born American pioneer of aeronautics and astronautics. During World War II, von Kármán directed jet propulsion research of the Guggenheim Aeronautical Laboratory at the California Institute of Technology, which in 1944 became the JET PROPULSION LABORATORY. Von Kármán developed solid-fuel rockets for assisting aircraft takeoff, and in 1941 was cofounder of Aerojet General, the first American rocket-engine manufacturer. He also developed hypergolic (spontaneous combustion) propellants, as later used in the Apollo command module and lunar module engines.

Voskhod
A modified Soviet VOSTOK spacecraft for carrying two or three cosmonauts. Voskhod contained permanent couches in place of the Vostok ejector seat; escape in an emergency was therefore impossible. The size of the Voskhod capsule was the same as that of Vostok. Voskhod 1 carried the first three-man crew into space; because of weight problems and the lack of space, the cosmonauts wore no space suits. On Voskhod 2 the third couch was removed and replaced by a concertina-like airlock for a space walk; both cosmonauts wore space suits on this mission. When extended, the airlock measured 6 feet by 3 feet (1.8 × 0.9 m) and was jettisoned before reentry. Voskhod had small braking rockets added to cushion touchdown; the cosmonauts remained inside the craft as it landed.

The Voskhod rocket was the VOSTOK LAUNCHER with a more powerful upper stage added. Because of this extra power, Voskhod entered a higher orbit than Vostok, and thus could not rely on atmospheric drag to pull it down if the retro-rockets failed; therefore a reserve retro-rocket was added at the front of the spherical crew compartment.

Mission	Launch date	Results
Voskhod 1	October 12, 1964	Cosmonauts Vladimir Komarov, Konstantin Feoktistov, and Boris Yegorov made day-long, 16-orbit flight in 11,728-lb. (5,320-kg) first multi-man craft
Voskhod 2	March 18, 1965	Alexei Leonov made first space walk from craft piloted by Pavel Belyaev during day-long mission. Manual entry on 18th orbit after automatic control system failed on previous orbit brought 12,527-lb. (5,682-kg) craft down over 1,000 miles (1,600 km) off course

Vostok
The first Soviet manned spacecraft, designed to carry a cosmonaut in orbit for up to 10 days. Vostok was basically a hermetically sealed sphere, 7½ feet (2.3 m) in diameter weighing about 5,290 lb. (2,400 kg). It contained normal air at atmospheric pressure. The sphere was coated with a honeycomb ablative heat shield, surfaced with metal foil strips to reflect sunlight in orbit. The capsule had no steering rockets, but relied on an offset center of gravity to align itself automatically as it encountered air resistance during reentry. Vostok went into orbit attached to the cylindrical last stage of its carrier rocket; during launch it was covered by a conical fairing. Joining the rocket stage to the capsule was a cylindrical instrument section, which contained air bottles, communications equipment, and a retro-rocket; this brought the craft's total weight up to about 10,400 lb. (4,717 kg). Before reentry, the retro-rocket was fired and jettisoned. However, Vostok was put into a sufficiently low orbit that atmospheric drag would

Mission	Launch date	Results
Vostok 1	April 12, 1961	First manned spaceflight; Yuri Gagarin made one orbit of Earth
Vostok 2	August 6, 1961	Gherman Titov made day-long flight
Vostok 3	August 11, 1962	Andrian Nikolayev made 64 orbits, landing on August 15
Vostok 4	August 12, 1962	Pavel Popovich made 48 orbits simultaneous with Vostok 3, landing on August 15
Vostok 5	June 14, 1963	Valery Bykovsky made 81 orbits, longest-ever individual flight, landing on June 19
Vostok 6	June 16, 1963	Valentina Tereshkova became first spacewoman, making 48 orbits simultaneous with Vostok 5, landing June 19

pull the capsule back to Earth after 10 days even if the retro-rocket failed. Inside the capsule, the cosmonaut sat in an ejector seat, with a porthole in front of him. After reentry, the hatch cover was blown off and the cosmonaut ejected to parachute to Earth independently. He could also eject in case of a launch emergency. On-board control was minimal and most functions occurred automatically. Despite its greater size and weight as compared to the American MERCURY, the Vostok was little more than an automatic satellite with a passenger. Vostok capsules were tested with animal passengers as part of the SPUTNIK series. Unmanned Vostok-type capsules have continued to be used for biological payloads, and for reconnaissance satellites in the COSMOS series. Vostok was launched by a modification of the rockets that launched Sputnik, called the VOSTOK LAUNCHER.

Vostok launcher

Soviet space rocket, developed from the first Soviet intercontinental military missile. This rocket was first shown to the West in 1967 in the version that launched the VOSTOK manned spacecraft. However, the same first stage was used to launch the first Sputniks, and with different upper stages it also launched the first Soviet lunar and planetary probes, and the VOSKHOD and SOYUZ manned craft. In its basic configuration, as used for Sputnik, the launcher is known in the West as the A type. It contains a central core, 92 feet (28 m) long and 9.7 feet (2.95 m) in diameter, to which are attached four tapered strap-on boosters 62 feet (19 m) long. Maximum diameter across the base is 33.8 feet (10.3 m). All engines ignite at launch, and the strap-on boosters fall away at altitude while the central core (which the Russians term the second stage) continues burning. The central core and four strap-ons each contain similar engines, producing a total first-stage thrust of 1,124,350 lb. (510,000 kg). Vostok launches were made with the addition of the so-called Luna top stage, introduced for the first three Luna flights, which gave a thrust of around 200,000 lb. (90,000 kg); this version of the launcher is termed the A-1 type. With a longer and more powerful top stage—the Venus stage, first used for launching Venus probes—the rocket became the A-2 type, as used for VOSKHOD and SOYUZ manned craft. This top stage, 25 feet (7.5 m) long, produces about 270,000 lb. (122,000 kg) of thrust. For Vostok launches, the overall height of the rocket was 125 feet (38 m); for Soyuz launches (including the escape tower) the height is 162 feet (49.3 m).

Vulpecula (the fox)

A faint constellation of the northern hemisphere of the sky next to Cygnus, best seen during the northern summer. It was introduced by Johannes HEVELIUS. Its brightest star, Alpha, is only of magnitude 4.63. However, the constellation contains the famous planetary nebula M27 (NGC 6853) known as the Dumbbell, 1,250 light-years distant. The first PULSAR, CP 1919, was discovered in Vulpecula.

W

Wallops Flight Center

A NASA installation at Wallops Island, Virginia, for aeronautical and astronautical research. It was founded in 1945 as a field station of the LANGLEY RESEARCH CENTER, but is now a separate facility. The center's headquarters and experimental research airport are located on the mainland, along with tracking instruments and the Range Control Center. On the island itself are rocket assembly buildings and launchpads. Spacecraft components have been tested on short flights from Wallops Island, but the center's main responsibility is for atmospheric and space environment research, as well as ecological sensing. Rockets fired from Wallops Island range from small sounding rockets to the SCOUT satellite launcher. The first satellite to be launched was Explorer 9, on February 16, 1961. Scout rockets have since launched several more small scientific satellites from the island.

wavelength

A term applied to the measurement of wave motions, particularly of ELECTROMAGNETIC RADIATION (such as light waves or radio waves). If the waves are visualized like sea waves, the wavelength is the distance from crest to crest. Optical wavelengths are often measured in ANGSTROMS (10^{-10} m); the wavelength of yellow light is about 5,000 angstroms. Wavelength is the wave motion's velocity divided by its FREQUENCY.

weather satellites

Earth satellites which monitor the Earth's atmosphere and surface, assisting meteorologists in their understanding and prediction of weather patterns. Weather satellites send back pictures of cloud, snow, and ice cover. They also have sensors which record the temperature of the atmosphere and oceans, thus measuring the overall heat balance of the Earth that determines weather systems. Weather satellites give advance warning of impending natural disasters, such as hurricanes forming at sea and floods from the melting of heavy snows, producing savings in lives and property that are said to have paid back the cost of the entire space program. The first weather satellites were the TIROS series; Tiros 1 was launched on April 1, 1960. The Tiros series was superseded by the ESSA satellites of the Environmental Science Services Administration, while new equipment was tested in the NIMBUS series. An advanced series, initially called the Improved Tiros Operational System (ITOS) was introduced early in 1970; the series designation was changed to NOAA on the formation of the National Oceanic and Atmospheric Administration in late 1970. NOAA 1 was launched on December 11, 1970; NOAA 2 in October 1972; NOAA 3 in November 1973; and NOAA 4 in November 1974. These satellites move in near-polar orbits for worldwide coverage, orbiting every 115 minutes at an altitude of around 900 miles (1,500 km).

In May 1974 and February 1975 a series of synchronous meteorological satellites (SMS) went into service in stationary orbit above the Earth's equator. These scan both the Earth's cloud cover and the Sun's activity, and relay data from remote automatic weather stations. From their high orbit, the satellites can cover a large area of the Earth in each picture, which they transmit at 30-minute intervals. The change of the system's name to GOES (Geostationary Operational Environmental Satellite) with a new launch in October 1975 indicates that the system became operational. These satellites provide the American contribution to the International Global Atmospheric Research Program (GARP). The European Space Agency, Japan, and the Soviet Union are also cooperating in this program.

The Soviet Union has its own weather satellite system, called Meteor. Meteor satellites are cylindrical in shape, 16 feet (5 m) long by 5 feet (1.5 m) in diameter, with wing-like solar panels. They move in similar orbits to the NOAA satellites, and return equivalent data. By international agreement, all nations exchange weather satellite results. Eventually, the Meteor system will be supplemented by low-altitude observations from Salyut space stations, and high-altitude monitoring from synchronous weather satellites. The Meteor system officially began on March 26, 1969, with the launch of Meteor 1, although there has been a series of experimental weather satellites launched previously in the Cosmos series. Several Meteor satellites are launched each year.

Weizsäcker, Carl Friedrich von (b. 1912)
German astrophysicist, who in 1945 proposed a theory of the origin of the solar system which suggested that the planets formed by the aggregation of dust particles from a disk orbiting the primeval Sun. Weizsäcker's view was derived ultimately from the ideas of Immanuel KANT and the marquis de LAPLACE, but provided a plausible mechanism for the formation of planets, which the Kant-Laplace theory lacked. It was the first of the modern theories of the solar

system's origin, from which the currently accepted views were developed (see PLANETS). In 1938 Weizsäcker proposed the carbon-nitrogen-oxygen cycle as the origin of stellar energy, independently of Hans BETHE.

Westar satellites
American domestic communications satellites, similar to the Canadian ANIK series, owned and operated by Western Union. The drum-shaped satellites, 6.3 feet (1.9 m) in diameter and 11.8 feet (3.6 m) tall to the top of their 5-foot (1.5-m) mesh antenna, weigh 1,265 lb. (573 kg) and can carry 7,200 two-way telephone circuits or 12 color television channels. The Westar satellites are placed in geostationary orbit, and are the first U.S. domestic communications satellites. They were launched in 1974. The system is planned to be extended as traffic grows.

Westerbork Radio Observatory
A radio astronomy observatory near Groningen, the Netherlands, operated by the Netherlands Foundation for Radio Astronomy, which also operates the 82-foot (25-m) dish at Dwingeloo. Westerbork contains a 1-mile (1.6-km) APERTURE SYNTHESIS telescope consisting of a line of ten fixed and two movable dishes of 82 foot diameter. The telescope, which began observations in 1970, works at wavelengths of 49, 21, and 6 centimeters. It is one of the world's most sensitive radio telescopes. Two more dishes are planned to be added about 1978 to extend the telescope's total aperture to 2 miles (3 km).

Nine of the twelve dishes of the Westerbork Synthesis Radio Telescope; the other three continue the line to the right. The two most distant dishes can be moved on a 300-yard-long track. Each dish is surfaced with fine wire mesh, and is counterbalanced by the large cylindrical weight.

Western Test Range
A missile firing range based at Vandenberg Air Force Base, Lompoc, California, previously known as the Pacific Missile Range. The range stretches southward for more than 5,000 miles (8,000 km) into the Pacific, and is used for test firing Minuteman and Titan II ballistic missiles. It is also used to launch sounding rockets, and to put satellites into polar orbit. The first satellite launch from the Western Test Range was Discoverer 1, on February 28, 1959. Facilities are now being installed for polar launches and landings of the SPACE SHUTTLE.

Whipple, Fred Lawrence (b. 1906)
American astronomer, who in 1949 suggested that a comet's nucleus can be compared to a dirty snowball, in which dusty and rocky materials are cemented into a ball a few miles across by frozen gases such as methane and ammonia; this is now the accepted view. Whipple discovered a total of six comets, the last in 1942. He devised a system of special cameras with rotating shutters to photograph meteors, and found from their orbits that they were debris from comets. From 1955 to 1973 Whipple was director of the Smithsonian Astrophysical Observatory, where he organized the optical tracking system for Earth satellites, setting up the worldwide system of Baker-Nunn cameras and the amateur Moonwatch teams.

White, Edward Higgins II (1930–1967)
First American to walk in space. White made a 21-minute spacewalk, maneuvering at the end of a tether with a hand-held gas gun, on June 3, 1965, during the flight of Gemini 4, commanded by James McDivitt. White was chosen as a crew member of the first manned Apollo flight; but all three crewmen were killed by a fire in the spacecraft during a simulated countdown.

white dwarf
A star whose diameter is only 1 percent that of the Sun, and whose luminosity is 10,000 times less. Although called "white" dwarfs, they can in fact be any color, depending on their surface temperature; the hotter white stars (temperature around 10,000°C) were generally the first to be observed. Best known is the companion to SIRIUS (Sirius B).

They represent the final stage of stellar evolution—dying stars, taking billions of years to cool to black globes. Their light originates in a thin hydrogen atmosphere made to shine by heat leaking away from the interior, and unlike other stars, they have no nuclear energy source. Consequently, there has been nothing to prevent them from collapsing until they are of planetary size. Their matter is in what is known as a degenerate state, in which atoms are stripped of their electrons and the electrons packed tightly together. White dwarfs are degenerate in their central regions, and become increasingly degenerate as they cool. Their density, as a result, is very high, about 1 million times that of water.

The gravity at the surface of a white dwarf star is tens of thousands of times greater than the Earth's, a consequence of its small size and relatively high mass. Even light loses energy and suffers a RED SHIFT when leaving these stars (see RELATIVITY). There is a limiting mass, called the Chandrasekhar limit (about 1.5 times that of the Sun), above which a white dwarf cannot exist. It is still not fully known how heavier stars lose enough mass to enable them to become white dwarfs.

White Sands Proving Ground
A U.S. Army rocket testing range at White Sands, New Mexico, used for launches of sounding rockets and missiles. White Sands was activated in 1945 as a testing ground for the American Wac Corporal sounding rocket; later, it was used for launching captured V-2 rockets by Wernher VON BRAUN, who was then at Fort Bliss, Texas. In 1950 the von Braun group moved to what is now the MARSHALL SPACE FLIGHT CENTER at Huntsville and their test launches switched to CAPE CANAVERAL, although White Sands remained in use for launches of smaller missiles and sounding rockets.

Wolf, Rudolf (1816–1893)
Swiss astronomer who confirmed the discovery by Heinrich Schwabe of the sunspot cycle, and more correctly determined its period as an average of about 11 years. Wolf showed that the sunspot cycle was correlated with aurorae and disturbances in the Earth's magnetic field, thus initiating the study of solar-terrestrial effects. Around 1850 he devised the system of using sunspot counts as a measure of solar activity; Wolf's system of sunspot numbers is continued to this day.

Wolf-Rayet stars
A rare group of small, hot stars surrounded by luminous clouds of ejected material. The first three such stars were discovered in 1867 by the French astronomers Charles Joseph Etienne Wolf (1827–1918) and Georges Antoine Pons Rayet (1839–1906). The spectra of Wolf-Rayet stars show broad emission lines caused by atoms of helium, carbon, and nitrogen in a hot atmosphere surrounding the star. A typical Wolf-Rayet star has a temperature in excess of 30,000°K and emits an abundance of ultraviolet energy which ionizes the carbon and nitrogen in its atmosphere. The star's energy also makes its atmosphere turbulent, and it streams off the star and into space at thousands of miles per second, thereby blurring the emission into broad bands. Wolf-Rayet stars may be the central cores of massive red giants that have been exposed to view by the stripping off of their outer layers. About 200 Wolf-Rayet stars are known. They are also called WC and WN stars, depending on whether they show carbon or nitrogen in their spectra.

Woolley, Sir Richard van der Riet (b. 1906)
Eleventh English astronomer royal (1956–1971). Woolley made fundamental analyses of the movements of stars around our Galaxy from their observed radial velocities and proper motions; this clarified knowledge of the Galaxy's structure. His studies helped distinguish several groups of stars which formed at different times during the Galaxy's evolution. These investigations led to the modern view that our Galaxy originated from a large collapsing cloud, forming first a widespread "halo" of old stars and finally flattening out to produce a

disk of young stars, including the spiral arms. In 1972 Woolley became director of the SOUTH AFRICAN ASTRONOMICAL OBSERVATORY.

Woomera

A rocket range in South Australia, 280 miles (450 km) northwest of Adelaide, set up in 1947 by Britain and Australia for the development of missiles such as Blue Streak. Woomera was also used to launch upper-atmosphere sounding rockets, and later became the launch site for the ELDO Europa 1 launcher, based on Blue Streak, which unsuccessfully attempted to orbit a European satellite. Two satellites have been orbited from Woomera: WRESAT 1, of Australia's Weapons Research Establishment, launched by a modified REDSTONE rocket on November 29, 1967, to study the upper atmosphere and solar radiation; and the British technology satellite Prospero on October 28, 1971, by a British Black Arrow rocket.

X

X-ray astronomy

The study of the Universe at X-ray wavelengths, comprising ELECTROMAGNETIC RADIATION between 0.1 and 300 Å in wavelength. It began on June 18, 1962, when an Aerobee sounding rocket was launched to study the X rays that were expected to arise from the interaction of solar radiation with the Moon's surface. No such X rays were found, but to the surprise of all astronomers a powerful source emitting X rays of a few angstroms in wavelength was detected in the direction of the constellation Scorpius. There was also evidence for a background of diffuse X-radiation coming from all parts of the sky, which has not yet been completely explained.

Throughout the 1960s other X-ray sources were discovered by rocket and balloon flights, but the real growth of this branch of astronomy occurred only in the early 1970s, with the launching of a succession of satellites carrying X-ray telescopes and detectors: Uhuru, Copernicus, ANS-1 and Ariel V. The Sun has also been extensively studied by rockets, and by the Orbiting Solar Observatory satellites (OSO). Solar X rays arise primarily in the very hot gas of the CORONA over active sunspot regions, but very small bright patches are also observed near solar flares.

X-ray surveys. Uhuru, launched in December 1970, provided the first X-ray sky surveys. These allowed many new sources to be identified, and showed that some are not long-lived. Before it ceased to function effectively, Uhuru completed three surveys, the most complete being the 3U catalog of 160 sources.

The British Ariel V satellite, launched four years after Uhuru, has shown that there are many extremely short-lived X-ray sources occurring all the time, flaring up briefly and dying away after a few days or weeks. These "X-ray novae" occur at the rate of at least one a month in our Galaxy.

Scorpius X-1. The first X-ray source discovered, named Sco X-1, was the most intensively studied and best-known source for a decade, because it was the brightest source known, and because it was also identified at optical and radio wavelengths. The optical counterpart emits strongly in ultraviolet wavelengths, shows "flickering" changes in brightness and occasional large flares, and its spectrum has many emission lines. These characteristics are in some ways similar to those of old NOVAE, and since these occur in binary-star systems it was suggested that Sco X-1 is also a double star.

The energy generated in Sco X-1 as X rays has been roughly constant for more than 10 years, and is estimated at 100,000 times the *total* luminosity of the Sun. Such a massive amount of energy can be liberated in a binary system when matter from the extended atmosphere of a giant star flows onto a small companion such as a NEUTRON STAR. According to this model, as material falls onto the companion, its rotation forms it into a disk (the *accretion disk*). The gas in the disk spirals inward and heats up to tens of millions of degrees; X rays are naturally emitted from gas at this temperature by the BREMSSTRAHLUNG process.

Such a model suggests that we should see periodic variations in the intensity of the radiation as the two stars orbit one another. Because there is very little evidence for such regular variations in the apparently random fluctuations of Sco X-1 at any wavelength, it may be that we are viewing the system from above the plane of its orbit, so that the two stars never eclipse one another as seen from Earth.

X-ray binaries. Two-thirds of the identified sources in the 3U survey are stars in our Galaxy, and in many cases their optical spectra show them to be binaries. Optically, we usually see the giant star of the pair, while the X rays comes from the compact companion, too faint to be seen in visible light. (Sco X-1 is one of the rare exceptions where the compact star outshines the giant.)

The X-ray intensities from the best-studied sources vary periodically, and the periods fall into two main groups: hours to days, and seconds to minutes. The longer periods represent the time taken by the compact star to orbit its companion (like the terrestrial year); while the short periods represent the time taken for the compact object to rotate on its axis (analogous to the terrestrial day).

The rotational periods are slightly longer than those of the radio PULSARS, which are probably younger neutron stars.

An X-ray binary evolves from a normal close DOUBLE STAR by what is known as mass exchange. The originally more massive star (the primary) evolves more rapidly at first, and expands into a giant star. The gravity of its lighter and smaller companion (the secondary) pulls matter out of the primary, and the secondary eventually grows to become the more massive of the pair. The original primary then explodes as a SUPERNOVA, whose compact remnant, a neutron star, is left orbiting the original secondary. When matter from this star flows back onto the neutron star, X rays may be produced. At this stage, theory predicts that the visible star must be a blue supergiant, and this is confirmed by observation in most cases. Emission of X rays should thus be a short-lived phase in the evolution of many binaries.

Transient sources. Of the transient sources detected by X-ray satellites, only one has so far been identified optically. This source was discovered by Ariel V in August 1975, in the constellation Monoceros, and it

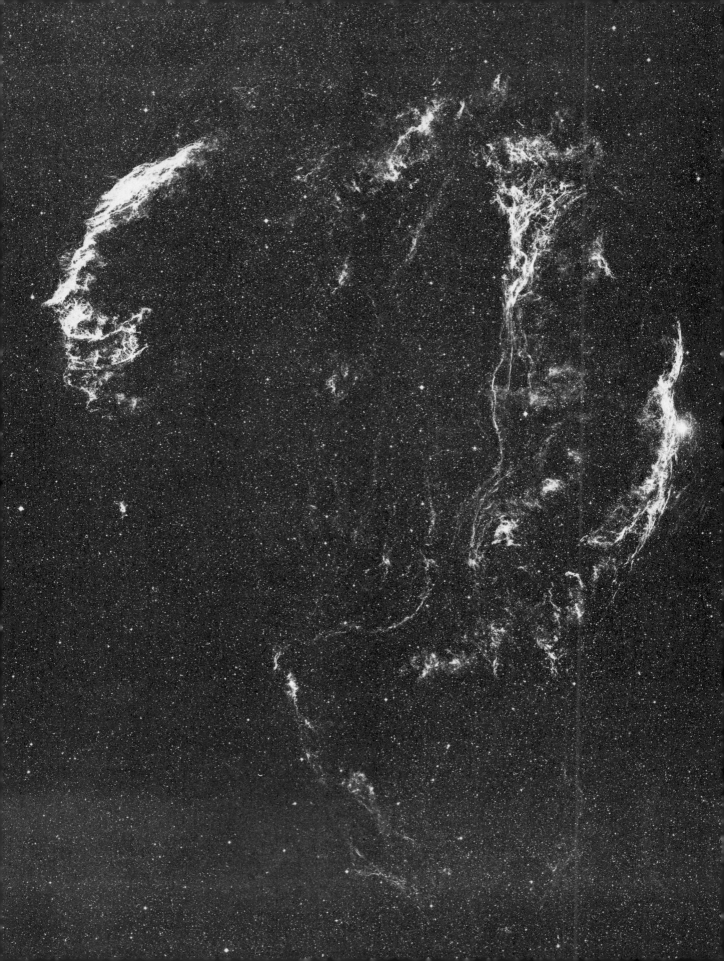

quickly became an X-ray source more intense than even Sco X-1. Optical astronomers identified it with a star of apparent magnitude 11, which had brightened from magnitude 20. Old photographs showed a previous brightening in 1917, and the star may thus belong to the class of VARIABLE STARS known as recurrent novae, which are in fact binary stars. Since all the X-ray binaries vary erratically to some extent, the transient sources may be no more than extreme examples in which mass transfer occurs spasmodically.

Black holes? Much attention has recently been focused on X-ray astronomy by the claim that it has revealed a BLACK HOLE in space, an object so dense that even light cannot escape from its gravity. Observations of binary stars can provide an indication of the masses of the two components, and in the case of the star seen at the position of the source Cygnus X-1, this technique suggests that the mass of the compact star is at least four times that of the Sun. Although an ordinary star can maintain its structure through the pressure produced by its nuclear reactions, theory predicts that any compact object more than three times the Sun's mass must collapse through its own gravity into a black hole. The observations of Cygnus X-1 are widely taken as showing a black hole orbiting a normal star, with X rays being emitted by gas falling toward the black hole, but not yet inside it. There is, however, an alternative explanation. Cygnus X-1 could be a triple system, in which the bright, optically visible star is orbited by a dim binary pair. The latter could be a conventional X-ray source, with the two individual stars each less massive than three Suns.

Other X-ray sources. The 3U catalog also contains supernova remnants, the expanding shells of gas from exploded stars. These move outward so fast that they form a shock wave in the interstellar gas, heating it to enormous temperatures and causing the bremsstrahlung emission of X rays. In a category of its own is the CRAB NEBULA, formed in a supernova observed to explode in the year 1054. This remnant contains a powerful radio and optical pulsar, and X rays are generated both in the pulsar and in the nebula by the interaction of very fast electrons from the pulsar with a magnetic field (SYNCHROTRON RADIATION).

A few extragalactic sources of X rays are known. Some powerful binary-type sources have been detected in the MAGELLANIC CLOUDS, the nearest external GALAXIES to our own Galaxy, the Milky Way. Much stronger X-radiation, however, is produced in the centers of galaxies with disturbed, energetic nuclei, such as SEYFERT GALAXIES, RADIO GALAXIES, and QUASARS. Extremely powerful extended regions of X-ray emission occur between galaxies at the centers of giant clusters of galaxies. This emission comes from gas heated to tens of millions of degrees, either because it has fallen into the strong gravitational field of the cluster, or because it is "stirred up" by the galaxies moving through it.

The Cygnus Loop, or Veil Nebula, photographed here in the red light of hydrogen, is a relatively powerful emitter of X rays. X-ray telescopes show that the radiation comes from the same circular arc as the visible light, a shell of gas ejected by a stellar explosion (supernova) 50,000 years ago.

Y

year
The time taken for the Earth to make one orbit of the Sun. The year on which we base our calendar lasts 365.2422 days; it is termed the *tropical* year. It is shorter than the orbit of the Earth with respect to the star background (the *sidereal year,* lasting 365.2564 days) because of PRECESSION, which effectively brings the Sun back to its starting point slightly earlier each year. A third type of year is the interval between two successive perihelion passages of the Earth. This, the so-called *anomalistic year,* lasts 365.2596 days, and arises because gravitational perturbations by the planets advance the perihelion of the Earth's orbit.

Yegorov, Boris Borisovich (b. 1937)
Soviet physician who became the first doctor in space when he flew on the Voskhod 1 mission in October 1964. Yegorov, an expert on the balancing mechanism of the inner ear, joined the team of Soviet space doctors in 1961, and joined the cosmonaut corps in the summer of 1964 purely for the Voskhod 1 flight.

Yeliseyev, Alexei Stanislovovich (b. 1934)
Soviet cosmonaut who flew on the Soyuz 5 flight in January 1969, which made the first docking between two manned spacecraft; Yeliseyev then walked in space to Soyuz 4. In October 1969 he flew with Vladimir SHATALOV on the Soyuz 8 mission, which performed joint maneuvers in space with Soyuz 6 and 7. On April 23, 1971, he made his third flight, on the Soyuz 10 mission, which docked briefly with the Salyut 1 space station. Yeliseyev, one of the Soviet Union's most experienced cosmonauts, was flight director for the Apollo-Soyuz mission.

Yerkes Observatory
The astronomical observatory of the University of Chicago, situated at Williams Bay, Wisconsin, at an altitude of 1,096 feet (334 m) on the shores of Lake Geneva. Yerkes Observatory was founded by George Ellery HALE with a grant from the businessman Charles Tyson Yerkes (1837–1905). It opened in 1897 with the famous 40-inch (102-cm) refractor, which remains the world's largest refracting telescope. In 1901 a 24-inch (61-cm) reflector was added, made by G. W. RITCHEY; in 1968 this was removed to make way for a new 41-inch (104-cm) reflector. The University of Chicago also shares in the operations of the Mcdonald observatory.

Young, John Watts (b. 1930)
Commander of the Apollo 16 mission, which landed on April 21, 1972, in the Descartes highland region of the Moon. Young and fellow astronaut Charles M. Duke (b. 1935) made three lunar-surface excursions, totaling 20 hours 15 minutes, during a stay on the Moon of 71 hours 2 minutes. Young, a former Navy test pilot, made his space debut on March 23, 1965, in Gemini 3, the first American two-man mission. He commanded the Gemini 10 flight in July 1966. Young was also command-module pilot of Apollo 10 in May 1969, the dress rehearsal for the first lunar landing mission.

Z

Zeeman effect

The splitting of SPECTRAL LINES by a magnetic field, named for the Dutch physicist Pieter Zeeman (1865–1943), who discovered the effect in 1896. The amount of splitting varies with the strength of the field. The lines split, usually into three, because the magnetic field changes the energy levels of electrons in an atom (see SPECTROSCOPY). The outer two lines are polarized in a different direction from the inner one (see POLARIZATION); this distinguishes the Zeeman effect from other line-splitting or broadening effects. A device called a magnetograph, developed by H. D. and H. W. BABCOCK at Mount Wilson in the early 1950s, compares the intensity and polarization on alternate sides of suitable lines in the Sun's spectrum. This produces an accurate value of the magnetic field at a single point on the solar surface. To build up a picture of the entire solar disk, the magnetoheliograph is used. This employs the same principle as the spectroheliograph (see SPECTROSCOPE). Magnetic fields have also been detected on some stars by noting the Zeeman effect (see MAGNETIC STARS).

Zelenchukskaya Astrophysical Observatory

An astronomical observatory at an altitude of 6,800 feet (2,070 m) in the Caucasus Mountains of Russia, containing the world's largest optical telescope, the 236-inch (600-cm) reflector, due to begin operation about 1976. At the same site is the RATAN 600 (Radio Astronomical Telescope of the Academy of Sciences), consisting of 900 aluminum reflectors mounted in a circle 1,890 feet (576 m) in diameter, capable of working at wavelengths from 0.4 to 21 centimeters. Each of the individual panels can be pointed up and down, and the instrument can work as a unit or as four separate sections.

zenith

The point in the sky directly above an observer and opposite the NADIR. *Zenith distance* is the angle between an object and the zenith.

zenithal hourly rate

The rate of a meteor shower that would be recorded by an experienced observer with a clear dark sky and the meteor radiant in the zenith (abbreviation ZHR). The altitude of the radiant in the sky has a major effect on the observed meteor rate. To derive the ZHR multiply by 1.1, 1.3, 1.5, 1.7, and 2.0 for radiant altitudes of 66°, 52°, 43°, 35°, and 27° respectively.

zodiac

A belt of constellations about 9° wide over which the Sun, Moon, and planets (except Pluto) appear to move. The constellations are Aries (The Ram) ♈, Taurus (The Bull) ♉, Gemini (The Twins) ♊, Cancer (The Crab) ♋, Leo (The Lion) ♌, Virgo (The Virgin) ♍, Libra (The Balance) ♎, Scorpius (The Scorpion) ♏, Sagittarius (The Archer) ♐, Capricornus (The Goat) ♑, Aquarius (The Water Bearer) ♒, and Pisces (The Fish) ♓. Of the twelve constellations, seven were said to represent animals, and the Greeks called the zone "zodiakos kyklos" or circle of animals; from this the word zodiac was derived. The zodiacal signs given above appear in manuscrips of the late Middle Ages, and their origin is obscure.

The twelve signs of the zodiac are taken by astrologers to cover a belt of the sky 30° wide, and the Sun is supposed to pass through them at specific, equally spaced dates. However, the constellations do not, in fact, occupy equal spaces, and PRECESSION (a movement of the Earth's axis) has now altered the dates at which the Sun is in a particular constellation. The Sun also now passes through Ophiuchus (The Serpent Bearer), which is not a zodiacal constellation at all.

zodiacal light

A diffuse pointed cone of faint light rising obliquely into the sky above either the last trace of twilight in the evening sky or the first trace of twilight in the morning sky. The zodiacal light lies along the ecliptic or zodiac, and is caused by sunlight reflected from a disk-shaped cloud of dust particles in the solar system. The cloud has a total mass of about 30 million million tons and contains particles down to about 10^{-12} gram in mass. These particles can enter the Earth's atmosphere to produce MICROMETEORITES, eventually falling to ground level. The solar wind sweeps about 100,000 tons of fine dust a year out of the solar system; larger particles spiral in toward the Sun (see POYNTING-ROBERTSON EFFECT). However, fresh dust is continually scattered by the periodic comets. The zodiacal light is not difficult to observe, but it does require a dark, haze-free sky. For observers in northern temperate latitudes, it is best seen after sunset in spring or before sunrise in autumn.

Zond spacecraft

A series of Soviet space probes that act as tests for future missions. Zonds 1 and 2 were launched toward Venus and Mars, respectively, after communications failures with the first Soviet Venus and Mars probes; however, radio contact was also lost with the Zond probes. Zond 3 took lunar farside pictures which it transmitted as it receded into space, to test picture transmission systems for future planetary probes. Subsequent Zonds were unmanned test flights of Soyuz craft (minus the spherical orbital module), in preparation for an intended manned circumlunar mission. However, this was postponed indefinitely after the manned circumlunar flight by Apollo 8 in December 1968.

Zond spacecraft

Probe	Launch date	Remarks
Zond 1	April 2, 1964	Launched toward Venus; communications failed
Zond 2	November 30, 1964	Launched toward Mars; communications failed
Zond 3	July 18, 1965	Flew behind Moon, photographing area not covered by Luna 3. Headed toward Mars, retransmitting lunar photographs in communications test
Zond 4	March 2, 1968	Unsuccessful test of circumlunar Soyuz

Probe	Launch date	Remarks
Zond 5	September 15, 1968	First flight to vicinity of Moon and back. Carried biological specimens to assess radiation hazard and tape recording to test voice transmission between capsule and Earth. Splashed down in Indian Ocean September 21
Zond 6	November 10, 1968	Photographed lunar farside. Skip-glide reentry using aerodynamic lift; landed in Soviet Union November 17
Zond 7	August 8, 1969	Repeat of Zond 6 mission. Landed August 14
Zond 8	October 20, 1970	Similar to previous Zonds, with modified reentry trajectory; splashed down in Indian Ocean October 27

Zwicky, Fritz (1898–1974)

Swiss astrophysicist, who in 1934 first clearly distinguished with Walter BAADE that supernovae are different in nature from the less bright novae, and proposed that a star is largely destroyed in a supernova explosion but may leave its compressed core as a NEUTRON STAR. Zwicky began an international program to search for supernovae in distant galaxies, personally finding many new examples that helped clarify the nature of these enormous outbursts. His observational studies led to the discovery of many galaxies and clusters of galaxies, which he studied in detail. Zwicky discovered that some nearby galaxies seem to be linked by bridges of stars, as if the galaxies were interacting. From 1943 to 1949 Zwicky worked with the Aerojet Engineering Corporation on development of jet engines and jet-assisted takeoff for aircraft.

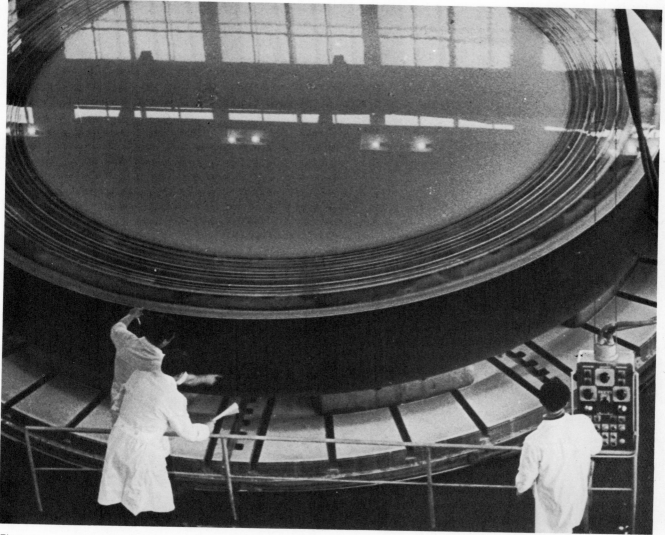

First stages in the grinding of the mirror of the Zelenchuk 236-inch (6-m) telescope, the world's largest. The 42-ton borosilicate glass disk is here being milled to a rough concave shape with a diamond wheel before polishing in a Leningrad optical workshop.

Index

Page numbers in italic refer to the main references.

Index

Index

Index